SOLAR ENERGY HOUSES

STRATEGIES • TECHNOLOGIES • EXAMPLES

Second Edition

Routledge
Taylor & Francis Group
LONDON AND NEW YORK

earthscan
from Routledge

First published 1997 by James & James (Science Publishers) Ltd
Second edition published 2003

Published 2022 by Routledge
4 Park Square, Milton Park, Abingdon, Oxon OX14 4RN
605 Third Avenue, New York, NY 10017

Routledge is an imprint of the Taylor & Francis Group, an informa business

A catalogue record for this book is available from the British Library

ISBN 13: 978-1-8497-1057-2 (pbk)
ISBN 13: 978-1-138-47144-3 (hbk)

Acknowledgements

Co-Editors
Anne Grete Hestnes
Faculty of Architecture, NTNU
N-7034 Trondheim, Norway

Robert Hastings
Architecture, Energy & Environment GmbH
Kirchstrasse 1, CH-8304 Wallisellen, Switzerland

Bjarne Saxhof
Department of Buildings and Energy
DTU, DK-2800 Lyngby, Denmark

Production:
Robert Hastings
&
Susan Rice
Solararchitektur, ETH-Hönggerberg
CH-8053 Zürich, Switzerland

Graphics:
Lydia Straubel & Hans Erhorn
Fraunhofer-Institut für Bauphysik
D-70569 Stuttgart, Germany

Contents

Foreword

Over the last few years considerable progress has been made in reducing the energy consumption for space heating. This is mostly achieved by the use of energy conservation and traditional solar technologies. The total energy consumption for all end uses is, however, still large and warrants considerable effort. To obtain a significant further reduction in the energy consumption for space heating, and to also reduce the consumption for cooling, ventilation, lighting, and the production of hot water, it has become necessary to develop new building concepts. Such new concepts require the use of new materials, components, and systems.

Recent advances in several areas of research have resulted in concepts and products that can be used. For instance, developments in material technology, with products such as low-E coatings on glass, and transparent insulation, are having a considerable impact. The most promising materials, components, and systems need to be analysed, tested, and developed for the purpose of integrating them in whole building concepts. IEA's Solar Heating and Cooling Programme's Task 13 was started for this purpose.

Its official objective was to 'advance solar building technologies through the identification, development, and testing of new and innovative concepts which have the potential for eliminating or minimising the use of purchased energy in residential buildings while maintaining acceptable comfort levels'. It dealt with the application of both passive and active solar technologies for space heating of residential buildings. The use of passive and active concepts for cooling, ventilation, lighting, and the production of hot water, as well as advanced energy conservation measures that reduce heating and cooling loads, were addressed.

In this process, most countries participating in the Task constructed one or more experimental buildings. These buildings were designed in part as a team effort by the Task participants. At Task meetings each expert submitted a design for a building project in his or her country. The designs were extensively reviewed and discussed in special design review sessions. After the meetings, the designers revised their plans based on the input from the other participants. Each design therefore benefited from the knowledge and experience of the group as a whole.

In general, the strategy used in the design of the buildings was to introduce energy conservation strategies first, and then to introduce passive and active solar strategies. In most cases all three strategies had to be used in order to reach the energy consumption targets of the Task. The strategies are described in Part 1 of this book.

The emphasis in the Task was on innovation and long-range (after the year 2000) cost-effectiveness. The materials, components, systems, and concepts considered for the experimental buildings did therefore not need to be economical, or on the mass market, allowing the participants to look at totally new solutions. However, most designers were required by the builders to consider cost. This did in the end limit the degree of innovation in the built projects.

As part of the concept development, a number of materials, systems, and components were studied and documented. Those that were of most interest to the participants were advanced glazings, transparent insulation materials, thermal storage materials, integrated mechanical systems, intelligent control systems, solar water collectors, and solar cells. The technologies analysed are described in Part 2.

The strategies and technologies are applied in the experimental buildings, described in Part 3. These building designs indicated that it is realistic to expect the total energy consumption can be reduced to a small fraction of the typical consumption today. The average total energy consumption calculated for the experimental buildings is 44 kWh/m^2 per year – about a 75% reduction of the typical consumption.

Part 4 gives an overview of the lessons learned, mainly based on the experience during design. The activities of Task 13 were extended in order to collect the results of the evaluation of most of the projects. The evaluation showed that the actual reduction in energy consumption was 60% compared to typical houses. This should be considered as a very successful outcome, bearing in mind the experimental character of the projects. Many interesting experiences were gained concerning the low energy concepts, the technologies and the user aspects. The evaluation results are described in Part 5. It was quite possible to reduce the energy consumption further, but that would require more extended use of technologies that were at that time not cost-effective and that the builders were not willing to finance.

Regional practices and traditions resulted in the need to include technologies and strategies that for participants of some countries seemed obvious or common knowledge, as the same technologies and strategies seemed totally revolutionary to others. For instance, mechanical ventilation techniques and heat recovery systems are common practice in the Nordic countries and were therefore, by the participants from those countries, not considered worth mentioning in a book of this type. However, for some of the other participants this was new knowledge that helped them reduce energy consumption. Therefore, a wide range of technologies and strategies have been included and the monitoring activities showed that sound concepts were developed. Many interesting conclusions were drawn from the project evaluation with great relevance for practice.

1

Strategies

J. Douglas Balcomb

(National Renewable Energy Laboratory,
1617 Cole Boulevard, Golden, CO 80401, USA)

Strategies

1.1 INTRODUCTION

This part of the book summarizes how energy consumption has been reduced by the IEA Solar Heating and Cooling Programme Task 13 houses (Task 13 house). The strategies that proved to be the most effective for the houses are highlighted and discussed quantitatively, both as a group and individually. The focus is on this group of houses as they were finally designed. Conclusions are drawn based on studying and comparing features common to all or most of the houses.

The reader should note that the numbers quoted in this part are all based on computer simulations made by the individual countries. Thus they are predicted, and not measured, numbers. Initial indications are that measured values are somewhat higher than predictions. Nonetheless, valuable inferences can be drawn from the predictions.

How much energy is predicted for these houses? The smaller pie chart on the right in Figure 1.1.1 shows the average purchased energy for the 14 Task 13 houses. The larger pie chart on the left shows typical values for a contemporary house, averaged over the same countries. The Task 13 houses use 44 kWh/m^2 of heated floor area compared to 172 kWh/m^2 for the typical house, a 75% reduction. The total solar contribution to these houses averages 37 kWh/m^2, including the useful passive solar gains, active solar, and photovoltaics.

Note that throughout this book, energy reported is the actual energy brought into the building. Bear in mind that the source energy required to produce and deliver energy to the dwelling can vary greatly. For example, electricity from a thermal power plant typically requires about three units of source energy to produce one unit of delivered energy compared to gas for which the ratio is more typically 1.3 units of source energy per unit of delivered energy. Another important consideration is that some energy is renewable, including solar, wood, and electricity from hydropower, whereas energy from fossil sources is not.

Although space heat accounts for the largest end use of energy in a typical house in these cold climates, space heat falls second behind electric energy for fans, lights, and appliances in the houses, requiring only 14 kWh/m^2 out of the total 44 kWh/m^2. On average, these houses use less than 15% of the heating energy of typical houses in each of their countries.

The houses were designed as total systems in which all energy uses were addressed. For example, appliance and lighting consumptions were reduced by selecting efficient units and water-heating consumption was reduced by using either a solar system or by heat recovery. The houses are a most remarkable group offering many instructive lessons for designers in any climate.

1.2 SPACE HEATING REDUCTION

The strategies used to minimize energy needed for space heat are:

- reduce transmission losses by designing a compact, well insulated, and tight envelope
- recover heat from exhaust air
- use passive and active solar gains
- satisfy the remaining heating requirement by producing and using auxiliary heat efficiently.

Reduce transmission losses

The Task 13 houses addressed reduction in the gross heating requirement as the first priority. Many of the houses use two or three times more insulation than a typical house. A good indicator of heat transmission is the overall building loss coefficient (BLC). This is the

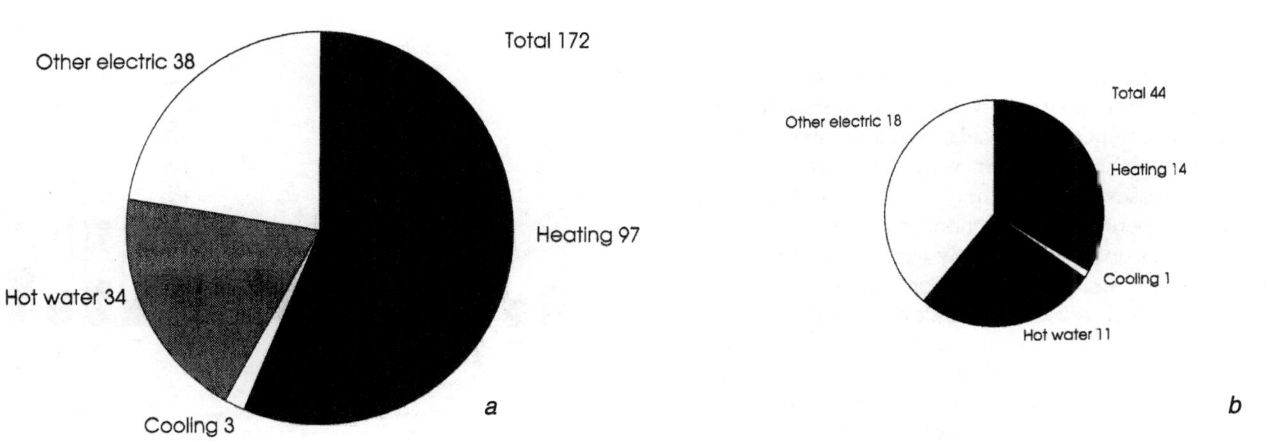

Figure 1.1.1. Energy use in (a) typical houses and (b) the Task 13 houses, expressed per m^2 of heated floor area

power, in units of watts, required to maintain a temperature difference of 1K between the inside and outside of the house.[1] The BLC values for the Task 13 houses are shown in Figure 1.2.1. To allow houses of different sizes to be compared, the values are listed per m² of heated floor area.

The two houses that stand out as having high BLC values, the Japanese house (J) and the American house in Yosemite (US-Y), are in warmer climates. The average BLC of the remaining 12 houses is 1.0 W/m²K, a good target value for a low-energy house in a cold climate. The variation between the BLC values of these 12 houses, from 0.68 to 1.24 W/m²K, is not particularly large.

There are two ways to achieve a low BLC – design a compact building form to minimize exposed surface area and use high levels of insulation and windows with low U-values. Both are effective and both have been used in the Task 13 houses. The average U-values (in W/m²K) are as follows:

walls	0.18
roof	0.13
windows	1.24

For the U-values achieved in the individual houses, refer to the technology chapters on super insulation (Chapter 2.2) and windows (Chapter 2.3).

Reduce infiltration and ventilation loads

Standards vary, but the amount of fresh air required to maintain adequate indoor air quality ranges between one third and three quarters air changes per hour. The approach to ventilation taken in nearly all of the Task 13 houses is to build the house to be very airtight. Air leakage is held to one tenth or less air changes per hour and then controlled ventilation is provided. There are two advantages to this. First, the ventilation is constant instead of varying from twice the required value to zero, depending on temperature and wind conditions. Second, exhaust air is collected in a single location, allowing heat recovery by either an air-to-air heat exchanger or a heat pump.

Required heat

Sharply cutting the BLC results in a much reduced annual requirement for heat. Figure 1.2.2 shows the average seasonal heating load of the Task 13 houses. This is total heat from all sources required to maintain 20°C room-temperature comfort from 1 October through 31 March. Note that contributions from internal gains (people, lights, and appliances), heat recovery, passive solar gains, and auxiliary heat are each about one fourth of the total required heat. However, the bars for the Norwegian house (N) and the German house in Berlin (D-B) illustrate that this distribution varies significantly from house to house.

Recover heat

The effectiveness of heat recovery in the Task 13 houses is shown in Figure 1.2.3. The values were determined by running the simulation program twice, first without heat recovery and again with heat recovery, and noting the difference.

On average, heat recovery meets 24% of the space heating requirements of the houses. The high value for the Norwegian house is due to a high ventilation rate of one air change per hour and a very high (theoretical) recovery rate of 90%. Recovery can be done by a heat exchanger which allows cold intake fresh air to be

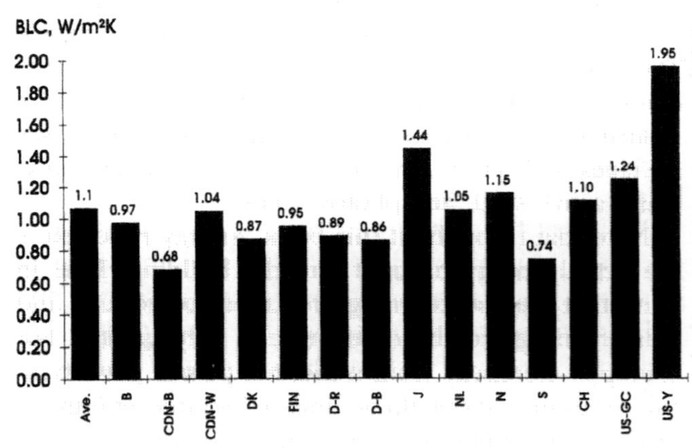

Figure 1.2.1. Building loss coefficients of the Task 13 houses, normalized by heated floor area

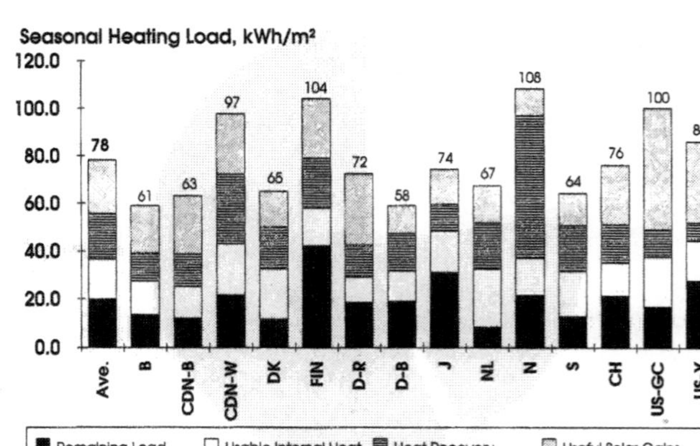

Figure 1.2.2. Space heating requirements of the Task 13 houses, 1 October through 31 March

[1] Sometimes the BLC is called the 'building UA' because the primary constituents are (U-value) x (area) products of the individual components (walls, windows, roof, etc.) all summed together. However, we use BLC because it also includes losses due to other effects, including natural infiltration, perimeter conduction, and losses to the ground. These losses don't scale directly with surface area but are in the same units of W/K. For this study we have computed an *equivalent* BLC index by dividing the gross heating energy (watt-hours) computed to maintain an inside temperature of 20°C from 1 October through 31 March by the base-20°C degree hours computed over the same time interval. For this particular calculation, solar gains, forced ventilation air, and heat recovery were not taken into account.

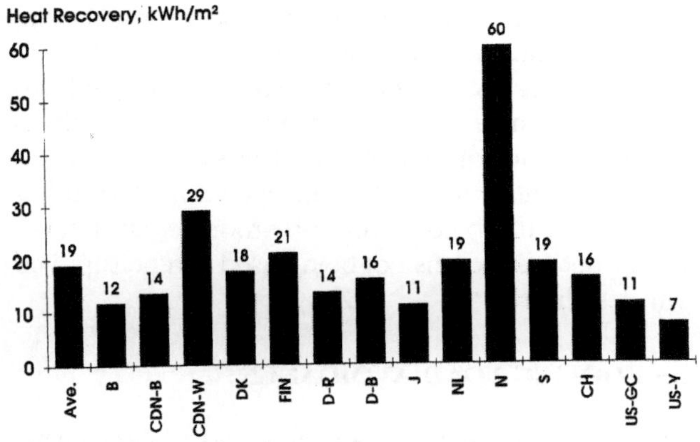

Figure 1.2.3. Annual heat recovery in the Task 13 houses

warmed by heat transfer through a partition separating it from the exhaust air. In three of the houses, a heat pump is used to recover heat from exhaust air, and the heat is used to warm hot water or the space, as needed. These techniques are discussed in the respective chapters of Part 2 on *Innovative Technologies*. A principal advantage of heat recovery is that it works day and night and is effective in any climate.

Use passive solar gains

Passive solar heating is an obvious answer in a sunny climate like that of the American house at the Grand Canyon in Arizona (US-GC), a cold climate at 2130 m elevation. It is less obvious at the high latitudes and in the cloudy winter climates of many of the Task 13 houses. The effectiveness of passive solar gains was determined by running the simulation program twice, first without solar gains and again with solar gains, and noting the difference. The results, shown in Figure 1.2.4, indicate a surprisingly high solar contribution, averaging about one third of the total required heat. This exceeds the contribution of the auxiliary heating system. The high values for the houses in the Southwest US are not surprising, but the high solar contribution in the German house in Rottweil (D-R) is impressive in a much less sunny climate.

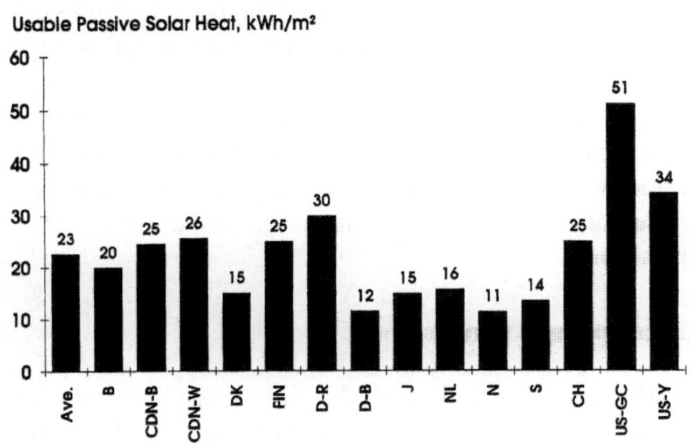

Figure 1.2.4. Usable passive solar heat in the Task 13 houses

The reason that passive solar heating is effective in the houses at high latitudes is primarily because of the extended length of the heating season. Solar gains are minimal in mid-winter, from December through mid-February, but make a major contribution in the fall and spring seasons, when heating requirements are still substantial.

Four strategies are used to make use of passive solar gains in the Task 13 houses:

- Direct solar gains through windows are an important aspect of all the houses. Unfortunately, in the colder north-latitude climates, window losses are greater than solar gains (taken over the whole heating season) for any type of currently available glazing. But in the range of U-values below 0.8 W/m^2K, windows are close to neutral.
- To increase comfort and the usefulness of solar gains, many of the house interiors include massive materials to store the heat, such as concrete floors and brick walls. The massiveness of the Task 13 houses varies greatly, from the foam insulation panel construction of the US Grand Canyon house to the masonry and concrete construction of the German houses. The Swiss house (CH) is a combination of light exterior walls to maximize insulation and massive interior walls and floors to absorb solar gains. The often-misunderstood role of thermal mass and the recommended minimum amount and placement are further discussed in Chapter 2.8 on *Thermal storage*.
- Sunspaces are included in five of the houses, the largest being the atrium of the Dutch apartment building (NL). Sunspaces have an amenity value, serve as a buffer space between the living area and the outside, and can be used to preheat ventilation air as described in Chapter 2.7 on *Sunspaces*.
- A Trombe–Michel wall was included to make up for the low mass of the US Grand Canyon house and balance daytime direct gain. In a Trombe–Michel (or Trombe) wall masonry below the windows is directly warmed by sunlight. In the Task 13 building, heat loss to the outside is hindered by using a selective surface on the wall exterior surface and by two layers of low-iron, high-transmission glass in front of the wall. Solar heat that accumulates in the masonry during the day, is at night conducted through the wall and radiates to the house interior. A Trombe wall thereby complements solar gains coming through the windows in the daytime.

Use active solar heating

Solar energy can be used actively in many ways ranging from heat-pump boost to seasonal storage. For example, the Japanese house uses a hybrid PV collector to improve the performance of their auxiliary heat-pump system in the winter. Warm air heated by the PV panels is used as the heat source for the heat pump.

The German Berlin house employs active solar heating to achieve the objective of zero auxiliary heat. A fluid, circulated by natural circulation through an array of solar panels, transfers heat to a tall 20 m^3 water heat-storage tank extending from the basement up through the centre of the house; thus heat losses from the tank contributes to heating the house. The large tank is required to balance the mismatch between the availability of solar in the summer and the need for heat in the winter. Heat is distributed to the house via radiators. In the summer this system provides domestic hot water for this house and two neighbouring houses.

Produce and distribute auxiliary heat efficiently

After the heat from solar gains, heat recovery, and lights and appliances has been taken into account, very little auxiliary heat is needed to maintain a 20°C room temperature. The average heating demand that remains is 20 kWh/m^2, which is only 26% of the total heating required. The challenge was to produce and distribute the heat as efficiently as possible. Figure 1.2.5 shows the remaining heat required and the auxiliary heating plant input energy required to meet this need. On average, only 14 kWh/m^2 are required to meet the 20 kWh/m^2 space-heating demand.

In those houses where the remaining load and the auxiliary are identical, electric resistance heating is used, for which the efficiency is essentially 100%. In the German Berlin house, as already mentioned, the remaining load is met by active solar. The Swiss (CH) house uses wood heat. In the Finnish (FIN), Norwegian (N), Japanese (J), and the two US houses (US-GC and US-Y), heat is supplied using a heat pump, effectively leveraging the electric input. In the Dutch (NL), Belgian (B), Canadian Waterloo (CDN-W), and Swedish (S) houses the heat is from fossil fuel at a much lower cost than electricity. In the Danish (DK) row houses, the heat is delivered from a local district heating plant.

In all of the task 13 houses attention was paid to the location of the heating plant and the design of the distribution system, especially to ensure that any losses are retained within the house thermal envelope and that the heat delivery is effective.

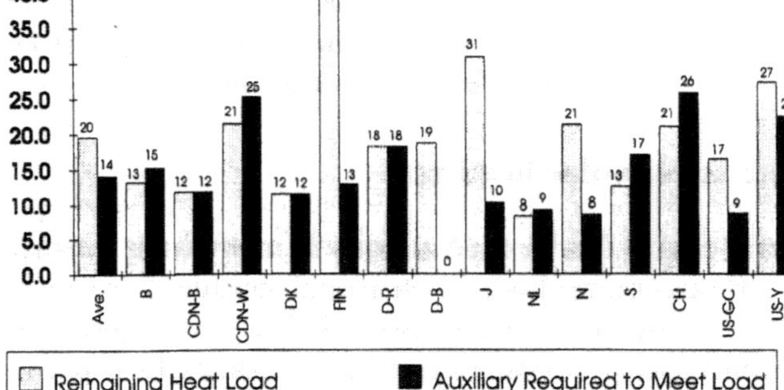

Figure 1.2.5. Auxiliary space heating requirements of the Task 13 houses

In general, the maximum heating power required in these low-energy houses is very much less than in typical houses. Accordingly, the size of auxiliary heating system can be reduced, often to a level that allows a change of the equipment type. This saves capital cost and can result in more efficient operation. A distribution system that allows room-by-room control saves energy because rooms not being used can be supplied with less heat.

1.3 COOLING LOAD AVOIDANCE

Any building, even one located in a cold winter climate, can overheat if care is not taken to both avoid and remove heat gains during warm or hot summer weather. Passive solar gains, so desirable during cold months, can cause discomfort at other times.

Only four of the Task 13 houses require any active cooling. These are the Japanese house, the US Yosemite house, and the two Canadian houses. Even in these cases the cooling loads are quite small except for the Japanese house, which is in a particularly hot and humid climate. All of the designers paid attention to cooling load avoidance. The strategies are to reduce solar gain, to use ventilation, and to use thermal mass.

Reduce summer solar gain

The noon elevation angle of the sun changes by 47 degrees between the winter solstice and the summer solstice. At lower latitudes, an overhang on a south window can be dimensioned to admit the winter sun all day long and totally shade the beam sunlight during mid-summer.

Solar-gain avoidance on orientations other than south requires different shading strategies because overhangs are not nearly as effective. Minimizing the glazing area, especially on the west, east, and roof, is the easiest. The Belgian, Norwegian, and Danish NS row[2] houses avoid the problem with their east and west party walls. In other Task 13 houses, removable screens have been used and trees have been planted on the east and west sides. A sun chart is particularly useful to assist in locating shading that will be effective when needed but not block the sun in winter when solar heat is desirable.

The Swedish 'magic window' design provides a desirable degree of solar gain control by adjusting a panel within the window assembly that can serve as a reflector, a shade, or an insulator, depending on its position.

Incorporate ventilation

Most houses, including all the Task 13 houses, rely on ventilation to get rid of unwanted heat and control

[2] In the UK the term 'terrace house' is normally used instead of 'row house'.

temperatures at some time of the year. Opening windows to allow ventilation when outside temperatures are cool usually solves the problem. Window placement should facilitate good cross ventilation, profiting from prevailing breezes and using the stack effect to move air.

Use thermal storage

Thermal mass alone cannot prevent overheating over an extended period, such as a day. However, the heat capacity of thermal mass tends to level out within temperature variations, helping to prevent discomfort due to temporary overheating. In the US Yosemite house, thermal mass is used in conjunction with a night-vent cooling strategy to shift the time when cooling is available at night until it is needed in the afternoon, as described in Chapter 2.8.

Sixty square metres of panels containing a phase-change material are used to store heating and cooling energy in the Japanese house. The phase-change temperature is 25°C, midway between the heating and cooling comfort zones. The panels are heated in the winter and cooled in the summer by means of a heat pump using off-peak electricity during the night.

Use other cooling techniques

The German Rottweil house uses ground ventilation cooling. The house ventilation air is drawn beneath the basement slab for pre-cooling in the summer.

The integrated mechanical system (IMS) in the Canadian house in Brampton (CDN-B) uses an ice storage tank to achieve a very high latent-heat storage density in the cooling mode. Water in the small tank is frozen by pumping heat out of the tank and into the exhaust air (or into the domestic hot water tank). A brine solution transfers heat from cooling coils in the house to the ice bath.

In the US Yosemite house, a heat pump heats the domestic hot water by cooling the house air in the summer.

A buried cistern is used to pre-cool air used to ventilate the Canadian Waterloo house.

1.4 WATER HEATING SAVINGS

The net heating load for domestic hot water is greater than the space heating load in these houses (23 vs 20 kWh/m²). Accordingly, it is important to apply to water heating some of the same principles that were used to save space-heating energy: first reduce the load and then use active solar or a heat pump system to supply the load. The hot-water demand and energy use of the Task 13 houses is shown in Figure 1.4.1.

Reduce demand

The load can be reduced by reducing hot water consumption. Examples include low-flow fixtures and shower heads. A heat recovery system was used to reduce demand in the Canadian Brampton house. The water drain pipes are in contact with the water supply pipe, making a long counter-flow heat exchanger. Thus some of the remaining heat in the drain water is transferred through the pipe walls to the incoming cold water.

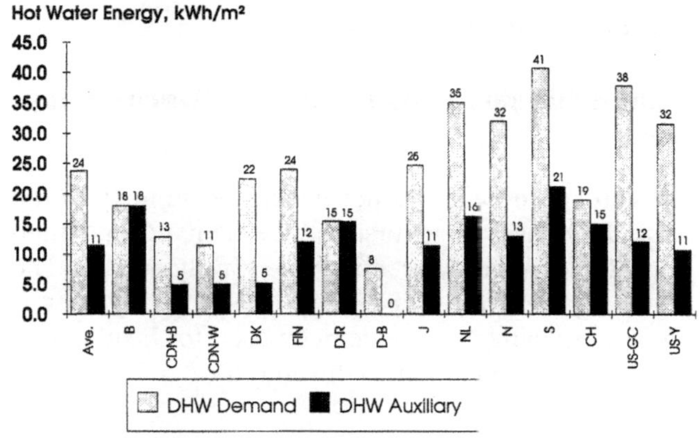

Figure 1.4.1. Hot-water demand and energy use in the Task 13 houses

Use active solar heating

Seven of the Task 13 houses use active solar hot-water heating systems for the domestic hot water. The solar contribution of these systems is better than that of active solar space-heating systems, because the demand for hot water occurs over the whole year, so that the long sunshine hours in the summer can be put to full use.

The solar water heater in the Canadian Waterloo house carefully avoids tank mixing, and this improves performance through stratification. Thin tubing from tank to collectors, associated control wiring, and insulation are bundled together in a flexible conduit to simplify installation.

Hot-water heating is a mature technology, and in most countries an industry exists that manufactures and installs reliable, proven systems. It is estimated that there are 6.2 million hot-water systems in use in the world today.

Use heat recovery

Four of the Task 13 houses, the Norwegian, Canadian Brampton, and the two American houses, use an integrated mechanical system to heat all their hot water. Heat is pumped out of the exhaust air (or out of the house air in the summer) and into the hot water tank. The result is to heat water electrically but with a high coefficient of performance.

The exhaust-air heat-pump hot-water system is a well developed technology; the industry is particularly well established in Scandinavia.

1.5 ELECTRICAL SUPPLY

Reduce fan, pump, lights, and appliance energy use

As with space heat and hot water, the strategy in the Task 13 houses is to first reduce demand and then consider low-energy or renewable supply options. The electricity demand and energy use of the houses is shown in Figure 1.5.1 (electricity used for heating and cooling is not included in this graph).

Use low-consumption lights and appliances

All of the Task 13 houses employ energy-efficient lights and appliances, which are usually available from local sources. In the Canadian Waterloo house energy is supplied at a lower cost by using gas appliances. In several of the houses particular attention was paid to low-energy refrigerators because this is the single largest energy-consuming appliance. New lights and appliances that dramatically reduce electric energy consumption are being developed and marketed throughout the world. The average predicted household electric use in the Task 13 houses is 17 kWh/m^2 per year.

The Belgian house achieves a household electric energy use of only 7 kWh/m^2 through the use of a home automation system.

Use care with fans and pumps

Electricity required to operate fans and pumps, often called 'parasitic energy', can be excessive if the systems are not well designed. This can obviate the apparent energy efficiency of mechanical systems. In the Task 13 houses, fan and pump energy averages 6 kWh/m^2. The values in the graph vary from 1.3 kWh/m^2 (in the Belgian house) to 14 kWh/m^2 (in the Norwegian house, which makes extensive use of heat recovery). This energy is generally a good investment because much more energy can be saved than is used. Note that the energy reduction due to heat recovery, which would not work without the fans and pumps, is 19 kWh/m^2, which is three times the parasitic energy.

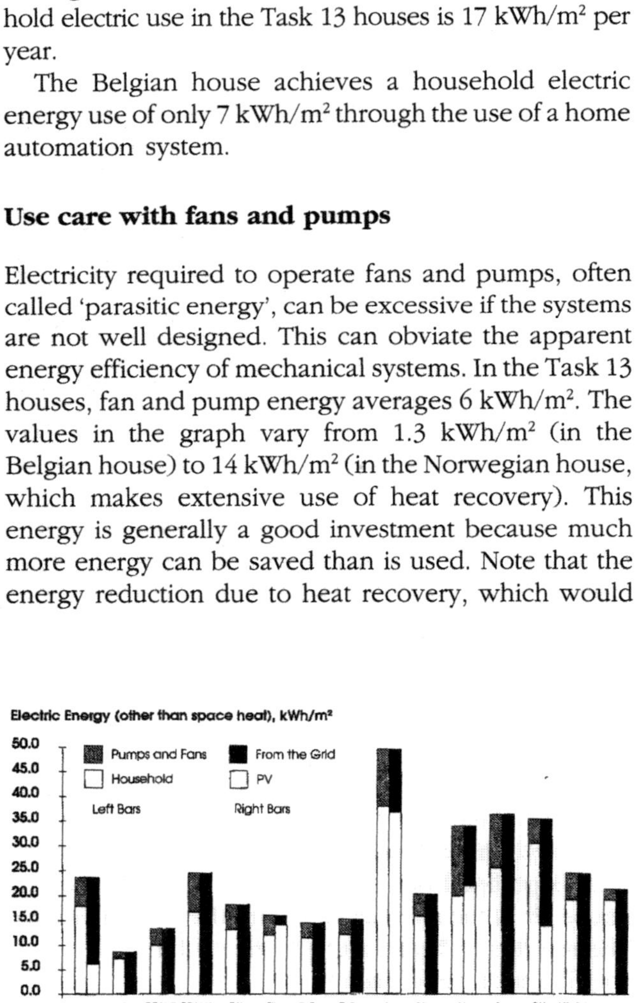

Figure 1.5.1. Other electrical use and sources for the IEA houses

Use photovoltaics

Four of the Task 13 houses (Japan, Finland, Switzerland, and Norway) use photovoltaic (PV) panels on the roof to generate electricity on site. The largest system is on the Japanese house. A 44.8 m^2 array generates 4.2 kW of peak power and has a predicted annual output of 5500 kWh. This is 36.7 kWh per m^2 of floor area. In Japan, the excess electric power from PV can be sold to the electric power company at the same rate as the purchase price of electricity.

The solar hot-water system on the Canadian Waterloo house and the Danish row houses use a PV-powered pump, eliminating the need for a controller, simplifying the system, and improving performance.

1.6 DESIGN INTEGRATION

The design of a low-energy house is an exercise in system integration. Each of the IEA design and analysis teams has gone through this exercise. There are priorities and conflicts that each team has addressed along the way, eventually converging on a design solution that minimizes energy. The solutions account for a myriad of important issues, from aesthetics to economics, from public acceptance to reliability.

Each design team relied heavily on computer simulation tools to quantify the complex trade-offs between design choices and design parameters. The analysis tool used was different from country to country, each team using a program that its members were familiar with. (For a discussion of the computer tools and the activities of the IEA Task 13 Simulation Support Group, see Appendix A.) Some of the final simulation results are presented in the individual house descriptions in Part 3.

An example of a trade-off that was investigated is the benefit of increasing glazing area. The benefit of additional solar gain to offset heating requirements must be evaluated relative to the added heat loss through the glazing. This depends not only on climate and on the specifics of the glazing properties, especially U-value and transmittance, but also on the particular house design, its need for heat in the daytime, and its thermal-mass heat-storage characteristics. These calculations have been done not just for one or two days but hourly for an entire year using climate data typical of the locality.

The end products of these evaluations are the individual house designs and the associated predictions of annual performance. The overall results are given in Figures 1.6.1 and 1.6.2. In Figure 1.6.1, energy use is presented per unit floor area disaggregated by heating use and all other uses. In Figure 1.6.2, energy use is presented as total kWh disaggregated by source – electricity, gas, and wood.

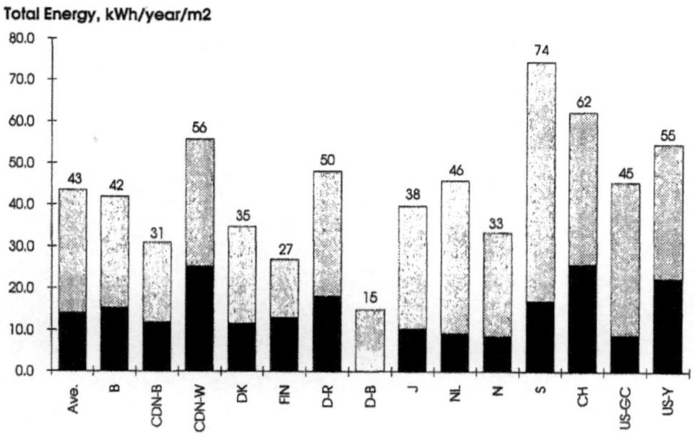

Figure 1.6.1. Total annual energy use of the Task 13 houses expressed per m² of heated floor area

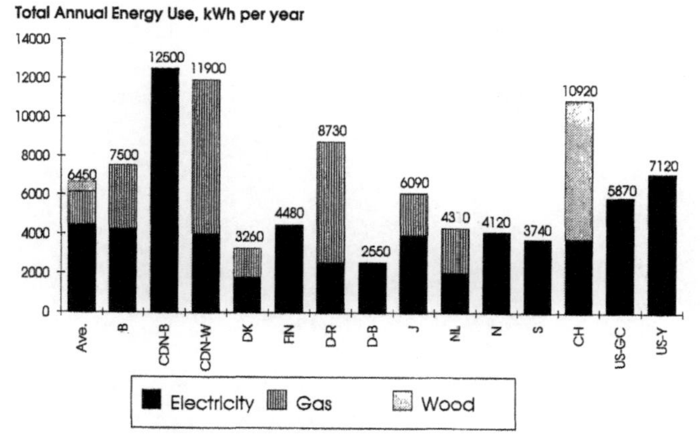

Figure 1.6.2. Total annual energy use for the Task 13 houses

The goal set for these houses was to achieve very low total purchased energy without compromising comfort. The energy requirements of all the houses is one third or less of the levels common for new construction in the respective countries. Electrical consumption for other than heating is the largest consumer at 48%. If the energy of thermal power stations and transmission losses were to be considered, it would be proportionally even larger. Space heating is in second place at 28%. The energy needed for fans and pumps is significant (12%). Finally, water heating represents a major consumption, arguing for solar or other energy-efficient water heating in very low-energy houses.

Inevitably, because heating loads are so large a part of the residential energy picture in most of the IEA countries participating in Task 13, heating energy became the primary focus of most of the design efforts. As a result, these houses are remarkable in their low use of heat, averaging only 14 kWh/m². This is one quarter or less of the levels common for new construction in the respective countries. Thus the lessons to be learned from these house are particularly relevant with respect to their minimal space-heating requirements. This part has highlighted the most important issues. The reader is advised to study the individual houses to learn more about the specifics and particular approaches taken in each case. Many of these features are not just appropriate to their climate or particular situation but applicable over a wide range of conditions.

1.7 OUTLOOK

Passive solar design, better insulation, tighter construction, and ventilation heat recovery have reduced space-heating demand to values that are so low that electricity and water heating are the next targets for savings. In the Japanese house, electricity use exceeds the total for all other energy demands. Future decreases in the energy consumption in houses will involve conservation and solar systems that serve multiple functions, such as space heating, water heating, and electrical needs.

2

Innovative technologies

Bjarne Saxhof (editor)

2.1 Introduction

Part 1 has described the overall strategies used to reach the very low levels of energy consumption aimed at in the advanced solar low-energy buildings designed in Task 13. This part supplies a closer description of the 12 major advanced technologies that were included in the designs and shows examples of how the technologies were applied in the Task 13 houses. An overview of advanced technologies reviewed for the Task 13 design process has previously been published in two Working Documents.

Although the focus during the design phase was on advanced technologies that might not be commercially viable within the decade, insufficient funding for the extra cost of prototype solutions and liability problems for the builders has limited the number of new constructions and systems in the buildings, so that many of the theoretically long- and high-performance solutions of the original designs were replaced by today's best commercially available products.

The technologies are described separately in this section, but it should be stressed that a major part of the design process consists of combining the proper technologies in the best way, giving several widely different design possibilities to reach very low energy consumption.

The technologies used to achieve the desired low total energy demands are a mixture of energy conservation and solar technologies. The 12 technologies are: super-insulation, high-performance windows, transparent insulation, ventilation heat-recovery systems, ground coupled heat exchangers, sunspaces, thermal storage (in building construction, i.e. thermal mass), active solar water systems, photovoltaic systems, integrated mechanical systems, home automation systems, and energy efficient lights and appliances.

The use of the different innovative technologies is summarized in Table 2.1.1. A chapter has been devoted to each of the 12 technologies listed and the ways they have been used in the Task 13 houses.

It should be noted that the order of the chapters does not indicate a priority of the technologies.

Super-insulation (in this context defined as insulation levels significantly in excess of the local building code or standard practice) is the most widely used technology, having a major part in all buildings but one, where passive gains trade off against it. Chapter 2.2 on super-insulation stresses the importance of airtightness and avoidance of thermal bridging, calls for careful design of building assemblies and generally for re-design of traditional building assemblies.

High-performance windows (in this book defined as windows with a total U-value of up to 1.5 W/m^{2}K, i.e. half the value of conventional double-glazed windows) are equally widely used.

In high-performance windows, there is a trade-off between very low U-values and a high solar transmittance. The heat balance calculations for the Task 13 designs called for highly insulating windows in all heating-dominated climates. Chapter 2.3 on high-performance windows stresses the importance of careful analysis of the window system heat balance in winter and in summer, and of paying special attention to the window perimeter (spacers and frames, and window-wall connections), because, for example, traditional frames have higher U-values than the new super-glazings.

Transparent insulation materials were investigated for inclusion in some of the designs and used for solar collectors, external insulation on south-facing walls, or for 'daylight walls', i.e. translucent rather than transparent areas, but they generally lost out to super-glazings, for cost or liability reasons. The Norwegian house, however, has a 10 m^2 'daylight wall' in the sunspace.

Another technology, used in most of the Task 13 houses, is heat recovery on exhaust air, either in air-to-air heat exchangers for pre-heating of supply air to the building, or in integrated mechanical systems where

Table 2.1.1. The technologies used in the Task 13 houses (i = inherent in construction; I = limited to active solar system pumping; x = yes)

	B	CDN-B	CDN-W	DK	FIN	D-R	D-B	I	J	NL	N	S	CH	US-GC	US-Y
Super-insulation	x	x	x	x	x	x	x	x	x	x	x	x	x	x	
High-performance windows	x	x	x	x	x	x	x	x	x	x	x	x	x	x	
Transparent insulation											x				
Ventilation heat recovery systems	x			x	x	x	x	x	x	x	x	x	x		
Ground coupled heat exchangers			x		x	x					x				
Sunspaces		x			x	x	x			x	x				
Thermal storage (building mass)	x			i		i	x	i	x			x	i	x	i
Active solar water systems			x	x	x		x	x		x		x			
Photovoltaic systems			I	I	x				x		x		x		
Integrated mechanical systems		x	x								x			x	x
Home automation systems	x				x				x	x		x			
Energy-efficient lights & appliances		x	x	x	x		x				x				x

heat may be recovered for space and/or domestic hot-water heating. The chapter on ventilation heat-recovery stresses the importance of very airtight building envelopes for the heat-recovery efficiency in houses with balanced ventilation systems, and the importance of using energy-efficient fans to limit parasitic energy consumption. Integrated mechanical systems provide several residential functions, such as ventilation, heat recovery, space heating and cooling, and domestic hot-water heating. Chapter 2.5 on these systems shows how this was achieved in five buildings, four systems including heat pumps (i.e. electrically heated units) and the fifth a combined gas furnace and heat-recovery ventilator. For the three systems providing space cooling, the optimum performance is achieved when the space-cooling load reasonably matches the domestic water-heating load.

Four buildings have ground-coupled heat exchangers in the form of air-duct or water-pipe loops in the ground under or next to the buildings. Two of the systems have been designed for heating and cooling, the other two for heating or cooling respectively. One system directly cools or pre-heats ventilation air, two systems use the ground loop to supply a heat pump (one of them for heating only), and the last system uses water from an underground cistern, circulated through a ground loop, to supply a water-to-air coil in the ventilation system and thus cool the house.

Six of the Task 13 houses have one- to six-storey sunspaces as part of their energy design, used in three different ways, as simple sun-tempered buffer zones, for pre-heating of ventilation air, or for supplying a multi-purpose heat pump. Chapter 2.7 on sunspaces illustrates the energy benefits as well as the amenity values of four of the six sunspaces.

Thermal storage (in thermal mass within the building thermal envelope) plays an important role in limiting excessive temperature swings, and this is essential since solar gains provide a significant input to all of the Task 13 houses at some time during the heating season. Thermal storage has been an issue in the design of most of the houses, although sufficient mass in several buildings was inherently present in the chosen construction. Chapter 2.8 on thermal storage gives examples of building thermal storage simply by use of added mass in the building structure, by use of Trombe walls, and by use of phase change materials.

Active solar water systems were used in seven of the Task 13 houses, mainly for domestic hot water heating, but in one case also for the total space heating, with a 20 m^3 seasonal storage (water tank) in the house. In another house, the solar system supplies a part of the space heating in combination with a heat pump – in this case there is a 3 m^3 storage in the house, also in form of a water tank.

Four of the buildings have grid-connected 2.3–4.2 kW PV systems, and two buildings have small systems operating and controlling active solar domestic hot-water systems. The expected annual generated power from the four large systems is 2300–5500 kWh, i.e. a quite substantial solar power contribution. However, not all generated power is used in the house, and surplus power has to be sold to the utilities at fairly low rates.

Five of the buildings are equipped with home automation systems, intelligent control systems that operate various energy functions in the building, e.g. control of the heating system or solar shading systems, to increase the thermal comfort and decrease the energy consumption. In addition, such systems may also provide amenity benefits, e.g. increased security in the dwelling (integrated fire and/or burglar alarms). Chapter 2.12 on home automation systems describes three of the large systems and also gives an overview of stand-alone controls in the Task 13 houses.

The consumption of purchased electricity in the Task 13 houses is reduced by local production in PV systems, by controls including intelligent controls of lights and appliances described in the chapter on home automation systems, and finally by selection of very efficient technical equipment and lights and appliances. Chapter 2.13 on energy-efficient lights and appliances deals with the possibilities of achieving very low electricity demands in the households and gives some examples of equipment chosen in some of the Task 13 houses where special attention was paid to reducing the electricity demand. Finally, it summarizes the Task 13 experience on electricity saving.

- *Author.* Bjarne Saxhof

2.2 Super-Insulation

Introduction

The building envelope consists of opaque and transparent parts, and in low-energy building design there is often a trade-off between super-insulation and solar gains – this chapter focuses on the opaque (insulated) envelope components. A well-designed building envelope will allow very little heat transfer, minimize air leakage and drafts, and protect building materials from moisture damage. Unfortunately, many conventional building assemblies do not meet these requirements. Low insulation levels contribute to high heating and cooling bills. Thermal bridging through building framing and junctions can result in condensation and mould growth. Air leakage through building components and junctions between components causes occupant discomfort. Accumulation of moisture in walls, either from air movement, diffusion of occupant-generated moisture, or transport of rain, reduces the effectiveness of insulation and ultimately causes irreparable damage to the structure.

A super-insulated building shell can solve the above problems and is an integral part of any low-energy building design. 'Super-insulation' means insulation to levels significantly in excess of the local building code or standard practice. As a result of super-insulating a building, the heat losses are significantly reduced such that much of the remaining heat load can be met by passive solar and energy-efficient technologies. The building assemblies presented in this chapter have been carefully designed to achieve maximum total performance and occupant comfort.

Concept definition

There are two key components to a super-insulated building shell: high levels of insulation with minimum thermal bridges, and airtight construction.

High levels of insulation are accomplished by constructing a thicker than normal wall and filling it with an insulation material. However, simply adding more insulation does not turn a conventional assembly into a high-performance assembly. The wall system and junctions between building components have to be carefully designed to be airtight and avoid thermal bridges or discontinuities. As more insulation is added, the thermal discontinuities become more important. The building assemblies have to be designed so that all non-insulating building materials (including wood, steel and concrete) are thermally protected by insulation.

The first layer of insulation is the most effective in reducing heat loss. The law of diminishing returns dictates that each additional layer of insulation is less effective than the previous layer. The amount of insulation used depends on the severity of the climate. Mild climates require a wall U-value of 0.2 W/m²K or less, whereas harsh climates might necessitate a value of under 0.13 W/m²K. As the insulation level of the wall is increased, the thickness of the wall also increases. The loss of interior floor area due to thicker walls is an issue that should be considered.

In most IEA countries wall systems are either wood frame or masonry construction. In wood-frame housing super-insulated wall systems are most commonly achieved using either single-stud and double-stud wall construction, although other systems can be adapted to achieve high insulation levels. In double-stud walls (where a 'cavity' is left in the middle, allowing continuous insulation of the wall between the interior and exterior studs) the studs may be staggered for optimum insulation efficiency.

Solid masonry or concrete walls provide sound insulation and thermal mass, but would become very thick (and expensive) to achieve extremely high insulation levels. In such systems loose-fill or batt insulation is sandwiched between masonry or concrete leaves (here, thermal bridging is an inherent problem), or rigid insulation is placed on the outside of a solid masonry or concrete wall to which the exterior finish is attached.

Super-insulating flat ceilings is relatively easy. Loose-fill insulation is blown into the attic space to the desired depth, but in very thick layers of low-density (and coarse) materials convection may slightly increase the thermal conductivity of the insulation. Batt insulation can also be used, but care must be taken to avoid air gaps, e.g. by using several layers with staggered joints. Better support, e.g. thicker gypsum boards, may be necessary to support the weight of the insulation. It should be emphasized that a good ceiling insulation is also very important to decrease cooling loads.

Concrete basement walls can be insulated on the inside, the outside, or both. On the inside, the interior walls are strapped and filled with batt insulation. The procedure for insulating on the outside is similar to that for masonry walls, but with further demands related to moisture protection and drainage. Basement and slab-on-grade floors can be insulated with rigid insulation installed underneath the slab. Wooden floors over crawlspaces can be insulated like single-stud walls. Concrete floors over crawlspaces can be insulated with rigid insulation glued or mechanically fixed under the floor panels.

The assembly of a wooden construction is made airtight by incorporating a continuous, sealed air barrier into the wall. It is an advantage to place the barrier a certain distance, e.g. 50 mm, from the interior finish-

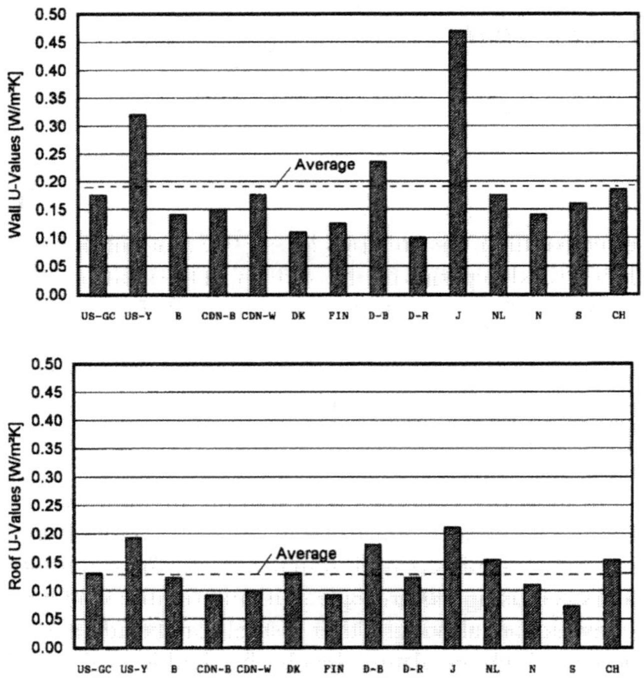

Figure 2.2.1. U-values for roof and walls in the Task 13 houses

ing for protection and for easy electrical wiring. Junctions between air barriers, e.g. polyethylene sheets, should be squeezed lap joints – taping is effective only if carried out at a support, e.g. a lath.

Some building materials, e.g. concrete and stuccoed masonry, with or sometimes without a painted finish, offer sufficient airtightness, but special attention should be paid to junctions with other building elements (including changes in geometry and in choice of materials) and window and door frames. That is true for wooden constructions also.

To ensure good airtightness, special attention during the planning phase should be given to areas where technical installations have to be taken through the envelope, and solutions for airtight lead-ins for cables, pipes, ducts, and flues should be designed.

When the building shell is made very airtight, mechanical ventilation of the house is usually necessary.

Variations

Super-insulation is a technology that has been used in most of the Task 13 houses. Figure 2.2.1 shows the U-values for roof and external walls in all the Task 13 houses.

Three generic construction techniques were used: wood or steel frame, masonry, and large panels. Wood (and steel) framing was the most popular wall construction system. The construction technique chosen was dependent on climate, availability of resources, and cultural traditions. In central and southern Europe, masonry construction is typical. In northern Europe, North America and Japan, wood-framed housing is most common. In countries with a large forestry industry, wood-framed walls are typically the least expensive construction technique, but require extra detail to eliminate thermal bridging and especially to ensure an airtight envelope.

Four approaches have been used to achieve airtight building shells: sealed polyethylene sheets, exterior air barriers, airtight pargings (e.g. stucco) and airtight panels. The air sealing approach used depends on the wall construction technique. Masonry housing can be inherently airtight, and concrete or stucco layers can, unlike wood framing, provide an effective air barrier. Often a combination of approaches is needed, e.g. sealing the polyethylene sheets used in a wood construction roof to concrete or stuccoed walls.

In most cases the designers of Task 13 houses had to redesign the conventional building assemblies to achieve the desired energy efficiency and performance. Typically, the key modification was to widen the building wall assembly to accommodate additional insulation. In addition, special details were designed to minimize thermal bridging and minimize air leakage. Each of the following sub-sections describes building assemblies used in one of the Task 13 houses.

Wood I-beam frame wall system – the Canadian Waterloo house

The above-grade walls for the Green Home are constructed using engineered wood I-beams spaced 600 mm apart (see Figure 2.2.2). This assembly allows for 240 mm of cellulose insulation with 19 mm of rigid insulation on the exterior for a thermal conductance of U = 0.16 W/m^2K. The engineered wood product is made up of two load-bearing members separated by a thin sheet of waferboard that resembles an I-beam. Thermal bridging is minimized over conventional wood stud framing by spacing the I-beams far apart (600 mm as opposed to 400 mm) and using an 11 mm-thick wood waferboard to separate the wood support members instead of 38 mm dimensional lumber. A polyethylene vapour barrier is placed behind the interior drywall and all joints are sealed with a non-drying sheathing tape. This wall system requires 30% less wood than conventional Canadian wood framing and does not require any large-dimensional lumber. Cellulose insulation was blown into the attic space to a depth of 400 mm.

The foundation system has low heat loss and requires only half the concrete of conventional poured basements in Canada. The below-grade walls are precast concrete panels that are flat on the exterior side but waffle-shaped on the interior side. 50 mm of rigid fibreglass is used on the exterior and the waffle cavities are filled with cellulose insulation. Wood stud walls are placed 63 mm inside the concrete wall, and another layer of cellulose insulation fills the entire cavity. 50 mm of expanded polystyrene is placed under the floor slab.

Double-stud wall – the Finnish house

The Finnish Pietarsaari house uses a double-stud wall system (see Figure 2.2.3). The two-storey wooden house is built of prefabricated wooden units. The wall units have 240 mm of mineral wool between two layers of gypsum board. On top of the outer gypsum board,

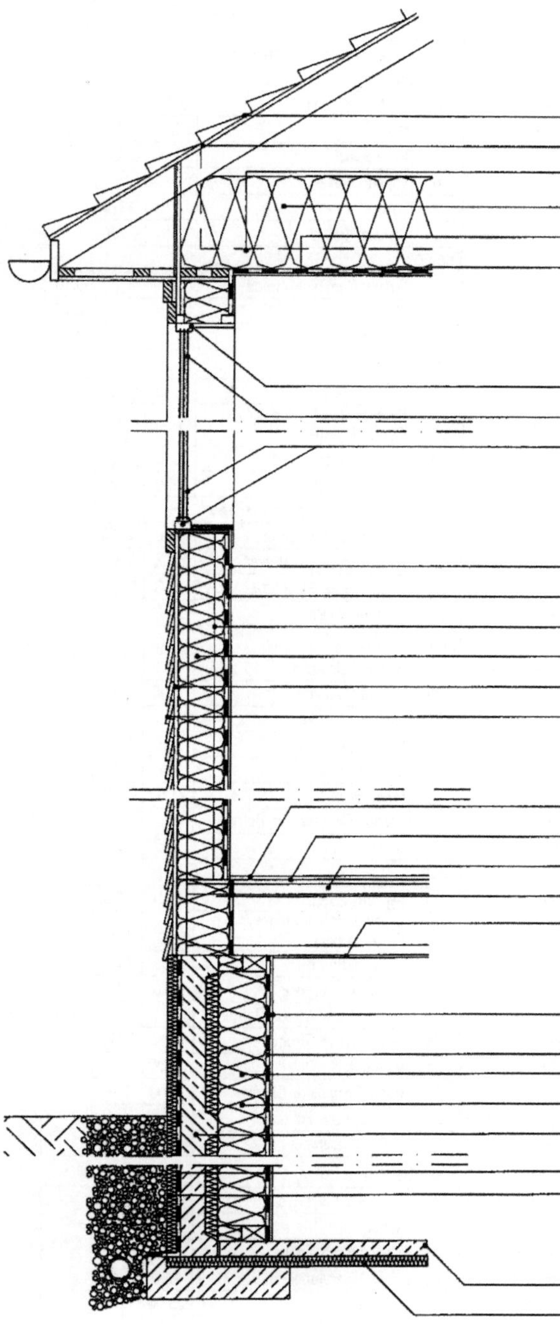

Layer	Description	Width (cm)	Thermal Conduct. (W/mK)	U-Value (W/m²K)
Roof				
1	Steel Roofing	–	–	0.094
2	Wood Waferboard	1.60	0.14	
3a	Wood I-Beam On 600 Centres	8.90	0.14	
3b	Cellulose Insulation	46.00	0.04	
4	Vapour Retarder	–	–	
5	Gypsum Board	1.20	0.21	
Window				
6	Fibreglass Frame			1.07
7	(Triple) Centre- Glazing			0.95
8	Glazing and Frame			1.02
Exterior Wall				
9	Gypsum Board	1.20	0.21	0.16
10	Vapur Retarder	–	–	
11a	Cellulose Insulation	24.00	0.04	
11b	Wood I-Beam On 600 Centres	24.00	0.14	
12	Insulated Sheating	1.90	0.06	
13	Wood Siding	1.90	0.14	
Ceiling				
14	Hardwood Flooring	1.90	0.14	–
15	Waferboard Subfloor	2.20	0.14	
16	Wood I-Beam On 600 Centres	36.60	0.14	
17	Polyurethane Foam	15.00	0.025	
18	Gypsum Board	1.20	0.21	
Below Grade Wall				
19	Gypsum Board	1.20	0.21	0.20
20	Vapour Retarder	–	–	
21a	Cellulose Insulation	24.00	0.04	
21b	Wood Studs On 600 Centres	6.40	0.14	
22	Reinforced Precast Concrete	5–20	2.00	
23	Moisture Barrier	–	–	
24	Semi- Rigid Fibreglass	5.00	0.04	
Below Grade Floor				
25	Concrete	7.60	2.00	0.61
26	Polystrene Insulation	5.00	0.035	

Figure 2.2.2. Wood I-beam frame wall system – the Waterloo house

rigid mineral wool coated with a plastic fibre fabric windproofing is installed using point-type fasteners. The idea of the additional insulation is to improve the U-value of the structure and to increase the temperature of the outer gypsum board to decrease the possibility of temperature-based natural convection inside the insulation cavity. The extra mineral wool layer goes continuously from the concrete footings to the top of the wall, forming an air barrier for the wall as well. It also acts as a thermal break for the wooden parts in the wall and upper floor structure. The insulation surface that is exposed to outdoor climate is rendered with thin rendering of about 5 mm.

The structure has three air barriers: inner gypsum board together with vapour barrier (with sealed joints), outer gypsum board, and the windproofing membrane on the external surface of the rigid mineral wool. Special attention has been paid to the sealing of the building envelope. Each joint between building elements was sealed during the installation.

The same procedure of several air barriers has also been used in the floor construction. The floor is built of prefabricated units with 300 mm mineral wool insulation. The total insulation thickness of the floor is 500 mm, since 200 mm of rigid mineral wool is fixed underneath the floor on site (crawl space foundation ventilated with outdoor air). The extra insulation is made of two layers with a plastic fibre fabric windproofing underneath each layer.

Figure 2.2.3. Double-stud wall system – the Finnish house

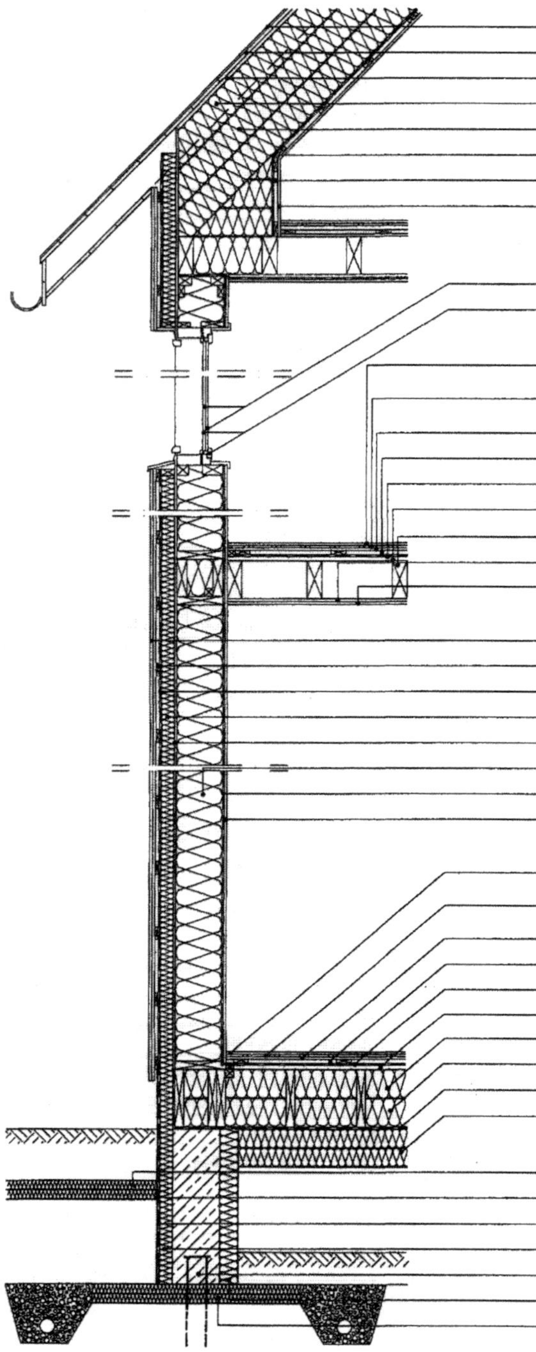

Layer	Description	Width cm	Thermal Conduct. W/mK	U-Value W/m²K
Roof				
1	Sheet Steel Waterproofing	–	–	
2	Boarding	1.90	–	
3	Ventilation Gap	5.00	–	
4	Wood	10.00	0.12	0.09
5	Mineralwool	45.00	0.035	
6	Vapour Barrier	–	–	
7	Lattice	2.20	0.120	
8	Gypsum Board	1.30	0.230	
Window				
9	Frame Aluminium/Wood			0.70
10	Glazing			0.70
11	Glazing and Frame			0.70
Ceiling				
12	Gypsum Board	1.30	0.230	
13	Gypsum Board Stripes/ Plastic Piping	1.30	0.230	
14	Gypsum Board	1.30	0.230	
15	Boarding/Airspace	2.20	0.120	–
16	Plastic Film Vapour Barrier	–	–	
17	Wood Frame Structure	2.20	0.120	
18	Wood Composite Beams	20.00	0.014	
19	Wood Frame Structure	2.00	0.120	
20	Gypsum Board	1.30	0.230	
Exterior Wall				
21	Wooden Cladding	3.00	–	
22	Lattice/Airspace	2.00	–	
23	Plastic Fibre Fabric	–	–	
24	Rigid Mineralwool	7.00	0.035	
25	Gypsum Board	0.90	0.210	0.12
26	Mineralwool	24.50	0.035	
27	Plastic Film Vapour Barrier	–	–	
28	Gypsum Board	1.30	0.230	
Floor				
29	Gypsum Board	1.30	0.230	
30	Gypsum Board Stripes/ Plastic Piping	1.30	0.230	
31	Gypsum Board	1.30	0.230	
32	Boarding/Airspace	2.20	0.120	
33	Plastic Film Vapour Barrier	–	–	0.11
34	Wood Frame Structure	2.00	0.120	
35	Mineralwool	30.00	0.035	
36	Wood Composite Beams	30.50	0.014	
37	Plastic Fibre Fabric	–	–	
38	Rigid Mineralwool	20.00	0.035	
Foundation				
39	Extruded Polystyrene	10.00	0.030	
40	Rendering	0.50	–	
41	Plastic Fibre Fabric	–	–	
42	Rigid Mineralwool	7.00	0.035	
43	Concrete	25.00	1.500	–
44	Rigid Mineralwool	10.00	0.035	
45	Extruded Polystyrene	10.00	0.030	
46	Steel Columns	11.00	60.000	

The roof of the house is insulated with 500 mm of mineral wool. Roof insulation was installed as batts at site.

Stress-skin panels – the American Grand Canyon house

An alternative to constructing a wood-stud wall is to use manufactured stress-skin panels as was done with the Grand Canyon 'Exemplary House' (see Figure 2.2.4).

These panels are a sandwich, comprised of two outside thin sheets enclosing a foam interior. The sheets are usually oriented wooden strandboard (sometimes referred to as chipboard) but can also be plywood. The inner and outer skins give the panels strength and rigidity. Almost any insulating value can be achieved by increasing the width of the wall. With this system the walls of the house are fabricated at a plant and shipped to the construction site.

The panels are an engineered system. Splines, pinned and glued, join the panels without the need for wood studs. Wood thermal bridges occur only at the top and bottom plates, at each corner and around openings, less than half the area of thermal bridging found in conventional wood stud framing. Electrical service is run through pre-cut holes. The construction is inherently airtight and rigid. Foam inserts are cut to fit into various odd openings, such as the joint between the top of the wall and the sloped roof.

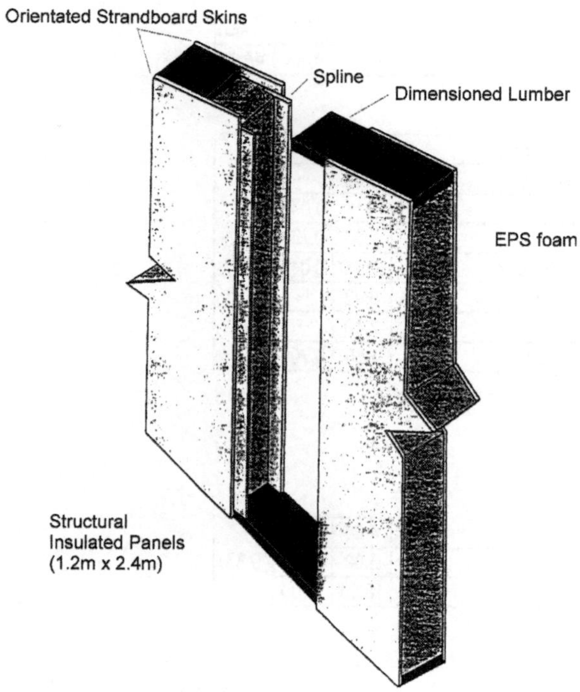

Figure 2.2.4. Stress-skin panels – the Grand Canyon

The large panels are constructed indoors from thin wooden studs, plywood, and mineral wool insulation under strict quality control. The wall elements are delivered complete with polyethylene vapour barrier and interior gypsum boards, windows and doors already in place. The joints between the wall panels are sealed with mastic sealants.

Similarly, the roof elements are delivered with roofing material so that the building may be quickly closed, and all airtightness operations can be performed under dry indoor conditions. The polyethylene sheets are put in on site with lap joints squeezed between laths, and also squeezed against the walls with wooden laths. An additional 50 mm layer of mineral wool insulation is installed under the vapour barrier, and this space is also used for running electrical wires for lamp outlets etc. (Figure 2.2.5).

The insulation thickness is 245 mm in the roof and 200 mm in the walls (thermal conductivity λ = 0.039 W/mK). This is an example of a redesign of an original Task 13 design into low-cost housing, where the combined considerations of cost and loss of interior usable floor area in these small dwellings caused a reduction of the original 300 mm wall insulation to 200 mm The flat roofs of the one-storey dwellings in the north–south running rows (house type B) have an additional layer of rigid 50–100 mm insulation on top. This layer is wedge-shaped to create the slight gradient required to drain the roof surface.

The slab-on-grade floor is cast on 100 mm expanded polystyrene (λ = 0.039 W/mK) on top of 200 mm expanded clay clinkers (λ = 0.13 W/mK). The slab is separated from the lightweight concrete foundation blocks by polystyrene insulation.

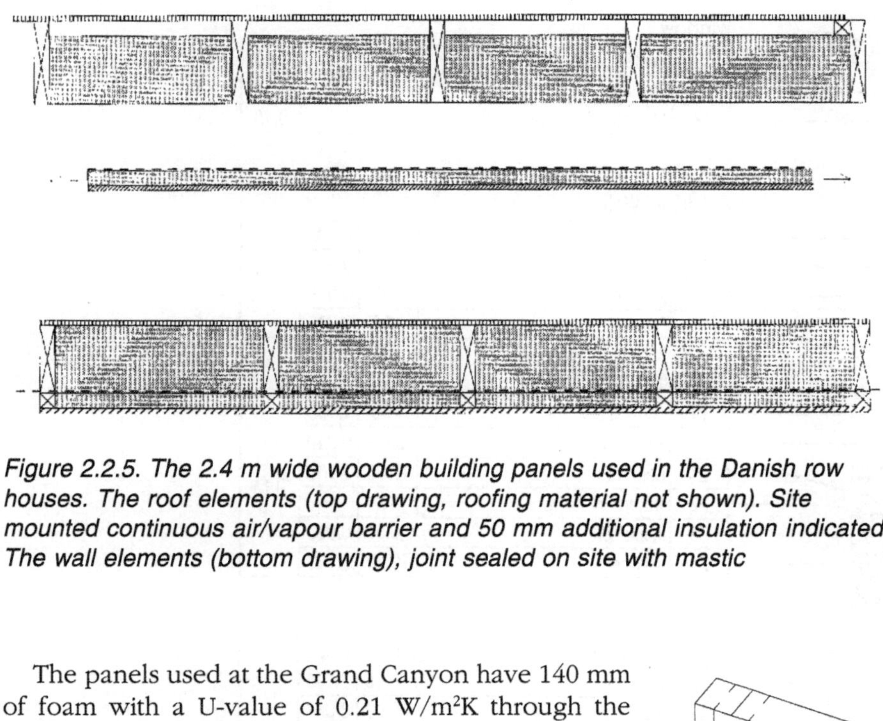

Figure 2.2.5. The 2.4 m wide wooden building panels used in the Danish row houses. The roof elements (top drawing, roofing material not shown). Site mounted continuous air/vapour barrier and 50 mm additional insulation indicated. The wall elements (bottom drawing), joint sealed on site with mastic

The panels used at the Grand Canyon have 140 mm of foam with a U-value of 0.21 W/m^{2}K through the foam. The roof system uses similar panels with 240 mm of foam.

Large wooden insulated panels – the Danish houses

This is another example of the use of large factory-produced building panels (Figure 2.2.5). In this case they are used for façades as well as roofs. The typical room element is 2.4 m wide and spans the room or dwelling – the wall elements are one or two storeys high and at least 2.4 m wide, some of the one-storey elements are up to 5.1 m wide. Most of the structural concrete walls are partition walls or party walls between dwellings, so the façades are not load-bearing.

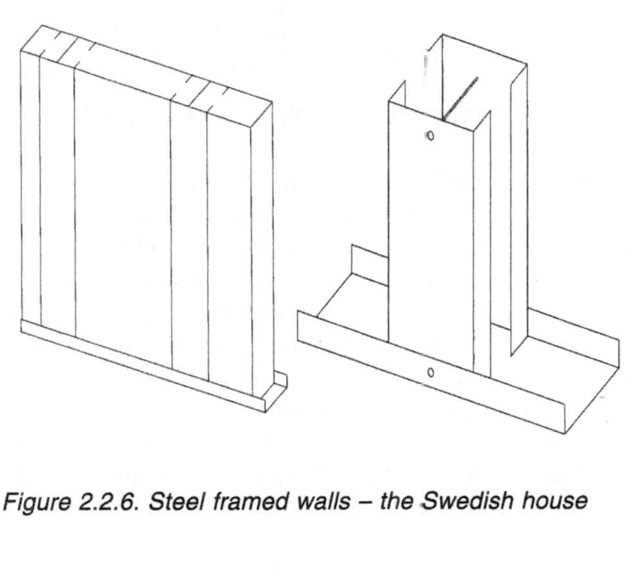

Figure 2.2.6. Steel framed walls – the Swedish house

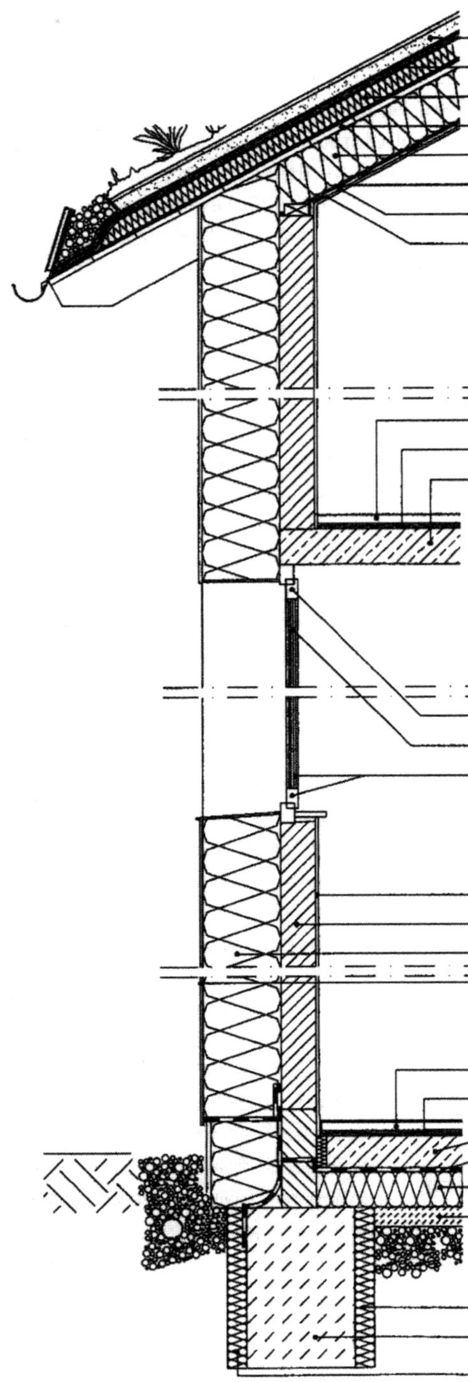

Layer	Description	Width cm	Thermal Conduct. W/mK	U-Value W/m²K
Roof				
1	Earth + Vegetation	11.50	-	0.12
2	Water Proofing	0.02	-	
3	Insulation	10.00	0.035	
4	Wood	2.00	0.130	
5	Insulation + Wood	24.00	0.049	
6	Vapour Barrier	0.02	-	
7	Lattice	2.40	-	
8	Gypsum Board	1.50	0.700	
Ceiling				
9	Screedfloor	4.00	1.400	0.83
10	Insulation	3.00	0.040	
11	Concrete	18.00	2.100	
Window				
12	Sealed Frame			1.40
13	Glazing			0.50
14	Glazing and Frame			0.75
Exterior Wall				
15	Plaster	1.50	0.700	0.11
16	Limestone	17.50	0.990	
17	Insulation	40.00	0.045	
18	Plaster	2.00	0.870	
Basment Slab				
19	Screedfloor	4.00	1.400	0.13
20	Insulation	3.00	0.040	
21	Concrete	18.00	2.100	
22	Water Proofing	0.02	-	
23	Insulation	20.00	0.030	
24	Concrete	10.00	2.100	
Foundation				
25	Insulation	10.00	0.035	-
26	Concrete	60.00	2.100	
27	Insulation	10.00	0.035	

Figure 2.2.7. Super-insulated masonry wall – the Rottweil house, Germany

Steel-framed walls – the Swedish house

There is a growing interest in using steel instead of wood as the house framing material (see Figure 2.2.6). The main concern with this substitution is thermal bridging through the steel members. The Swedish Röskär house uses an innovative system of steel channels and expanded polystyrene insulation to overcome this concern. The steel channels clip the foam panels together on the inside and outside but do not extend through the entire insulation slab. The system is strong, easy-to-assemble and has a high insulating value.

Super-insulated masonry wall – the German Rottweil house

The 'Ultra House' in Rottweil shows how super-insulation can be achieved with masonry construction (see Figure 2.2.7). The exterior walls are constructed with 175 mm limestone blocks and 400 mm mineral wool insulation on the outer side with the surface sealed with plaster. The mineral wool is glued to the blocks.

The roof is constructed with rafters. Between the rafters there is 250 mm thick mineral wool insulation, while above the rafters there is a layer of wood boards

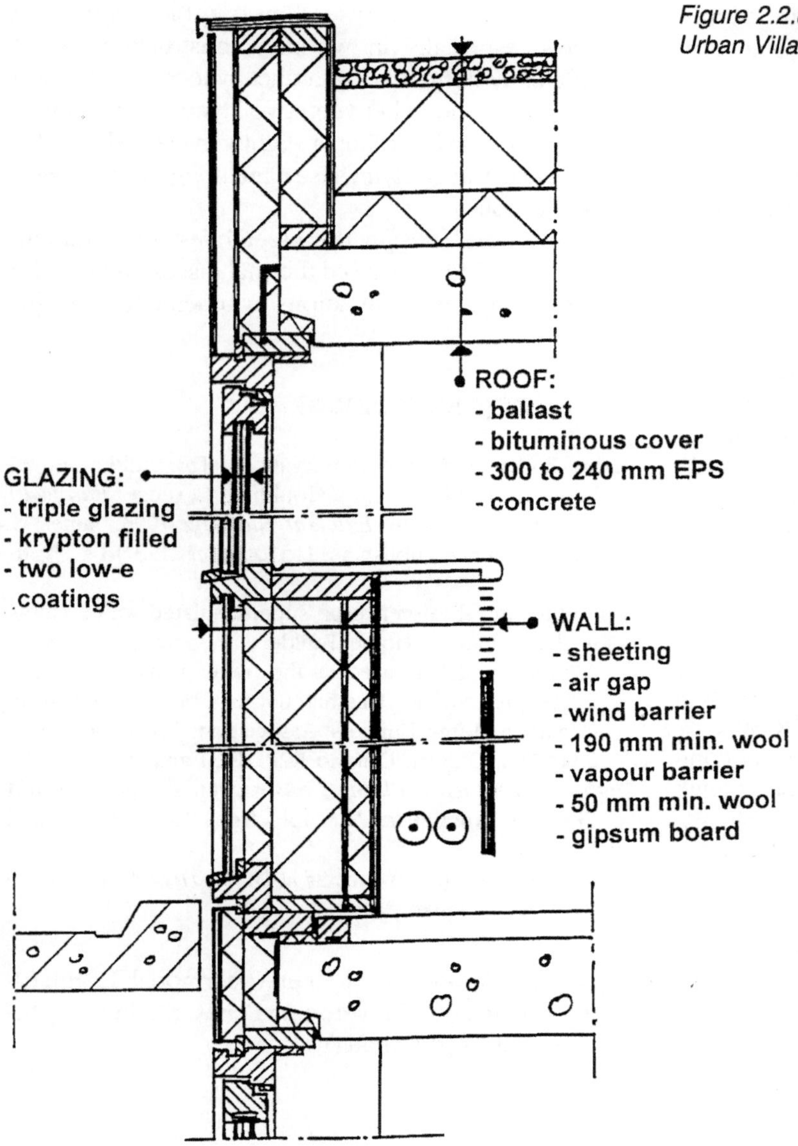

Figure 2.2.8. Construction assemblies o' the Amstelveen Urban Villa, the Netherlands

which supports an unbroken layer of 100 mm rigid insulation. Under the rafters there is an aluminium vapour barrier. The outer layer of the roof is finished off with vegetation and consists of (from top to bottom):

- vegetation with clawed webbing
- soil
- clawed webbing
- sealant against roots.

The floor slab has 200 mm of expanded polystyrene sandwiched between two layers of concrete. The foundation is insulated with 100 mm of extruded polystyrene insulation on both sides to minimize thermal bridging.

External insulation on a high-rise building – the Netherlands house

The 'Urban Villa' in Amstelveen is a good example of how super-insulation can be incorporated into high-rise residential buildings (see Figure 2.2.8). The building consists of a concrete load-bearing structure (floors, ceiling, and common walls insulated on the exterior with at least 200 mm expanded polystyrene). The north-facing exterior wall is wood framed with 38 x 140 mm studs and mineral wool. 50 mm of mineral wool is added on the interior side to increase the thermal resistance and minimize thermal bridging. The south-side façade is an integrated wall element of windows, doors, vents, and opaque sections. The opaque sections are insulated with 240 mm of mineral wool. An innovative system was developed for the balconies and stairtowers to eliminate thermal bridging through the floor. These elements have their own foundation with only minimal connections to the main building. This configuration allows insulation to be run between the floor and the balcony to maintain a highly insulated envelope.

Conclusions

The Task 13 experience of designing super-insulated building envelopes may be summarized as follows:

- Do not rely on standard practice. Design for efficiently insulated and airtight constructions and assemblies

- Use at least twice the amount of insulation prescribed in building codes or used in standard practice
- Use low-conductivity insulation in walls to limit the space needed (minimize loss of interior floor area).
- When using batt insulation (or rigid boards) use several layers with staggered joints to avoid air voids and make sure that the insulation fills out the available space
- Make wooden constructions airtight by incorporating into the interior part of the wall/ceiling a continuous sealed air (and vapour) barrier, e.g. a polyethylene sheet with squeezed lap joints or lap joints effectively taped with non-drying sheathing tape. Pay particular attention to 2D and 3D conditions at junctions, corners, window, and door openings etc. Place the barrier 50–75 mm under the surface finishing (even up to one third of the total insulation thickness) for protection of the barrier and for easier electrical wiring
- Use the inherent airtightness in masonry or panelled constructions, either in the material (e.g. concrete) or in airtight pargings (e.g. stucco), but be sure to establish airtight connections at changes in construction principles, e.g. to seal a ceiling polyethylene air barrier to a concrete wall, preferably by squeezing the polyethylene against the wall by slats tacked or screwed onto the walls, with a sealant to compensate for warping
- In panelled constructions (with inherently airtight panels) where sealants or mortar are used in the joints, the panels must be mechanically fixed to ensure long-life airtightness of the joints
- Plan for the necessary lead-ins of cables, pipes, and ducts, especially in wooden constructions, most efficiently done for, say, a duct by locally squeezing the polyethylene between two layers of wooden or gypsum boards, drilling a slightly oversized hole for the duct, and applying a sealant along the perimeter of the hole
- Analyze and design the assemblies for minimum thermal bridges to avoid thermal discontinuities (be sure to insulate foundations or basement walls, for example).

RECOMMENDED READING

Saxhof B (1992). Energy Conservation – The Building Envelope. In Lewis O. and Goulding J. (ed.), *European Directory of Energy Efficient Building* 1993. James & James (Science Publishers) Ltd, London. (ISBN 1-873936-14-1)

A good technical reference on super-insulated wall systems is the Canadian Home Builders' Association *Builders' Manual,* which is used as the training manual for the R2000 programme. This manual may be obtained from: Canadian Home Builders' Association, 200 Elgin Street, Suite 702, Ottawa, Ontario K2P 1L5, Canada.

High-Performance Building Construction Details: The IEA Task 13 Experience. Available from Task 13 National Contacts.

Reports from IEA *BCS Annex 24 HAMTIE (Heat And Moisture Transfer In Envelopes).*

- *Author.* Stephen Carpenter (Enermodal Engineering Limited, 650 Riverbend Drive, Kitchener, Ontario N2K 3S2, Canada)

2.3 High-performance windows

Introduction

High-performance windows are an integral part of an energy-efficient building envelope. High-performance windows can be defined as any window system that has a U-value of 1.5 W/m^{2}K or lower (half the value of conventional double-glazed windows). In addition, the windows should be designed for high solar and visible-light transmission and low air leakage. Low heat-loss values are achieved through a combination of multiple glazings, low-emissivity coatings, inert-gas fills, insulating edge spacers, low-conductivity frames and insulating shutters. High solar and daylight transmission is achieved by carefully selecting glazing systems and coatings, and by minimizing the frame area.

Concept definition

A high-performance window must have low heat loss and air leakage, but high solar transmission.

Several technologies are available to reduce radiative and convective heat transfer between glazings. Radiative heat loss is reduced by choosing glazings with a low-emissivity (low-E) coating. Low-E coatings can be applied to the inner surfaces of each glazing (on glass panes or thin plastic films). These coatings can reduce radiative heat transfer by up to 96% (i.e. $e = 0.04$). There are two types of low-E coatings: sputtered (soft) or pyrolytic (hard). In general, soft coats have lower emissivities (0.04 to 0.15), but also lower solar transmission. They are extremely fragile, and exposure to water and physical contact (even fingerprints) must be completely avoided during assembly. Hard low-E coatings have emissivities of about 0.2, but they are more durable, and some of them have almost no effect on solar transmission. Some hard coatings with lower emissivities, e.g. 0.12–0.15, as used in the Norwegian house, do reduce the solar gain appreciably, however.

Convective losses are reduced by replacing the air between the panes with an inert gas. Argon is the cheapest and most common. Krypton, and more recently xenon, are sometimes used. They are more expensive, but provide lower convective losses and have a much smaller optimum glazing spacing. The use of low-E coatings without gas fill is a false economy, because decreased radiative heat transfer is offset by increased convective heat transfer.

The most straightforward approach to achieving a high-performance glazing system is to use multiple panes of low-E films and gas fills. The number of radiative and convective transfers occurring in series is thereby increased, reducing the overall U-value. Each extra layer, however, reduces solar gains and adds to the bulk of the window unit. Nevertheless, this is a reliable route to a very low window heat loss. The bulk may be reduced by using transparent films for the inner layers. In warm countries, double-glazed windows are used, whereas in cold climates designers opt for triple and quadruple glazing systems.

There are other technologies to reduce glazing heat loss. Convective heat transfer can be reduced by filling the cavity with plastic or glass honeycomb-like structures, often referred to as transparent insulation. Aerogel can also be inserted between panes of glass to reduce heat loss. Aerogel is a combination of glass and very small air pockets to produce a very low-density, low-conductivity glass. The use of transparent insulation materials in windows or 'daylight walls' is discussed in Chapter 2.4. A promising emerging technology is evacuated windows. The space between the two panes of glass is evacuated and held apart by microspheres of glass. One or two low-E coatings reduce radiative heat loss. Sealing the edges in a durable way without creating a major thermal bridge is the main development problem.

Multiple layers of inert gas fills and low-E coatings are so effective that the glazing typically has a lower heat loss than the frame and edge components of a window. Considerable attention is now, therefore, being devoted to improving the insulating properties of frames, seals, and pane spacers. In conventional windows, aluminium edge spacers are used to separate the panes of glass and provide a sealed unit. While the aluminium provides a good seal, it transmits a great deal of heat. The band of condensation at the bottom of the windows on cold days is due in large part to the aluminium spacer bar. Many manufacturers are switching to insulating edge spacers to reduce heat loss and condensation. Common insulating spacers include silicone foam, butyl rubber, and thermally broken aluminium.

Frame thermal performance can be improved by using low-conductivity materials, minimizing frame area and reducing thermal bridging within the frame. There is a trend away from aluminium windows (with or without thermal break) towards lower-conductivity materials such as vinyl, wood and, most recently, fibreglass. Fibreglass frames are similar in appearance to vinyl frames but are made from fibreglass fibres in a pultrusion process. Window-frame heat loss can also be reduced by minimizing frame height (i.e. the projected frame area) and avoiding frame projections (which act like fins). Reducing the frame height also increases the amount of glass (for the same overall window size) and the solar heat gain. Thermal bridging in frames can be reduced by avoiding metal cladding and reinforcing. Recently, some manufacturers have been incorporating insulation into the frame. In vinyl and fibreglass frames, insulating foam is injected into the cavities. For wood windows, manufacturers incorporate rigid foam pieces into the non-exposed portions of the frame.

With this vast array of window technologies, it is difficult to gauge the thermal performance of the window just by looking at it. Several computer programs have been written to calculate the window thermal properties. For example, the VISION and FRAME programs are used in many countries to perform a one-dimensional calculation of glazing heat loss and a two-dimensional calculation of frame and edge heat loss.

Variations

Advanced window technologies are used in all of the Task 13 houses. Colder-climate houses generally opted for triple- and quadruple-glazed windows with inter-pane gas fills of argon or krypton and two low-emissivity (low-E) coatings. Windows in the warmer climates tend to be double-glazed with a low-E coating and argon gas fill. Figures 2.3.1 and 2.3.2 compare the window performance characteristics of the windows used in the Task 13 houses.

For the most part, the designers opted to achieve low heat loss through multiple layers of glass as opposed to aerogel, transparent insulation, or evacuated glazings. In general, multiple glass layers could achieve lower heat loss and higher solar transmission at a lower cost than the other glazing technologies (cf. Chapter 2.4). Nevertheless, there are many window innovations used in the case-study houses.

Although most of the houses have wood frames, three projects feature 'insulated' framing systems. The Canadian Waterloo house has pultruded fibreglass frames with the interior cavities filled with polystyrene insulation. A low-conductivity edge spacer made from silicone foam reduces edge-of-glass heat loss. The Finnish house has foam insulation on the interior parts of the frame to obtain the same insulation as in the quintuple glazing. A prototype foam-insulated sandwich frame construction in the Danish Task 13 design (cf. Chapter 3.5) was not used in the building project because of liability problems.

Four examples of efficient window systems have been chosen for presentation below.

The Canadian Waterloo house

The windows used in the Waterloo Region Green Home are triple-glazed with low-emissivity coatings on surfaces 2 and 5 (e = 0.16), counting the surfaces of panes and films from the outside. The 12 mm glazing cavities are filled with argon gas. The edge spacer is a

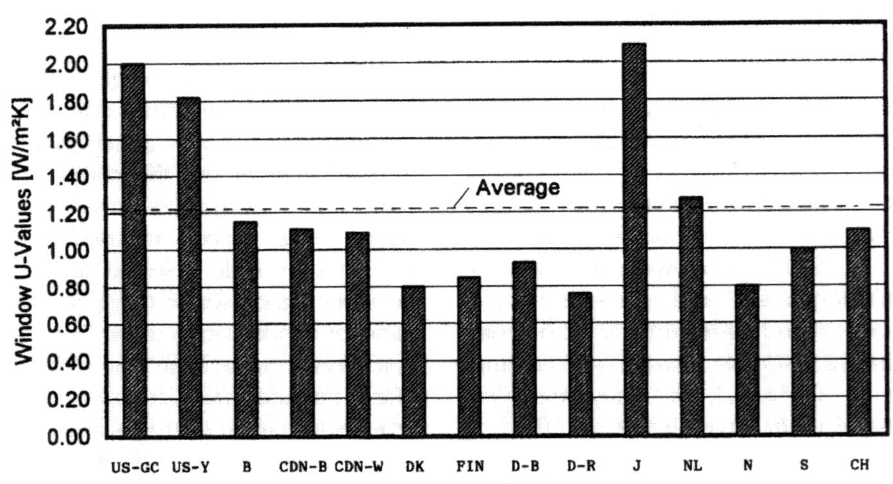

Figure 2.3.1. Overall U-values of the window systems in the Task 13 houses

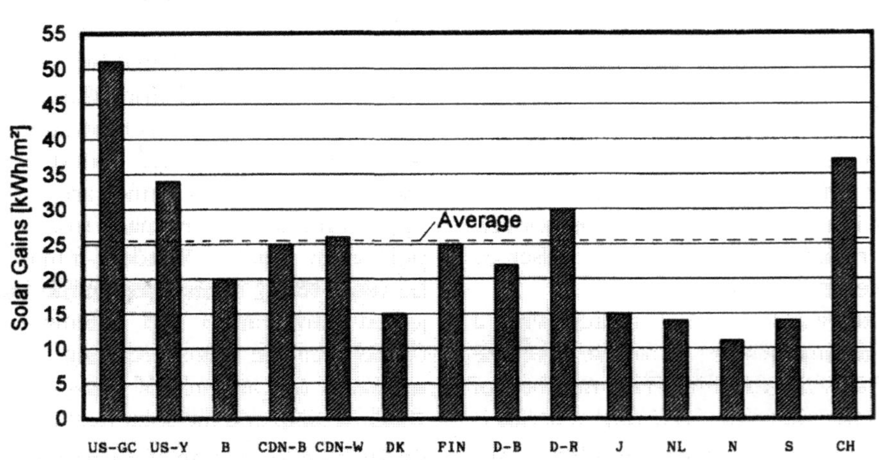

Figure 2.3.2. Calculated usable solar gains in the Task 13 houses

silicone foam. The framing system is made from pultruded fibreglass (wall thickness 3 mm). The cavities within the frame are filled with polystyrene insulation to reduce further window heat loss. The U-value of the frame is 1.2 W/m²K, and the total window U-value is just over 1.0 W/m²K.

The Finnish house

The windows in the Pietarsaari house are quintuple-glazed: quadruple-glazed sealed units (double-glazed, krypton-filled, with two low-E films) with an exterior storm glazing. The window frames are sandwiched with sections of foam insulation to reduce frame heat loss (Figure 2.3.3).

The German Rottweil house

All external glazing, including the sunspace glazing, in the German Task 13 house in Rottweil is triple-glazed with a xenon gas filling (Figure 2.3.4). The framing is pine and accounts for approximately 28% of the window area. The glazing has a U-value of 0.5 W/m²K, while the total window has a U-value of 0.75 W/m²K.

The Swedish house

The special window system concept used in the Swedish Röskär house has 1.1–1.2 m wide and 1.1–1.2 m deep glazed boxes attached to the external wall (Figure 2.3.5). The 2.1–2.7 m high boxes contain a movable reflector system. The concept is based on the following five strategies:

- standard glazing, i.e. non-low-E-coated, should be used in all but north-facing windows to allow high light transmission

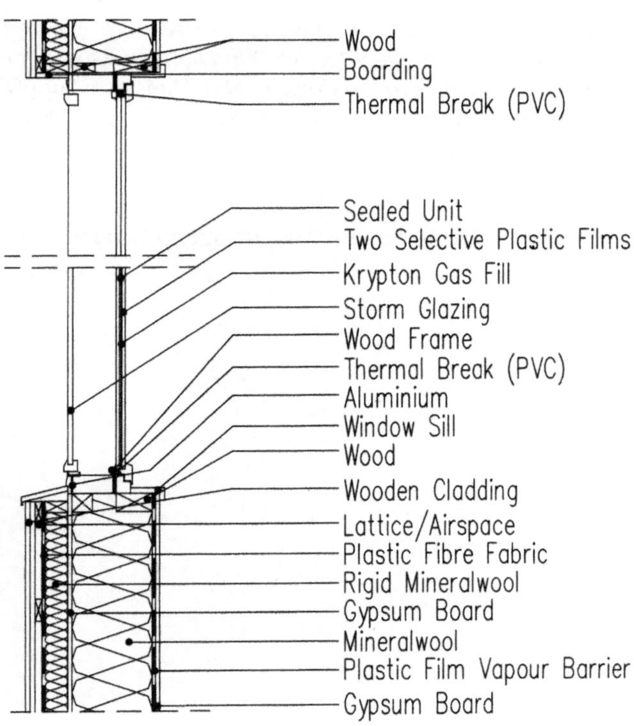

Figure 2.3.3. Multiple-glazing window system with insulated frames – the Pietarsaari house

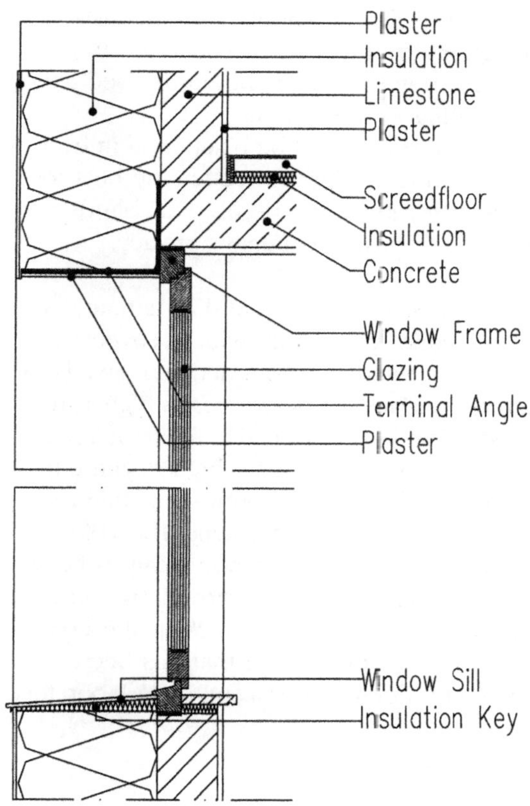

Figure 2.3.4. Window detail from the Rottweil Ultra House, Germany

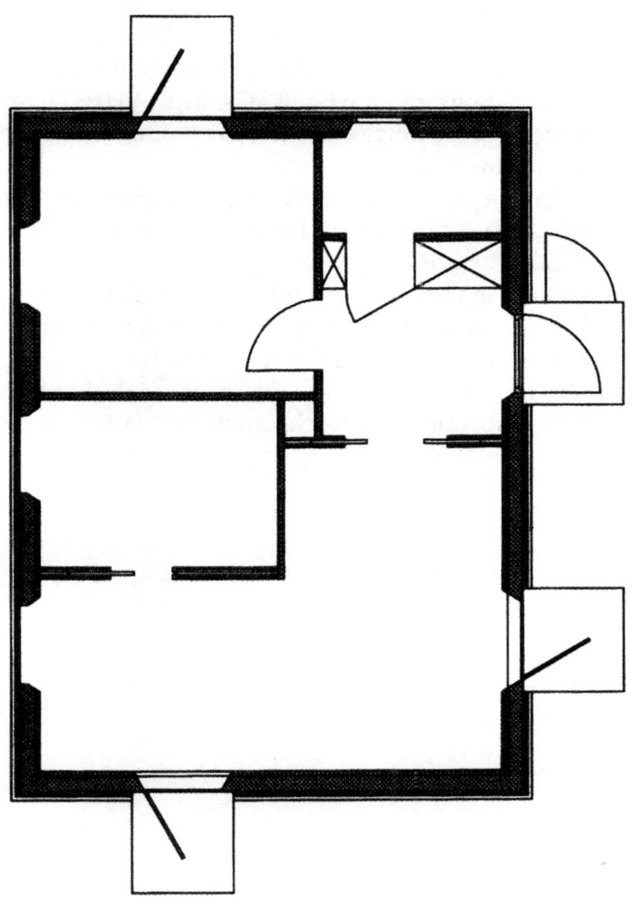

Figure 2.3.5. Floor plan of the Röskär house showing the special window system

- an outer reflector should be used to enhance daylighting in the winter
- the same outer reflector should be used to prevent overheating in the summer
- the reflector(s) should be positioned in front of the wall opening at night to reduce the heat loss
- the reflector(s) should be used 'as blinds' (parallel position) to ensure occupant privacy.

The reflector is protected from dirt, rain, and wind by the single-pane glass box which covers a normal double-glazed window positioned in the wall opening. The interior surface temperature is as high as for a low-E coated window because the reflector reduces the radiation heat transfer. A very high daylight transmission is possible because of normal glass being used and because of reflector enhancement. The colours of the light are only slightly distorted by two panes of clear glass.

The movable reflector is carried by motor-driven wheels running on a Velcro track in the box (Figure 2.3.6), a fairly simple system that has been tested for reliability and durability in a full-scale laboratory test.

Figure 2.3.6. Prototype of the glass box including the reflector with wheels on the top

Conclusions

The Task 13 experience of designing high-performance windows may be summarized as follows:

- Do not rely on standard practice, which focuses one-sidedly on the heat loss factor; the solar heat gain is also important. However, the Task 13 designs all called for highly insulating windows in most of the buildings, often with at least triple-glazed units with low-E coatings and inert gas fillings, albeit the solar transmittance is lower
- Calculate and analyse the heat balance for the windows – both in the heating season and in the cooling season
- Include in the analyses (at least) 2D calculations of heat loss at the window perimeter (sealed unit edge, frame, and surrounding wall assembly)
- Use large window units (low edge-to-glazing-area ratio) and 'warm-edge technology' (insulated spacers and frames) to minimize perimeter heat losses
- Consider special solutions using non-coated glazings, if the quantity and quality of, especially, winter daylighting is an issue (as it may be in some northern climates).

RECOMMENDED READING

Aschehoug Ø, Thyholt M, Andresen I (1995). *Frame and Edge Seal Technology – Design Guidelines.* IEA SHC Task 18 Working document T18/B9/WD3/95. Department of Architecture, Norwegian University of Science and Technology, N-7034 Trondheim, Norway.

Button D, Pye B (ed.) (1993). *Glass in Building. A Guide to Modern Architectural Glass Performance.* Butterworth Architecture, Oxford. (ISBN 0-7506-0590-1)

Lampert C M (ed.) (1993). *Glazing Materials for Solar and Architectural Applications. A Final Report of IEA SHC Task 10.* LBL Report 34436, Lawrence Berkeley Laboratories, California, USA.

Window Innovations '95 Conference Proceedings. A World Conference on State-of-the-Art Window Technologies for Energy Efficiency in Buildings. (1995). Minister of Supply & Services Canada, Catalogue No. M91-7/332-1995E. (ISBN 0-660-16085-4)

- *Authors.* Bjarne Saxhof and Stephen Carpenter

2.4 Transparent insulation

INTRODUCTION

Transparent insulation can be described as materials that have a high thermal insulation and a high solar transmittance. According to general practice, the term includes materials that technically speaking are not visually transparent, but translucent.

Transparent Insulation Materials (TIM) are normally classified according to their geometric structure. Figure 2.4.1 shows a classification based on four different geometric structures.

The parallel type consists of plastic films or glass panes, possibly with infrared-reflective coatings. Within the vertical or perpendicular structures we find the so-called honeycomb structures, capillaries, and parallel slats. A honeycomb material has angular plastic tubes. A capillary material consists of circular plastic or glass tubes placed perpendicular to the transparent cover or to the absorber. The cavity structure is a combination of the parallel and the perpendicular types. The convection loss is suppressed, as for the perpendicular structure, and the reflection loss is large, as for the parallel structure. The last type, the quasi-homogeneous structure, is characterized by its solar absorption and scattering properties. Silica aerogels and glass fibres may be attributed to this group.

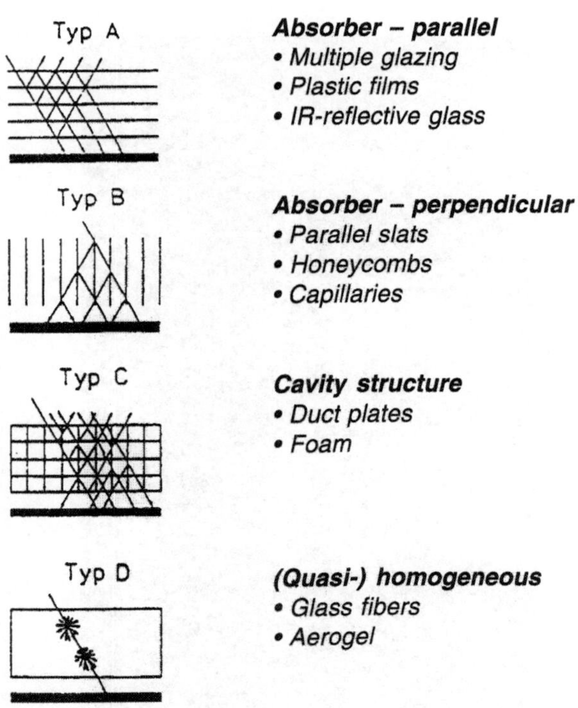

Figure 2.4.1. Four geometric TIM structures: the parallel structure, the perpendicular structure, the cavity structure, and the (quasi-)homogeneous structure

Solar and thermal properties for some TIMs are given in Table 2.4.1.

Concept definition

Transparent insulation materials are used in windows, on walls exposed to direct solar irradiation, and as insulation for absorbers in solar collectors.

Since none of the existing TIMs have a transparency acceptable where the view is an issue, TIM glazings are mainly used in clerestory windows and skylights, or where a clear view may even be a problem, as in bathroom windows. TIM glazings are particularly well suited for skylights where condensation and dripping may otherwise be a problem. Vertical translucent glazings are also called daylight walls.

Application of TIMs externally on walls, usually mass walls, instead of glass (as used in Trombe walls) lowers the heat loss from the walls, but also decreases the solar gains because of the lower transmittance. In northern climates, however, the use of TIMs increases the performance significantly. TIMs are also used on insulated walls, with an air gap between the TIM and the wall, ventilated to the building in an open or closed loop. In this context, these walls are referred to as solar walls.

In solar collectors, the use of TIMs in the transparent cover increases the efficiency significantly, especially at high system temperatures.

For some applications, high performance TIMs can replace the opaque insulation normally used on the

Table 2.4.1. Data for some typical constructions with transparent insulation materials (TIMs)

TIM construction	Total solar energy transmittance	Heat loss coefficient (W/m²K)
Eight layers of corrugated cellulose acetate sheets between two layers of glass	0.44	1.2
200 mm of capillary material between two layers of 6 mm glass panes	0.71	0.52
40 mm of granular aerogel between two layers of 6 mm low-iron glass panes	0.34	0.51
20 mm of evacuated monolithic silica aerogel between two 4 mm glass panes	0.73	0.51
Reference values		
Double glazing, 12 mm air space	0.75	3.05
Double glazing, with one low-E and argon	0.65	1.60

back and sides of the absorber and enable the system also to collect heat from diffuse and reflected irradiation.

Variations

During the early phases of Task 13, some design teams considered the use of transparent insulation materials for their low-energy buildings (Table 2.4.2) and investigated materials and systems experimentally, or through computer simulations. The ability of existing simulation programs to analyze transparent insulation was also assessed in the Task.

The following sections describe the main uses of TIM that were investigated and show examples of their implementation.

Windows (i.e. transparent components) and translucent components

The original design for the Danish houses at Kolding incorporated aerogel windows, i.e. double-glazed partially evacuated units filled with 20 mm monolithic silica aerogel, giving a centre of glass U-value of 0.5 W/m²K, and a total solar transmittance of 0.73 (Table 2.4.1). The prototype aerogel windows are not perfectly transparent, so the glazing units were designed for use in large skylights, a total of 6 m² transparent window area insulated with 0.6 × 0.6 m² aerogel panels (presently the largest available size). When the building project was re-designed to have smaller and cheaper dwellings, the two Task 13 units no longer had skylights.

Initially, the design for the Netherlands house included TIM windows of the honeycomb type between two panes of glazing in a wood frame. These windows were located in the upper part of the south façade on each floor. Simulations showed no significant difference in the heating load if TIM windows were replaced with super-glazing. The main reason to include TIM in a small strip of upper windows was the visual aspect of daylighting, where TIM windows were considered an interesting architectural feature.

Unfortunately, during the cost-reduction phase of the design process the TIM windows were replaced by super-glazing. During the component testing TIM windows were still included in the design. They were a part of the full-scale outdoor tests on the performance of the south façade (Figure 2.4.2). However, the tests focused on the total performance of the façade element and no explicit results for the TIM windows came out of the test.

Table 2.4.2. Studies and uses of TIM within the Task 13 work

	Considered use of TIM	Decided to use TIM	Ways to be used
Denmark	Yes	No	Solar walls Some windows
Finland	Yes	No	Solar walls Solar collectors
Netherlands	Yes	No	Some windows
Norway	Yes	Yes	Some windows

In the Norwegian house, a 10 m² south-facing TIM wall was installed as a daylighting element for the sunspace (Figure 2.4.3). The daylight wall was constructed from four 2.1 × 1.0 m² panels of 122 mm of capillaries between two 6 mm glass panes. This gives a total transparent area of 8.4 m² with a U-value of 0.8 W/m²K.

After one year of occupation the users of the house were interviewed about their experiences with the TIM wall. The occupants found the daylight in the sunspace comfortable, they did not have problems with overheating, and they liked the appearance of the daylight wall.

Norwegian studies showed that a daylight wall can replace some of the windows in a dwelling where visual contact is not wanted. A daylight wall can also replace part of an opaque wall if more daylight is needed. When a daylight wall with capillary materials or granulated silica aerogel replaces conventional windows, energy savings are achieved. When a daylight wall replaces a well insulated wall on the south façade, the heating demand is not affected. Replacing walls on

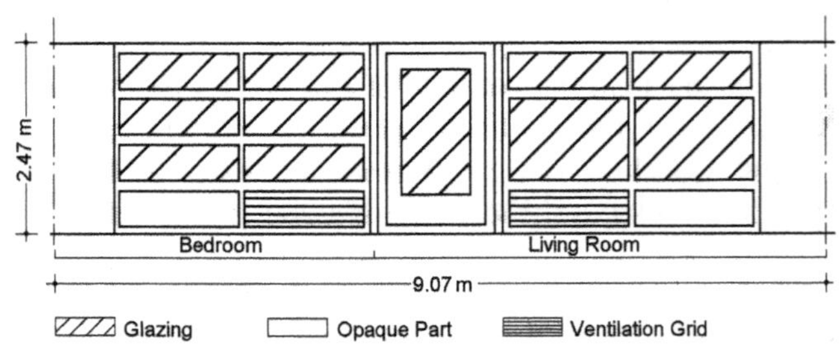

Figure 2.4.2. Façade element (south façade) for the Amstelveen Urban Villa. The four window panes along the top were planned as TIM daylight windows

Figure 2.4.3. The south façade of the Hamar house

the north façade, however, will increase the heating demand. From these results it can be concluded that daylight walls with TIM at high latitudes can offer both energy savings and increased daylight penetration into the dwelling.

Solar walls (TIM walls exposed to direct solar radiation)

A solar wall (mass wall) consists of three parts: a wall with high thermal capacity, e.g. brick or concrete, an absorbing surface, e.g. dark paint, and a transparent cover, e.g. a TIM (Figure 2.4.4). Solar radiation will pass through the cover and be absorbed by the dark surface, and the heat will accumulate in the wall. Ideally, the heat is stored in the wall during the daytime and released to the living spaces during the night-time. In contrast to windows, there is no need for transparency. There is, therefore, a wider selection in transparent insulation materials for solar walls.

In the analyses done for the Netherlands house, honeycomb materials applied on walls were found not to be able to compete with external opaque insulation, based on considerations of future cost versus benefits.

Some TIM walls have been tested in Danish indoor test facilities for passive solar components. 'Solar irradiance' from a solar simulator on solar walls was measured throughout one day and used for validation of the models and calculation programs. The validated models were then used for calculation of the annual saving in solar walls in real buildings. The performance was calculated for a dwelling with a floor area of 70 m^2 with 7 m^2 south facing solar walls, placed on a normal brick wall. The savings obtained in a northern climate were 65 kWh/m^2 with a single sheet of glass (reference savings), 168 kWh/m^2 with one layer of glass and 100 mm honeycomb material, and 211 kWh/m^2 with sealed partially evacuated units filled with 20 mm monolithic silica aerogel.

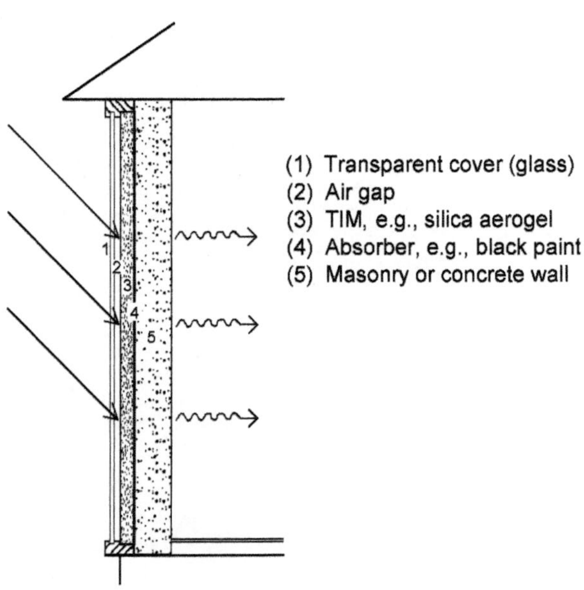

Figure 2.4.4. Schematic section in solar wall

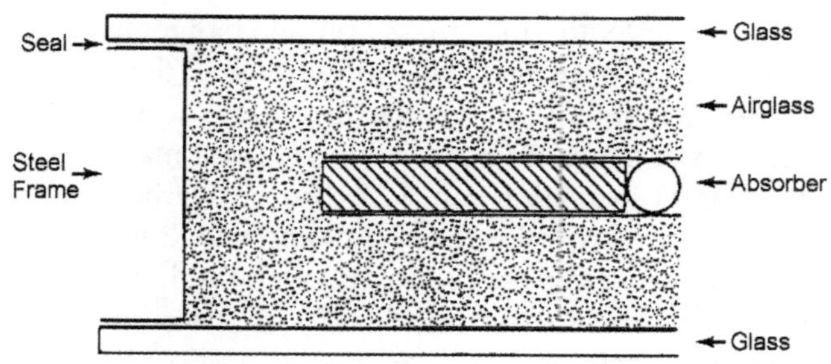

Figure 2.4.5. Cross-section of an aerogel solar collector

A sensitivity study for the Danish climate showed that even if the area of the solar wall was doubled, the output would be practically the same. The solar wall was competing with a super-insulated wall with 300 mm opaque insulation. In the design of the Danish houses, the use of approximately 5 m^2 south-facing TIM walls was considered.

Solar collectors

Transparent insulation has a considerable influence on the output of solar collectors. A new method of making solar collectors insulated with monolithic silica aerogel was evaluated for use in the Danish project.

The solar collector shown in Figure 2.4.5 consists of a black painted absorber, two tiles of aerogel and two pieces of glass. The whole construction is sealed hermetically and partially evacuated. The calculated efficiency and the measured efficiency for a 1.3 m^2 prototype was high, especially in the high-temperature range, where normal collectors have a fairly poor performance.

This type of solar collector is, of course, especially well suited for purposes where high temperatures are needed, e.g. in domestic hot-water systems and in district heating systems with or without seasonal heat storage. Calculations have shown that aerogel solar collectors under Danish conditions have an output of about 700 kWh/m^2 at a temperature of 60°C. This performance is twice that of the high-efficiency flat-plate solar collectors produced today.

In the Danish Task 13 house, this type of solar collector was considered for a combined domestic hot-water and space-heating system connected to the local district heating system, but the concept was abandoned for economic reasons.

Discussion and conclusions

Although the energy performance of transparent insulation used in different ways in laboratory experiments and computer simulations was quite good, transparent insulation materials were only used in one of the Task 13 houses, the Hamar House in Norway.

The main reasons for this were cost, durability (builders' liability), and risk of overheating. The combined effect of those concerns caused transparent insulation in walls to lose out to opaque super-insulation.

When transparent insulation is used in windows, the direct solar gains are very high, and full utilization requires large thermal mass, phase-change material storage, or even actively charged storage. If the storage is insufficient, severe overheating may occur. Even when used as exterior insulation on solid mass walls, transparent insulation can easily cause overheating in an otherwise super-insulated building with low-U windows. However, these problems can be reduced to an acceptable level if shading devices are used or if it is possible to achieve good ventilation on hot days.

Some of the materials and systems tested experienced partial or complete failure due to high system temperatures, for example stagnation in high-efficiency solar collectors insulated with plastic TIM.

Several of the systems were quite expensive and thus showed a poorer cost–benefit ratio than opaque super-insulation or more conventional high-performance windows. As well as the high cost, the aerogel windows described were still experimental prototypes and technically not ready for installation in real dwellings.

However, the Task 13 work demonstrated that the use of transparent insulation in buildings is both technically and practically viable. User satisfaction studies conducted on the Norwegian house showed that the occupants found the translucent wall pleasing with respect to daylight, comfort, and appearance.

RECOMMENDED READING

Goetzberger A (1991). *Transparent Insulation Technology for Solar Energy Conversion.* Fraunhofer-Institut für Solare Energiesysteme, Freiburg, Germany.

Jesch L F (ed.) (1986–1994). *Proceedings of the International Workshops and Meetings on Transparent Insulation Technology TI1–TI7.* The Franklin Company Consultants Ltd, Birmingham, UK.

Lien A G (1995). *Transparent Insulation Materials for Low Energy Dwellings in a Cold Climate.* PhD Thesis 1995:70, Norwegian Institute of Technology, Trondheim, Norway. (ISBN 82-7119-817-3)

- *Author.* Bjarne Saxhof

2.5 Ventilation heat-recovery systems

Introduction

In very well insulated and airtight dwellings, such as the those studied in Task 13, the heating of ventilation air may represent more than 50% of the total heating energy demand. Therefore, recovering heat from exhaust air is an important challenge. This normally implies the installation of a mechanical ducted system for exhaust or balanced ventilation, including a device in which part of the heat in the exhaust air is transferred to the incoming fresh air, or to the domestic hot water.

An airtight building envelope is a prerequisite for efficient heat recovery in balanced ventilation systems, i.e. systems where both supply and exhaust air are controlled, and usually balanced to a slight depressurization of the building. The guide on energy-efficient ventilation published by AIVC (the IEA Air Infiltration and Ventilation Centre), indicates that at a pressurization/depressurization of 50 Pa, an air change rate (n_{50}) below 3 per hour (a.c.h.) (in some countries, e.g. Switzerland and Sweden, even below 1.0 per hour) is recommended or required.

Concept definition

The following ventilation heat-recovery technologies were used in the Task 13 houses:

- air-to-air plate heat exchangers
- air-to-air thermal wheels
- heat exchange with the ground
- heat exchange in building components or rockbeds
- air-to-air or air-to-water heat pumps.

The heat-pump ventilation systems and a combined furnace flue heat exchanger/regenerative rockbed heat-recovery system used in Task 13 are parts of integrated mechanical systems and are described in Chapter 2.11. Heat exchange with the ground may be a part of a ventilation heat-recovery system and is described in Chapter 2.6. The special use of hollow-core concrete floor panels used in the Italian building is described in Chapter 3.11.

In all the ventilation heat-recovery systems, one or more fans are used to move the air. It is therefore important to design for minimum use of parasitic energy (electricity), as the systems are often operated all year (or at least all through the heating season), 24 hours a day.

Variations

Heat-recovery systems are used in all Task 13 houses (Table 2.5.1). All systems except one (in one of the Danish rowhouse types) are systems for individual dwellings. In most cases, the system is used for the whole dwelling, in one case (the German Berlin house) a separate range hood not connected to the heat-recovery system is installed in the kitchen, and in another case (the Finnish house) two separate systems are used, one for the house and one for the range hood. Use of separate range hoods or kitchen systems may upset the operation of a balanced system, but has the advantage of adding less dirt (and fire-hazardous grease) to the heat-recovery system, which can be important, especially to plate heat exchangers. Additionally, it offers the possibility of increased exhaust from the kitchen when necessary (which may, however, also disturb the balance during operation). System cleaning possibilities exist in all cases. In one system the total air flow can be increased by manual controls located in the bathrooms.

In balanced ventilation systems, the heat-recovery efficiency can be very strongly dependent on user behaviour, as open windows create an unintended bypass of the heat recovery. A ducted bypass for summer use in the balanced ventilation systems is installed in one case only, but in some projects the mechanical air supply (the Danish houses, the Swedish house) or the whole system (the German Berlin house) is designed to be stopped in the summer and the windows or special vents opened instead.

In two cases the sunspace is used for pre-heating the ventilation air, although in both cases it is mentioned that this is mainly done for comfort reasons and not so much for energy reasons. The system is sometimes

Table 2.5.1. Overview of heat-recovery systems in the Task 13 houses

	Building[a]														
System	B	CDN-B	CDN-W	DK	FIN	D-R	D-B	I	J	NL	N	S	CH	US-GC	US-Y
Plate, cross-flow	*X*				*K*	*B*	*X*		*X*	*2X*		*K*	*X*		
Plate, counterflow				*X*											
Thermal wheel					*X*						*X*				
Heat pump (in IMS)		*X*									*X*			*X*	*X*
Miscellaneous		*S*	*F*			*G*		*H*		*S*					

[a] *B* = With bypass; *F* = Furnace/rockbed; *G* = Ground pre-heating; *H* = Solar assisted hollow-core; *K* = Kitchen range hood; *S* = Sunspace pre-heating; *2X* = 2 units in series

installed in an unheated space with the ducts well insulated; in other cases it is installed in a heated space.

In most cases, standard products are used, although more or less modified. In the Netherlands house, two heat exchangers are used in series in order to increase the thermal efficiency.

Plate heat exchangers

For dwellings, cross-flow and counterflow plate heat exchangers are the most common type of heat-recovery systems (Figures 2.5.1 and 2.5.2). Plate heat exchangers are fairly simple so-called static devices, with no moving parts, and thus require minimal maintenance, chiefly cleaning of filters. Driven by one or two fans, supply and exhaust air pass through the unit on either side of a series of thin plates, across which heat is exchanged. When the unit is properly airtight, contaminated exhaust air (smell, moisture, etc.) is not transported to the fresh supply air. Small units suitable for application in dwellings (air flow capacities from less than 250 m^3/h) are commercially available.

Among the disadvantages are that balanced ventilation systems require a lot of ducting because the air supply and extraction need to be brought together, while the supply and exhaust grills need to be placed sufficiently far apart to avoid the re-entry of some of the exhaust air. Also, the systems are sensitive to occupant behaviour, e.g. opening of windows. Generally only sensible heat can be transferred, and therefore the efficiencies are not as high as for, say, thermal wheels. Usually the ventilation system works continuously and would also recover heat in the summer, so in order to avoid or limit overheating problems, a bypass must be provided, or the user must be able to switch off the ducted air supply and replace it with natural ventilation through windows or other openings – in both cases the user must remember to switch back to the winter operation mode.

In principle, it is possible to construct plate heat exchangers with a thermal efficiency of close to 80–85%. This means that the supply air leaves the exchanger with a temperature approaching that at which the exhaust air enters the exchanger. However, this would require very large devices and a significant amount of fan energy for driving the air through the exchanger. In practice, heat exchangers are designed to minimize the total energy consumption or the total energy cost, including the depreciation of the investment. Therefore, the total efficiency (= reduction of total energy demand for ventilation, including fan energy) is normally not more than 55–70%. Generally, counterflow heat exchangers are more efficient than cross-flow units, but often also need larger fans. Improved performance of cross-flow heat exchangers is possible by having two units in series.

To limit the parasitic energy consumption, efficient fan motors should be chosen and the fans should be carefully (over)sized (an oversized fan running at low speed generally has a better performance than a small fan operating at full speed). In most commercial units, fans and motors have been chosen solely on the criteria of cost.

In most cases, a drainage system is required for the water that condenses in the exchanger in cold weather. Moreover, in very cold climates the condensate may freeze in the unit and make defrosting necessary. In two-fan systems this is usually done by stopping the fresh air fan and letting the exhaust air thaw the ice.

It is very important that all ductwork between the heat exchanger and the heated space is well insulated.

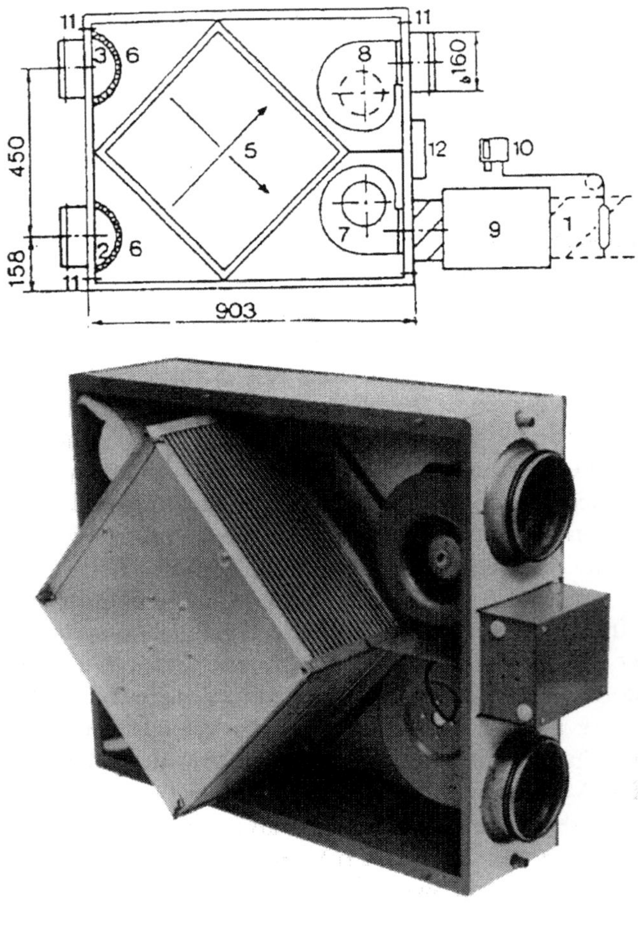

Figure 2.5.1. Example of cross-flow plate heat exchanger (diagram and photo of open unit)

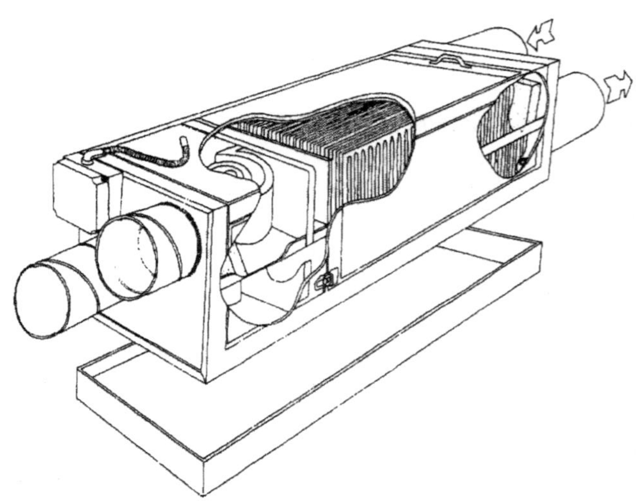

Figure 2.5.2. Example of counterflow plate heat-exchanger recovery unit

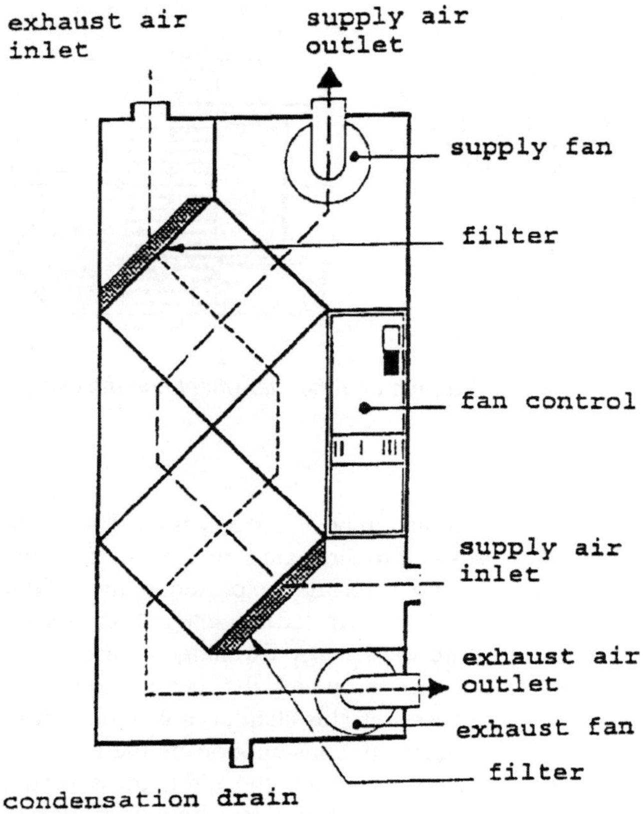

Figure 2.5.3.The heat-recovery unit in the Amstelveen Urban Villa

This is especially important if the heat exchanger is installed in an unheated attic space (which is not recommended). Also, the cold ducts inside the building envelope should be insulated to prevent condensation.

In addition to the cleaning of filters, which is easy, the plate heat exchangers themselves must be cleaned at regular intervals (about once a year). Some of the small cross-flow exchangers are designed so that they can be cleaned in a dishwasher – larger units must be washed-down with high-pressure equipment.

Heat-recovery system in the Netherlands house

In this apartment building, each dwelling is provided with a balanced ventilation system with heat recovery.

During the *heating season* the supply ventilation air is taken from the atrium (cf. Chapter 2.7). This preheated air enters a heat-recovery unit consisting of two cross-flow heat exchangers connected in series. In winter the residents are advised not to use the natural ventilation provisions. However, any opening of the windows will automatically shut off the heating element in the room.

Outside the heating season, based on in- and outdoor air temperatures, a red light on each room controller will indicate that ventilation by mechanical exhaust or by natural supply is more energy-efficient. When the occupant opens the special vents in the façade, this is detected by the sensors of the control system and the inlet ventilator is switched off. This reduces the electricity use for the fans.

The heat-recovery unit consists of two cross-flow heat exchangers, two fans, two filters, and a control unit (Figure 2.5.3).

A drain takes care of condensation. The two fans both create an underpressure, thus reducing internal air leakage. The supply fan contributes its heat to the incoming air. The efficiency of the unit (fan contribution excluded) for three airflow levels (135 m^3/h, 210 m^3/h and 325 m^3/h) depending on the outdoor temperature (the indoor temperature is 20°C), is given in Figure 2.5.4. Those efficiencies are the result of component tests. The yearly efficiency of the heat-recovery unit including the heat loss from the fan is conservatively estimated to be 80%. Normally an 80% efficiency heat-recovery unit needs additional heating for the supply air for comfort reasons (minimum inlet temperature is 15°C). Using the preheated air from the atrium means that the additional heating can be left out. This unit was a special assembly for the Netherlands Task 13 house.

The counterflow systems in the Danish houses

For economic reasons (based on first cost only, not on future trade-off with running cost), only the row houses with north–south-running rows are equipped with heat-recovery systems. However, in the row houses with east–west-running rows, the Task 13 house is also equipped with a heat-recovery system. This building has a single-dwelling counterflow heat exchanger with an exchanger efficiency of 80–85% and an overall efficiency (accounting for parasitic energy) of 70–75%. It is a balanced system exhausting air from the bathroom and the kitchen part of the living room, and supplying air to the living room and the three bedrooms. The range hood has a separate grease filter, but is connected to the heat-recovery system. In the summer the occupant can turn off the supply-air fan in the recovery unit and let fresh air in through the windows, which can be locked in a ventilation position, or through three special fresh air vents in each dwelling.

The houses in the north–south-running rows are equipped with a central heat-recovery system serving from four up to eight dwellings. The advantage is a slightly higher exchanger efficiency, and lower parasitic energy use per dwelling, set off against the inconvenience of no individual control. The central system is linked to a computerized remote control system from where, for example, the supply air fan can be turned off during warm weather and the conditions

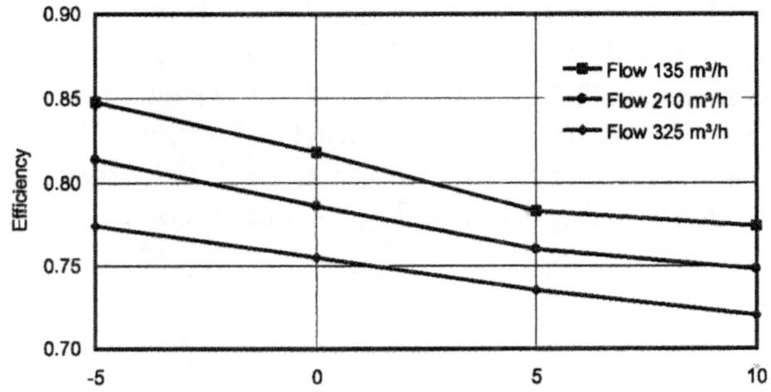

Figure 2.5.4. Measured efficiency of the heat-recovery unit in the Urban Villa

of the filters can be monitored. This is also a balanced system with a counterflow heat exchanger, with insulated ducts running under the slab-on-grade floors of the houses. The insulation thickness is 100 mm. In this case, the range hoods (each with its own grease filter) are connected to the central heat-recovery system. The living room and bedroom on the upper floor are heated with the supply ventilation air through a water-to-air heat exchanger supplied from the combined solar and (chiefly) district heating system, cf. Chapter 3.7.

Thermal wheels

A thermal wheel consists of a revolving cylinder divided into a number of segments packed with a coarsely knitted metal mesh or some other material (Figure 2.5.5). It operates by rotating between 10 and 20 revolutions per minute, picking up heat from the warmer exhaust air stream and discharging it to the cooler supply side. These wheels can also transfer latent heat and therefore have higher efficiencies than plate heat exchanger units, typically 80–90%.

Generally, thermal wheels are not used very much in dwellings, one of the major reasons being the risk of cross-contamination, another being cost. In Task 13, these high-efficiency units were chosen in the Norwegian and Finnish houses and operate part of the year on PV power. In very cold climates, where the outside winter air is extremely dry, the recovery of moisture from the exhaust air is an advantage and makes humidification unnecessary.

It is fairly easy to control the degree of heat recovery automatically from maximum down to zero by slowing down and eventually stopping the rotation of the wheel. Thus, with the proper controls, this heat-recovery system does not contribute to overheating of the rooms. It is a big advantage that the occupants do not have to work (and misuse) by-pass dampers, on–off summer/winter controls etc. As the units are much more complicated than the plate heat exchanger units, professional service maintenance is necessary.

Thermal-wheel heat-recovery system in the Finnish house

In the Finnish Pietarsaari house, the main heat-recovery system is equipped with a thermal wheel unit designed for an air flow of about 240 m^3/h and a thermal efficiency of 80%. As back-up and a specific kitchen ventilation system, a range hood cross-flow heat exchanger with a thermal efficiency of about 60% has been installed. The air flow in both systems can be manually controlled through switches for four different power settings for each fan. The control of the thermal wheel is on–off, giving full or no heat recovery.

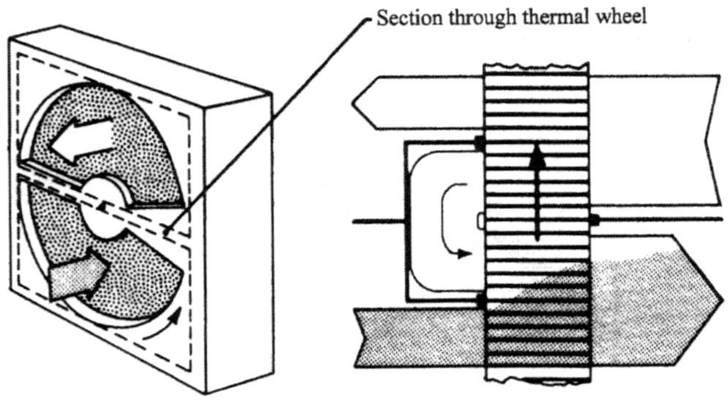

Figure 2.5.5. Diagram of a thermal-wheel heat-recovery unit

Conclusions

Some sort of ventilation heat recovery is used in all the Task 13 houses. Air-to-air plate-type heat exchangers are most commonly used and are present in nine of the Task 13 houses. It is clear that the integration of such systems in future low-energy dwellings is one of the important tools for reducing the heating demand. However, as stressed in this chapter, it is important to minimize the electricity consumption by the fans and pumps, to evaluate whether the airtightness of the building is sufficient, to design for summer as well as winter conditions, and to evaluate whether there are any practical problems with installation, e.g. to assign the necessary space for insulating the ducts or for recovery unit maintenance.

Finally, although not discussed in this chapter, it is also clear that optimizing, or rather minimizing, air-flow rates is important. One very attractive way of achieving this is the use of so-called demand-controlled ventilation systems. Such systems try to adapt the air flow rates to the ventilation needs at different times and locations.

RECOMMENDED READING

Irving S. *Air-to-Air Heat Recovery in Ventilation. Technical Note 45.* AIVC, University of Warwick Science Park, Sovereign Court, Sir William Lyons Road, Coventry CV4 7EZ, UK.

Liddament M W (1996) *AIVC Guide to Ventilation*', AIVC.

Orme M S (1994). *An Analysis and Data Summary of the AIVC's Numerical Database*'. Technical Note 44. AIVC.

AIVC also maintains a database on literature relevant to ventilation and airtightness, AIRBASE, which is available as a diskette package from the AIVC operating agent, Oscar Faber Consulting Engineers, Marlborough House, Upper Marlborough Road, St. Alban, Herts AL1 3UT, UK.

• *Authors.* Peter Wouters (WTCB/CSTC, Violetstrasse 21-23, B-1000 Brussels, Belgium) and Bjarne Saxhof

2.6 Ground-coupled heat exchangers

Introduction

The ground is a natural heat storage system. In the winter, the ground temperature is warmer than the air temperature, while in the summer, the ground temperature is cooler than the air temperature. Several of the countries participating in Task 13 have designed their houses to enable utilization of this natural heating and cooling source. While the concept behind the ground storage system is very simple, the applications in the Task 13 house are quite different.

Concept definition

Ground-coupled heat exchange systems are systems connected to pipes buried in the ground to make use of the ground temperature for the cooling or heating of slowly flowing fluids.

In the Task 13 houses, three types of ground-coupled heat exchange systems are used. The first is used in conjunction with a ventilation system where the fresh air is preheated, the second uses the ground as heat source for a heat pump, and the third is used for air conditioning. Most of the systems used in these projects have also been designed with a secondary purpose. The Finnish house, for example, uses the ground mainly as a source of heat in the winter, but also for cooling in the summer.

In the design of the system, the ducting from the ground storage system should be kept as short as possible, and if the system includes a heat-recovery ventilator, the connecting ducts should also be well insulated.

Variations

Ground-coupled pre-heat ventilation systems

Even the most efficient heat-recovery ventilators have a maximum heat-exchange efficiency of around 80% at –15°C, and the fresh air still requires auxiliary heating to raise its temperature to room temperature in the winter. A simple and cost-effective method is to pre-heat the air before it passes through the heat-recovery system by passing it through a ground storage system.

In areas where the soil has high concentrations of radon, there is the possibility that the ground storage system could be used as a transfer medium. Any condensation which reaches the heat-exchanger ventilator would be collected in drip trays while the large diameter of the pipes in the ground should ensure that there would be no significant restriction to airflow by condensation that collects in the pipe.

The layout of the system is also very important. The pipes' resistance to airflow must be small enough to ensure that the energy gains are significantly higher than the energy needed to run the fan. Another important factor is that, when windows are constantly open, the system should be switched off.

The German Rottweil house. The Ultra House in Rottweil has a central balanced ventilation system that is coupled to a ground storage system. The fan has an airflow of 90 m^3/h, i.e. an air change of 0.2 litre/h related to the heated volume of the house. The ground-coupled heat exchanger is used both for heating the fresh air supply in the winter and for cooling in the summer.

The ground heat-exchanger system consists of plastic piping buried approximately 1 m under the ground surface beneath the garage and carport (see Figure 2.6.1). It is 34 m long with an inner diameter of 0.2 m, giving a total heat exchanger surface area of 21.4 m^2. The piping enters the house through the basement and travels through a shaft in the living area to the heat-recovery ventilator in the roof. The air duct is insulated from where it enters the basement to where it reaches the heat-recovery ventilator so as not to draw any heat from the basement or the living space.

Besides the two fans in the heat-recovery ventilator, an extra fan is installed in the system so that the heat-recovery ventilator can be bypassed. The system is designed to be energy efficient and has four modes of operation selected automatically depending on the ambient air temperature:

<8°C	pre-heating in the ground storage system and the heat-recovery ventilator
8–15°C	ground storage bypassed, pre-heating in the heat-recovery ventilator only
15–25°C	no pre-heating, the fresh air is supplied directly to the living space
> 25°C	cooling through the ground storage system, heat-recovery ventilator bypassed.

Figure 2.6.1. The ground heat exchanger system in Rottweil

A schematic diagram of the total system with the various operation modes can be seen in Figure 3.10.5.

Ground-coupled heat pumps
In some of the Task 13 houses heat pumps are used in auxiliary space and water heating systems. In two of the houses using heat pumps, Hamar, Norway, and Pietarsaari, Finland., a major part of the heating energy is taken from the ground through a buried piping loop system. These houses have different secondary uses for the ground storage. In order to achieve high performance of the heat pump, the temperature difference between the ground and the heating system should be as small as possible. The house should also be equipped with a low-temperature heating system.

The Norwegian house. In the Norwegian house at Hamar, the main auxiliary space and domestic hot-water heating system is a heat-pump system drawing heat from solar collectors located in the sunspace and from a ground storage system. The storage system consists of a 20 m long looped pipe buried 2 m below the surface of the ground away from the house. Only the ground storage loop is directly connected to the heat pump. The solar collector is used to provide surplus energy for the ground storage system during the summer. A schematic diagram of the heating system, including the ground storage system, can be seen in Figure 3.14.3.

The Finnish house. The Finnish house at Pietarsaari is heated by a hydronic system with a central hot-water storage tank. Heat is provided by flat solar-plate collectors and a heat pump (COP 3.3). Approximately 400 m pipes are buried at a depth of 1 m in the form of two 200 m loops. The loops can be used separately or together. Calculations indicate that one loop is sufficient for the heating requirements of the house. A water/glycol mixture is used as the energy transfer medium from the ground to the heat pump. A schematic diagram of the heating system, including the ground storage system, can be seen in Figure 3.8.4.

The hot-water supply and the solar-collector fluid medium can be cooled in summer during periods of overheating. The system can be used for cooling the house as the floor heating system is also connected directly to the ground loop.

Cistern/ground-loop cooling
In this case, a small cooling system with a water-to-air coil is connected to an underground rain-water cistern through plastic tubes in the ground. When cooling is required, cistern water is circulated in the ground loop and supplied to the cooling coil. There is only one house in Task 13 using this system, the Canadian Waterloo house.

The Canadian Waterloo house. In this house, the required cooling load has been minimized. This means that only a small mechanical system using a ground-

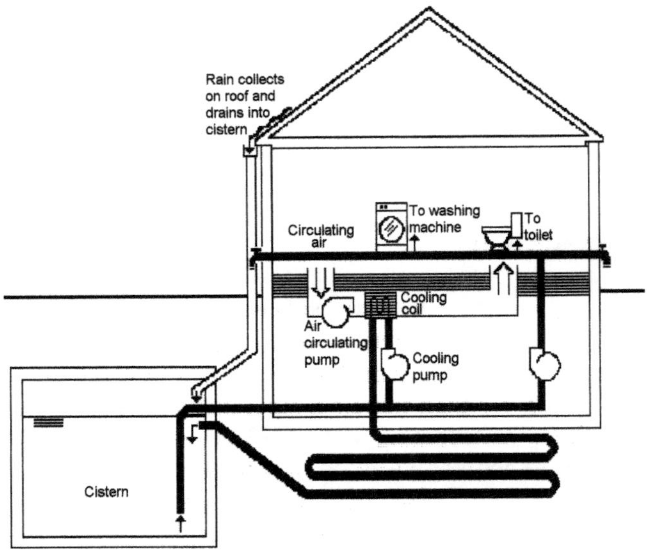

Figure 2.6.2. Schematic diagram of the ground storage system in Waterloo

storage system is required to meet the cooling load. A cistern collects rain-water as a supply for the washing machine, the toilets, and landscaping. This cistern is connected to the ground-storage loop. The loop is buried under the house and is made up of approximately 450 m plastic tubes (see Figure 2.6.2). When cooling is required, water from the cistern is circulated through the ground storage system and then through a retractable cooling coil in the furnace ducting system. The cool air is circulated in the house through the furnace heating system ducting. An advantage of this system is that no CFC-based coolants are required. It does, however, have limited cooling capacity.

Cross comparison

Table 2.6.1 shows a cross comparison of the ground-coupled heat exchangers used in Task 13. Regardless of the type of system used, the basic aim was to reduce either the space heating load or the space cooling load.

Table 2.6.1. Use of ground-coupled heat exchangers

House	Space heating	Space cooling
Rottweil, Germany	X	X
Hamar, Norway	X	
Pietarsaari, Finland	X	X
Waterloo, Canada		X

RECOMMENDED READING

Wittchen K B (1988). Air-Supply in Airtight, Highly InsulatedBuildings. In: Kronvall J (ed.), *Symposium Proceedings of Building Physics in the Nordic Countries*, pp. 241-246. Lund University and the Swedish Council for Building Research. D13:1988. (ISBN 91-540-4905-9)

• *Authors.* Michael Beckert and Hans Erhorn (Fraunhofer-Institut für Bauphysik, Nobelstrasse 12, D-70569 Stuttgart, Germany)

2.7 Sunspaces

Introduction

The effectiveness of a sunspace as an energy-saving feature depends very much on its energy concept, how it is constructed, and not least how the occupants use it. Often simple cost/benefit calculations for sunspaces, based on energy savings only, do not show a profit. However, in many cases a main or major reason to apply a sunspace is not energy savings, but amenity values, creating a pleasant living space. The Task 13 houses include three different types of sunspaces. They provide good examples of the energy contribution a sunspace can make.

Concept definition

A sunspace is a glass-covered space attached to a dwelling or integrated in a complex of dwellings. It is a separate thermal zone, insulated from the dwelling itself, and heated to some degree by solar radiation and to some degree by heat loss from the dwelling.

The main advantage of such a sunspace is that it both collects heat and provides additional living space. Because higher temperatures are acceptable in sunspaces, more solar energy can be collected than comfort or glare would allow in a sunlit living space. At night lower allowable sunspace temperatures provide the freedom to incorporate more glazing than would be sensible for fully conditioned spaces. While a sunspace is admittedly less efficient than a solar air collector, the fact that it also provides living space for many hours of the year helps justify the high kWh cost of the energy savings.

Variations

The Task 13 houses illustrate three ways sunspaces can be used as part of an energy concept:

- The most simple sunspace use is as a sun-tempered buffer space to passively reduce heat losses of the building envelope. This is illustrated by the German Rottweil house. This use is less effective with highly insulating walls between the sunspace and the fully conditioned spaces
- Alternatively, sunspace-warmed air can be extracted and serve as a pre-heated fresh air supply for ventilation of a house. Since in highly insulated houses heating ventilation air can be a major end use of energy, this sunspace use can be very effective. Two quite different examples occur among the Task 13 houses. The Canadian Brampton house incorporates a two-storey sunspace, while the apartment building in the Netherlands makes use of a six-storey atrium for the same purpose
- Another variation is to use a sunspace as a heat source for a heat pump to heat water to temperatures usable for domestic purposes or space heating. The Norwegian house illustrates this concept.

Sunspaces as buffer spaces

The German Rottweil House. This three-storey sunspace, inset in the building, is a prime space in this duplex house (Figure 2.7.1). Emphasis is accordingly on habitability. To achieve comfort for as many hours as possible, the outer glazing is triple pane with xenon gas filling. All the glazing faces south. Thermal mass is provided by a concrete slab floor. The air temperatures in the sunspace average 10 K above the ambient, according to monitoring data collected in 1994/95. In December temperatures ranged from 10–24°C. This sunspace saves heating energy primarily by acting as a buffer space.

Key design issues for sunspaces that mainly function as buffer spaces are:

- the glazing and framing for the sunspace should have a high insulation value to improve winter comfort, if this is required
- the glazing should be limited as much as possible to the south front, as glazing to the east or west and especially roof glazing are not as effective solar gainers in winter and tend to cause overheating in the sunspace in summer. If the roof is glazed, shading devices are required
- the common house/sunspace wall should ideally be exposed masonry (with insulation on the house interior side, if needed) to provide thermal mass to reduce sunspace temperature swings. A floor of

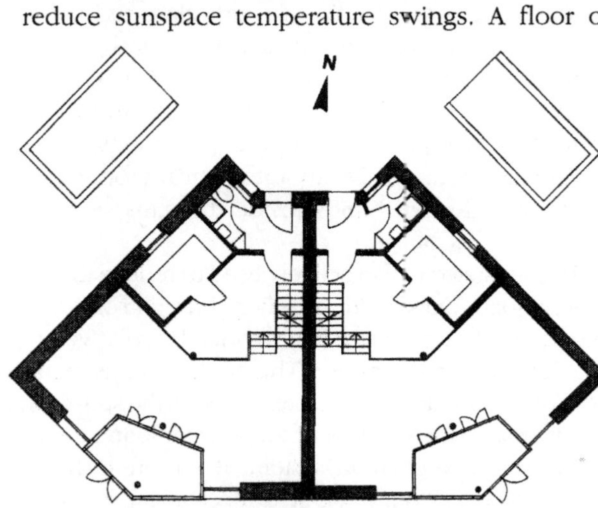

Figure 2.7.1. Plan of the Rottweil house showing the sunspaces

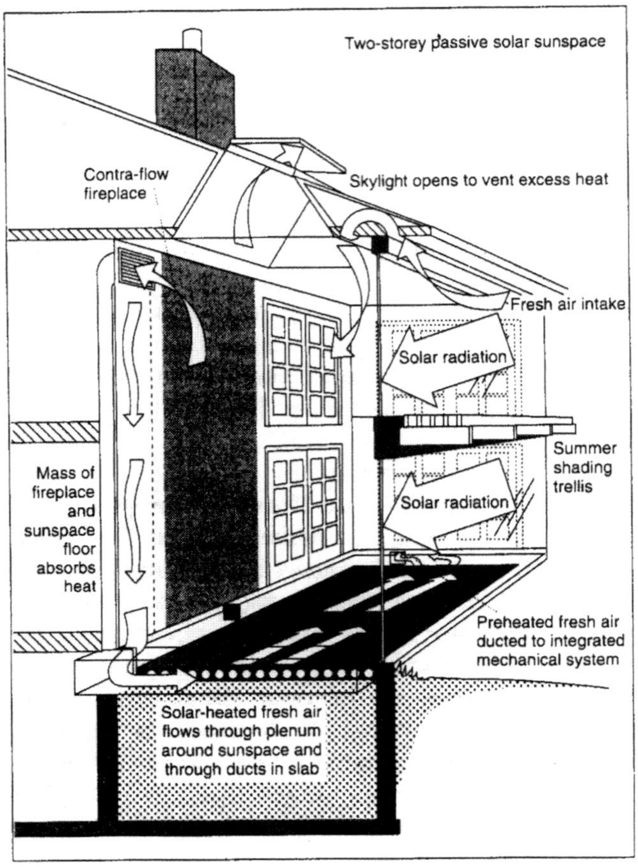

Figure 2.7.2. Cross section of the sunspace of the Brampton house

concrete or clinkers is also a good provider of thermal mass

- good venting possibilities are required for summer comfort - for multi-storey sunspaces vents should be placed high and low to efficiently make use of the stack effect.

Sunspaces/atria for ventilation pre-heat

The Canadian Brampton House. In this house, as in the previous example, a two-storey sunspace (Figure 2.7.2) is an important additional living space. The corner position and the largely opaque and well-insulated roof provide a good ratio of south glass to total glass area. To maximize comfort, a glazing with a low U-value was selected for the sunspace. A massive masonry chimney facing into the sunspace and the concrete floor dampen the temperature swings. Summer comfort is enhanced by an automatically opened roof window, reflective blinds inside the sunspace, and a shading trellis outside.

The sunspace is also actively used to reduce the house heating load. Outside air is drawn through soffit louvres into the sunspace, where it is warmed by solar gains and heat losses from the house. This warmed air is extracted from the top of the sunspace, fed through the sun-warmed hollow-core concrete floor of the sunspace, and delivered to the house ventilation system. If required, the air is further heated by an integrated mechanical system.

Key design issues for sunspaces used to preheat ventilation air include:

- outdoor air inlets should be offset from the air extraction grills at the top of the sunspace to avoid thermal short circuits
- down drafts may occur if the air inlets into the sunspace are placed in high positions
- cooler sunspace temperatures and a risk of freezing must be accepted because heat is actively extracted.
- Interior windows and doors connected to the sunspace should be well sealed to prevent short-circuiting of ventilation air flow
- the hollow core floor slab of the sunspace, which transports the hot air, should be well insulated from the ground to minimize heat losses and have a dark coloured surface to give the channelled air a solar boost in temperature
- the overheating risk should be addressed.

The Netherlands Apartment Building. An atrium with a glazed, south-facing roof links a four-storey apartment block to the south with a six-storey apartment block to the north. Solar radiation is absorbed by the large south wall of the north block, tempering the atrium temperature. Summer comfort is provided by movable shading suspended beneath the supporting structure and by venting with air drawn in through low vents, passing between the glazing and shading, and exiting from top vents. For economic reasons the 1060 m^2 glass area of the atrium (of which 540 m^2 is south-sloping roof glazing) is only single-glazed with aluminium framing without a thermal break. The atrium temperature averages 5 K above the ambient temperature. The air flow configurations for summer and winter operation are shown in Figure 2.7.3.

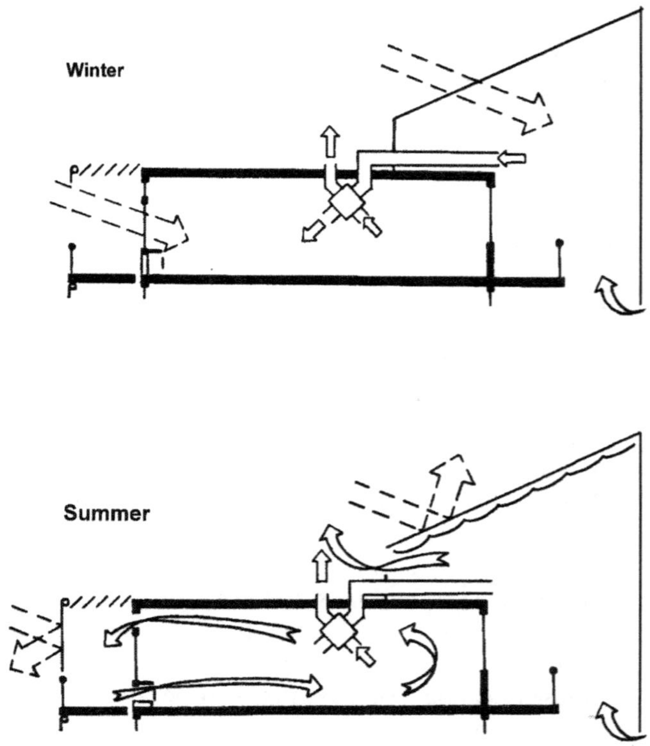

Figure 2.7.3. Air flow configurations of the Urban Villa with atrium, the Netherlands

In the active, winter mode, outside air is drawn through the vents at the base of the atrium, is sun-tempered and then fan-extracted from the upper part of the atrium. It is then ducted to the dwellings. Dwelling air is extracted through a heat exchanger to warm the incoming fresh air. The overall efficiency of this combined system is 85% with a 20% contribution estimated coming from the atrium. This translates to annual energy savings of c. 400 kWh per dwelling due to the atrium. The contribution of the atrium made it possible to omit auxiliary heating of the incoming fresh air, resulting in significant savings in capital costs.

During summer the ventilation air is not extracted from the sunspace, but is taken in directly from the outside through provisions in the façades to prevent overheating in the apartments.

Key design issues for atria used to pre-heat ventilation air include:

- there should be generous venting possibilities, with vertical separation between inlets and outlets
- there should be sunshading for protection against overheating and glare
- small radiant heaters can be placed in atrium areas where local comfort is required, with strong emphasis on 'local'.

Sunspace as heat source for heat pump

The Norwegian house. A two-storey sunspace extending over two thirds of the south façade of this house provides heat for a liquid solar collector which assists a ground coil in providing thermal input to a heat pump during periods of excessive solar gains (Figure 2.7.4). A black-painted aluminium room radiator located on the house wall at the top of the sunspace serves this function. As required, the sunspace is heated to maintain a minimum temperature of 15°C. It serves as an entry airlock and as a stairway as well as admitting daylight into the house. Because it is heated, the outer glazing is highly insulating, comprised of 12.5 m^2 triple, infrared-reflecting and argon-filled glazing and 9.5 m^2 145 mm thick capillary transparent insulation.

Key design issues to be learned from this project include:

- The set-point temperature for the sunspace back-up heating is critical. Above 15°C there is a sharply diminishing return between more sunspace heating and less house heating
- In northern climates, weak mid-winter solar gains do not offset the thermal losses of windows. Buffering windows with a sunspace decrease these losses, particularly in this case where the sunspace is heated to 15°C. Accordingly, windows with higher U-values can be used between the sunspace and the fully conditioned rooms of the house
- Because the sunspace is heated, a high-quality insulating glazing and frame construction is imperative to minimize sunspace heat loss to the ambient

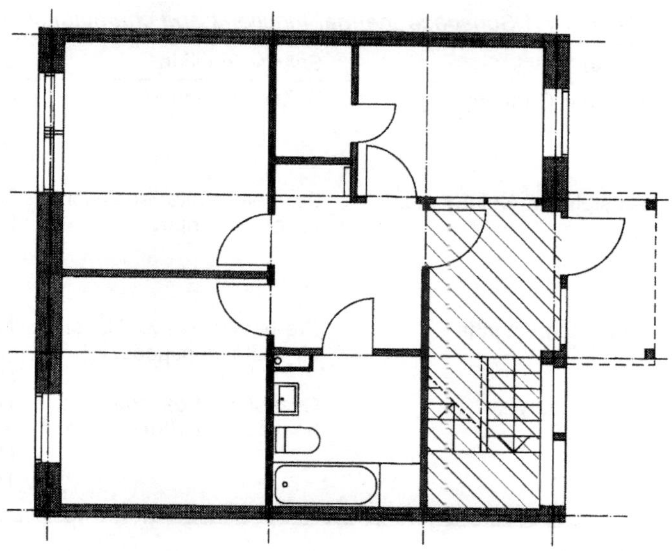

Figure 2.7.4. Plan showing the sunspace, a heat source for the heat pump (Hamar, Norway)

- The solar collector for the ground-coupled heat pump should be located high in the sunspace to profit from thermal stratification, and be irradiated by the sun. It might be configured as a finned sunshade providing good heat exchange and glare control
- Large vent openings to the outside are also important in northern latitudes where low sun angles on vertical glazing in autumn/spring can rapidly overheat a sunspace.

Cross comparisons

How well a sunspace performs depends on its geometry, the construction of the common house/sunspace façade and the outer glazing, and how the entrapped energy is put to use.

A key issue regarding sunspace geometry is to provide as large as possible a ratio of south-facing glazing relative to the total glazing area. For this reason, sunspaces embedded in the building may be at an advantage. Tall sunspaces are also advantageous because they allow stratification of warm air to the top where it can be extracted and put to use. An additional benefit is that the ground level is likely to be more comfortable on days of strong sunshine.

The glazing of a sunspace and of the wall(s) common between the sunspace and house should be constructed consistent with the energy concept for the sunspace:

- If comfort in the sunspace has priority, as in the Rottweil House, the sunspace glazing should be highly insulating to reduce heat losses. The common glazing can be conventional since its heat losses are beneficial to the sunspace
- If outside air is drawn through and pre-heated in the sunspace, as in the Canadian Brampton house or the Netherlands atrium, then the sunspace will generally be cool and the glazing to the house must be better insulating than the sunspace glazing

Table 2.7.1.Sunspace thermal functions and characteristics

House	Energy function	Glass to ambient[a]	Glass to house[a]
Rottweil, Germany	Buffer sunspace	Triple glazing, infrared relecting., xenon gas $U = 0.5$ W/m²K	Double glazing,air filled $U = 2.6$ W/m²K
Amstelveen, Netherlands	Pre-heat intake ventilation air, buffer atrium	Single glazing $U = 5.7$ W/m²K	Triple glazing + one pane for fire protection, infrared relecting, krypton $U = 0.6$ W/m²K
Brampton, Canada	Pre-heat intake ventilation air, buffer sunspace	Triple glazing, infrared relecting, argon $U = 1.05$ W/m²K	Double glazing, air filled $U = 2.8$ W/m²K
Hamar, Norway	Occasional heat source for heat pump, buffer sunspace	Triple glazing, infrared reflecting argon, $U = 0.85$ W/m²K, Transparent insulation $U = 0.8$ W/m²K	Double glazing, infrared reflecting, argon $U = 1.3$ W/m²K

[a] Glass U-values are at the centre of the glass

- If the sunspace is heated, as in the Norwegian house, then high-quality sunspace glazing is essential!

Table 2.7.1 summarizes the construction of the four examples of sunspaces presented in this chapter. All of the sunspaces have highly insulating glazing. Only the atrium has a low-quality glazing because of the large glass area, but it has good glazing for the apartments. It should also be noted that all of the examples here are multi-storey sunspaces.

The energy concept of each of the sunspaces presented in this chapter has been proven through monitoring to be effective, helping make possible the very low heating energy requirements of the houses. At the same time, the occupants greatly value the amenity the sunspaces provide.

RECOMMENDED READING

Hastings R (ed.) (1994). *Passive Solar Commercial and Institutional Buildings.* John Wiley & Sons, Baffins Lane, Chichester. (ISBN 0 471 93943 9)

- *Author.* Robert Hastings (Forschungsstelle Solararchitektur, ETH-Hönggerberg, CH-8093 Zürich, Switzerland)

2.8 Thermal storage

Introduction

Thermal mass inside the building thermal envelope increases comfort by providing heat capacity that prevents excessive temperature swings that might otherwise occur in lightweight construction. Traditionally, houses in some IEA countries have been constructed using heavy materials – block, stone, solid or hollow brick, or concrete – and therefore temperature swings were seldom a problem. The exceptions are Canada, the USA, Japan, and the Nordic countries, where wood-frame construction has been more common. However, in super-insulated walls there is a tendency in most countries to use lighter weight materials because the cavities in double-shell walls provide an inexpensive space for insulation. Another factor in the Task 13 houses is that solar gains are deliberately increased during the winter, spring, and fall seasons. This will increase temperature swings, compared to conventional construction. Thus the issue of adequate thermal mass can become an important consideration in low-energy houses.

Concept definition

Underlying principles

Thermal mass is important in low-energy houses because it decreases temperature variations that would otherwise cause discomfort or increase auxiliary heating or cooling requirements. Heat stored in the thermal mass during an overheated period is released later during an under-heated period. The mass, sometimes referred to as 'thermal inertia', resists changes in temperature by virtue of its heat capacity. However, thermal mass may prolong the period of thermal discomfort in the case of consecutive days of net heat input.

Figure 2.8.1 shows temperature variations over a period of two days inside two houses that are responding to solar gains. The graph marked 'very light' is for a house with little thermal mass, perhaps only sheathing inside the insulation; the graph marked 'very heavy' is for a house that has massive walls, partitions, floor, and ceiling. Note that the average temperature is the same for the two houses because they are assumed to have the same overall loss coefficient between inside and outside (same insulation) and the same solar gains. The temperature variation in the massive house is much lower than in the lightweight house, and it is delayed in time. In the optimal case, heat stored in the thermal mass during the day is released at night, reducing both the daytime peaks and the night-time lows.

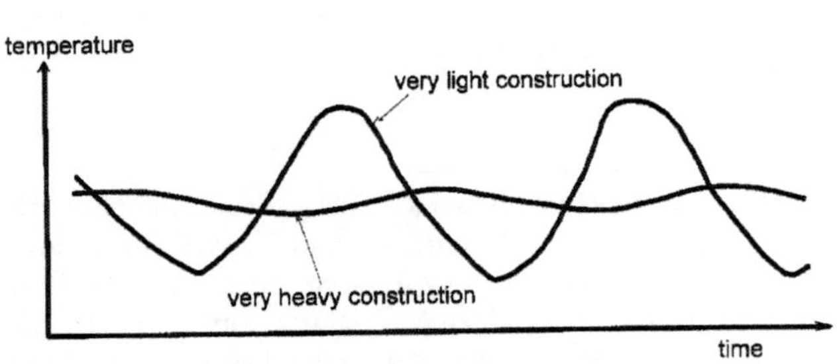

Figure 2.8.1. Effect of thermal mass on temperature swings

This thermal-mass effect can be important in any house that experiences significant variations in heat input or heat loss. Such variations can result not only from solar gains but also from internal gains, such as heat released from appliances, for example kitchen cooking heat that peaks in the late afternoon. Temperature swings are also driven by large variations in outside temperature. It is important to note that solar gains provide a significant input to all of the Task 13 houses *at some time during the heating season.* In the American Grand Canyon house, solar gains during the frequent periods of clear-cold mid-winter weather are large enough to heat the house totally. The same is true in the Nordic houses, but only during clear periods in the spring and fall.

Advantages/limitations of high thermal mass

With the possible exception of buildings in tropical climates, adding thermal mass will enhance thermal comfort. Whether it is desirable to add mass depends on how much is already present in the construction and on other factors discussed below. Adding mass can be expensive, but the mass may serve multiple purposes, for example, a brick wall might be an attractive amenity, and it might reduce sound transmission from an adjoining unit.

Thermal mass can shift the electricity load of a house from periods of electric utility system peak use to off-peak periods. Some utilities offer special off-peak electricity tariffs or other incentives that will result in lower electricity costs.

A limitation to adding thermal mass would occur when all the available surfaces have been covered. However, this limit is normally not reached in practical cases. Another limitation is that thermal mass tends to be difficult to engineer into highly insulated exterior walls owing to thermal-bridging effects. One way around this is to place thermal mass entirely within the building interior with both sides exposed, thus increasing its effectiveness.

If the house thermostat is set back at night, the effect of thermal mass will be to reduce the temperature drop. If the setback is large, the peak heating required to warm up the house again the following morning may be excessive. The simple remedy for this is to use a smaller setback and to reset the thermostat somewhat earlier.

Design issues

Both location and thickness of the thermal mass are important to its effectiveness. Increasing the mass thickness beyond about 100 mm does not increase its ability to reduce daily temperature swings because the surface layers insulate the deeper layers from the daily storage-and-release cycle. This means that spreading a given amount of mass over a large surface makes it more effective, provided that all of the surface communicates with the space.

A popular misconception is that for mass to be effective it must be located in direct sunlight. This is not true because long-wave (infrared) radiation effectively redistributes thermal energy from the initial point of absorption to other locations within line of sight. Direct sun through windows is most likely to strike the floor; the energy is then scattered or re-radiated to the walls and roof (but significantly, not to other, non-sunlit, parts of the floor because they are not in the direct line of sight). Thus, except for the floor, the mass does *not* need to be in the direct sun. To be effective, a mass surface should not be covered by any insulating material such as carpet.

A popular strategy for massive floors is to expose a narrow strip (about as wide as the height of the windows) along the south wall next to the windows where the sun will fall directly on it. The rest of the floor, which is not very effective anyway, can be carpeted without much loss of overall thermal-storage effectiveness.

For the reason cited above, floor mass is much more effective in the direct sun. Sunlit thermal mass floors should be dark in colour, to absorb and store energy low in the room. However, mass walls and ceilings should be light in colour to distribute the heat and light more evenly.

Adding one large heavy surface to a lightweight room makes a great impact on the temperature swings.

Table 2.8.1. Thermal storage characteristics of building construction in the Task 13 houses

Country and building	Construction	Is construction inherently massive?	Is mass an issue here?	(Net south glass area) ÷ (net floor area) (%)
B	moderate	no	yes	7
CDN-B	light	no	no	6
CDN-W	light	no	no	6
DK	moderate	yes	yes	11
FIN	light	no	yes	5
D-R	heavy	yes	yes	35
D-B	heavy	yes	yes	20
I	heavy	yes	yes	
J	light	no	yes	10
NL	heavy	yes	no	12
N	light	no	no	25
S	moderate	no	yes	11
CH	moderate	yes	yes	8
US-GC	light	no	yes	13
US-Y	moderate	yes	yes	5

Variations

Overview – cross comparison

The general characteristics of the IEA buildings are shown in Table 2.8.1.

Mass was considered to be required by the designers in 73% of the buildings. Mass was intentionally added in 33% of the houses. The construction technique used is inherently massive only for the houses in Germany, Denmark, the Netherlands, and Switzerland. The average net south-facing glazed area is 13% of the net floor area.

Note that there is a wide variation in the amount of south-facing solar glazing used in the various designs. The south glass area is given in the last column, expressed as a percentage of the net floor area. This single number largely explains the different attitudes of the various designers regarding the need for thermal mass.

Houses where added mass is needed

A key indicator for designing a solar house is the ratio of the south glass area to the floor area. Studies of buildings in the southwest of the USA (generally with double glazing) have shown that temperature swings inside lightweight buildings will be too large for comfort if the net south glass area exceeds about 6% of the net floor area.[1]

The last column in the table reveals that only about half of the Task 13 houses employ a south-facing glass area that is more than 6% of the house net floor area. Thus it is only in these houses that one would expect that added mass will be needed. The south glass area in the remaining houses is small enough that no added mass is required, over and above the inherent mass associated with lightweight construction. Thus it comes as no surprise that the designers of these latter houses find that thermal mass is not an important design issue in their climate. One should note, however, that a different designer using a larger area of south-facing glazing would surely come to a different conclusion – added thermal mass would be required. Thus it seems that the need for added thermal mass is more a function of the size of the south glazing than a function of the climate (however, in some – mainly northern – climates, overheating is often caused by solar gains through west-facing windows).

[1] Another way to state this is that construction materials, such as framing and internal sheathing, and internal furnishings provide enough heat capacity to store the *excess* solar heat collected by an area of glazing equal to about 6% of the net floor area. The heat capacity of a lightweight house is about 23 Wh/K (or 81400 J/K) per m^2 of floor area. This is the *effective* heat capacity, effective in the sense that it is thermal mass that is closely coupled to the building interior.

In the cases of the Belgian house, the American Grand Canyon house, the Swedish house, and the Japanese house, additional mass was required and was intentionally added to the design. In the German, Danish, and Swiss houses, and the American Yosemite house, thermal mass was required but was present anyway because of the massive construction. In the Netherlands building the mass was there but was not considered an issue. Four examples of the use of thermal mass in low-energy building design are described in more detail below.

Mass can also be beneficial in the summer. The particular case of the American Yosemite house, where added mass was required only for the summer, is discussed below.

The Netherlands house

Mass was employed in the Urban Villa apartment building. Mass is located in the walls separating the dwelling units. Roofs and floors, and most of the internal walls, are of concrete. Room temperature swings are predicted to be under 7 K, an acceptable range. Although this type of construction would have been used in any case, it is clear that the added mass is required to maintain comfort. This is expected because the ratio of south-facing glass to floor area is 11.5%, well above the 6% limit. Phase-change material (PCM) was evaluated during the design process, but it was decided not to incorporate it. In general, experience with phase-change materials in the Netherlands has not been favourable. Problems have been high cost, due to the packaging required, inadequate heat transfer required to take advantage of the high thermal capacity of the materials, and the poor longevity of incongruent-freezing materials such as salt–water mixtures which tend to precipitate over time.

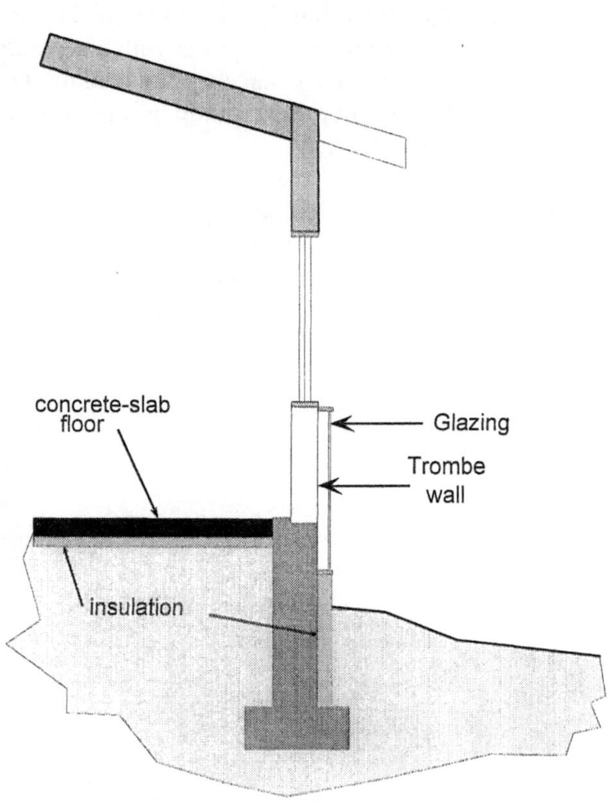

Figure 2.8.2. Section through the Trombe wall in the Grand Canyon house

The American Grand Canyon house

The construction technique used in the American house at Grand Canyon is very lightweight (structural insulated panels, consisting of layers of plywood enclosing a foam core). The design solution was to use as much direct gain as the construction would allow. This is about 7.2% of the house net floor area (the slight increase over 6% is due to the mass of the concrete-slab floor construction).

Besides the transparent glazing, the house has a solar glazing, in the form of a 1-m-high strip of Trombe wall across the entire south side of the house located beneath the windows (Figure 2.8.2). The Trombe wall is constructed of 300-mm-thick, high-density, solid concrete blocks. The Trombe wall delays the solar heat delivery to the room until the evening hours. Temperature swings inside the house are *not* increased by the Trombe wall because the heat flow from the inside wall surface into the house is completely out of phase with the direct gain solar heat.

At night, the house is primarily heated by the Trombe wall, and during the day primarily by direct gain. About one half of the net solar heat in the house is provided by the Trombe wall; the other half is from direct gain. The tile-covered concrete-slab floor is sufficient effective thermal mass for storing the direct-gain solar heat. The Trombe wall has no thermo-circulation vents to the inside or the outside – these would be counter-productive.[2]

It is worth noting that, per unit of mass, a Trombe wall makes the most effective use of the mass in reducing temperature swings. Another point to note is that the summer cooling load created by the Trombe wall is only about one sixth the cooling load that would be created by an equal area of direct-gain windows. This is seldom a problem, and it is certainly not one in the Grand Canyon location with its mild summer climate.

The American Yosemite house

Mass was added to the design of the second American house for a completely different reason. A key problem at this location, in a valley west of Yosemite National Park, is that afternoon peak temperatures are very high during the summer, averaging about 40°C in July. However, diurnal temperature swings are exceptionally large, about 26 K, with a cool downstream breeze through the valley, providing an opportunity to pre-cool the house at night by opening windows. The windows are closed during the hot afternoon so that the

[2] Such vents are frequently included in Trombe wall designs but they require additional effort to build, usually create a cleaning problem, are likely to degrade performance owing to reverse thermo-circulation at night, often overheat the house during the day, and tend to defeat the main advantage of the Trombe wall, which is delayed heating of the house.

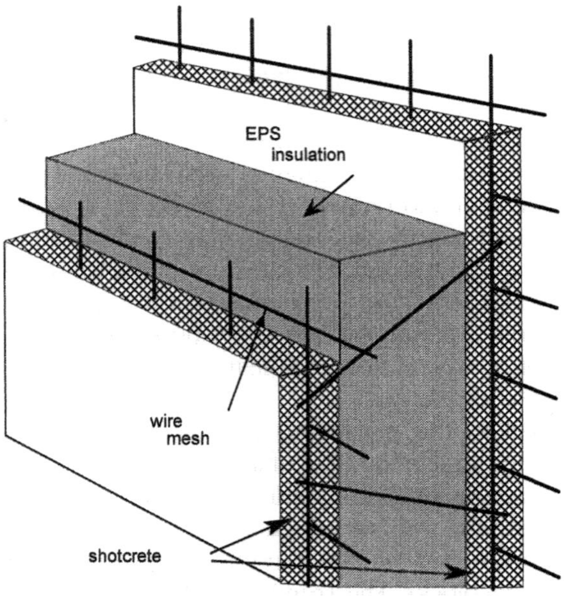

Figure 2.8.3. Mass sandwich construction in the Yosemite house

daytime excess heat is slowly stored in the walls. Mass was added to the design to make this strategy work.

The added mass consists of 50 mm of concrete sprayed on a structural wire mesh inside the foam insulation of the exterior walls (Figure 2.8.3). This concrete spray process is called 'shotcrete'. The wire cage with the foam inside is a commercial product manufactured in the USA.

The amount of south-facing transparent glazing (4.5% of the net floor area) does not require adding mass to control temperature swings in winter. However, temperature swings during sunny winter days will be less with the added mass.

The Japanese house

In the Japanese lightweight steel-framed for Task 13 house, where additional thermal mass is needed, thermal storage is used to decrease both the heating and the cooling load, with the primary aim of reducing the peak electricity load.

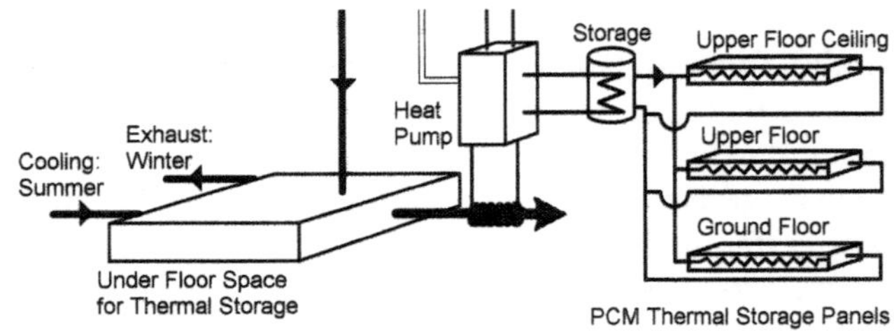

Figure 2.8.4. Storage system in the Japanese WISH house

The house uses an under-floor space for thermal storage (Figure 2.8.4). During summer nights, the under-floor space is cooled by ventilation with outside air. The thermal mass allows this cooling effect to benefit the heat pump efficiency on the following day, when air drawn through the heat pump to the fan coil unit is taken from the under-floor space. In the winter, this space is heated by cooling air drawn through a large PV system, which incorporates an air collector. The thermal mass of the under-floor space makes this heat available during the following night, when the space is used as a heat source for the heat pump.

The design also incorporates the use of a phase-change material. The PCM phase-change temperature is 22°C for both heating and cooling, which provides a benefit both winter and summer. The thermal storage panels (each 260 × 300 mm) are embedded in the floor and in the ceiling, 22 m^2 on the bottom floor, 12.5 m^2 on the upper floor, and 26 m^2 in the ceiling of the upper floor. The total thermal storage capacity is about 550 kJ. Water from the heat-pump heat exchanger is circulated through these panels to shift the electricity load to off-peak times. An air-conditioning system with a separate dehumidifying unit serves as back-up for cooling. For heating, a high-efficiency heat pump is operated at night on off-peak electricity.

- *Author.* J. Douglas Balcomb

2.9 Active solar water systems

Introduction

In locations with high solar irradiation throughout the whole year, very simple systems combining the functions of collecting and storing solar energy are feasible. Black painted solar-absorbing steel tanks placed on flat roofs, e.g. of residential bungalows, were the first active solar water systems. Smaller and more irregular availability of solar energy makes an effective separation of solar collecting and storing necessary. High utilization of solar energy thus requires distribution within the building. The more storage and energy distribution required, the more advantageous it is to use water systems.

Active water systems are a highly efficient way to collect, distribute, and store solar energy. The most common applications are solar (pre-)heating of domestic hot water, solar space heating (with short-term and/or seasonal storage) and finally, pre-heating of inlet air. Combinations of the three applications can increase the yearly solar use significantly, as each works at different temperature levels.

Concept definition

Active solar water systems provide three functions:

- collection of solar energy
- distribution of collected solar gains into storage (storage charge mode)
- distribution of stored solar gains to the place of usage (storage discharge mode).

In water (or more correctly, liquid) systems, once heat has been collected, all heat transfer takes place between liquids. In most countries, antifreeze additives are needed in the collector loop in permanently water-filled systems. With some additives, including some glycols, double separation (i.e. an extra heat exchanger or tank) is required between the collector loop and the domestic hot water.

Advantages/limitations of the concept
Non-renewable auxiliary energy is directly displaced by active solar systems. In locations with only limited available solar gains in the winter, replacing auxiliary energy for space and domestic hot-water heating calls for seasonal storage. Today, such extended time shifting of solar gains can only be achieved with water storage systems. Limitations are given by the balance of solar gains and the amount of parasitic energy necessary to store (charging and discharging of the storage) and distribute the collected solar energy.

Geometric/location variations
To optimize the collection of solar gains (e.g. minimizing the shading), collectors should normally be placed on the roof. In a multi-storey building, an optimal solution can be vertical façade collectors at each storey, as short distribution lines become the dominating system design issue. For thermosiphon systems, an optimum geometry is given by a maximum height between collector inlet and outlet in order to create temperature-driven buoyancy.

Variations

Several of the Task 13 houses have active solar water systems. Examples of three system variations are given here.

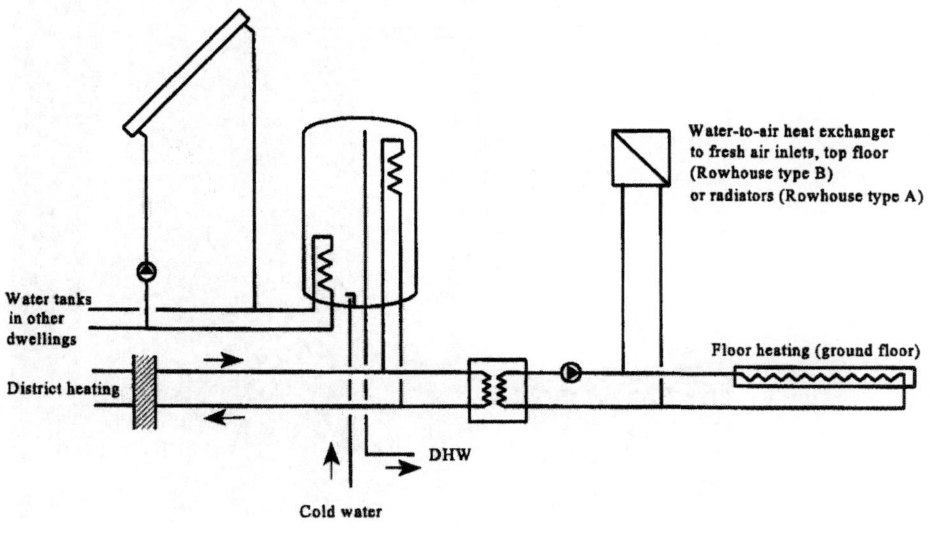

Figure 2.9.1. Active solar water system diagram in the Danish house (domestic hot-water and space heating)

Solar domestic hot-water system equipped with a circulation pump powered by a PV panel – The Canadian Waterloo House

In this house, solar energy transfer from the collector to the storage tank is controlled and regulated by a PV-powered circulation pump which starts operating only when the solar irradiation level produces sufficient power. Such a configuration eliminates the need for a controller and simplifies the installation (no temperature sensors and digital controls for optimized running of the circulation pump). Better performance is provided because the pump flow rate will vary with the incident solar irradiation. The system does not require any grid-supplied electricity for operation.

Solar domestic hot-water system with domestic hot-water tank acting as buffer in the space heating system – the Danish row houses

The south-facing solar collectors of this low-flow system (Figure 2.9.1) are mounted inclined at 45° on racks on the roof. The distribution system is minimized (short distances from collector to storage and from storage to place of usage). A collector array serves up to four dwellings, each with a 200 litre domestic hot water tank.

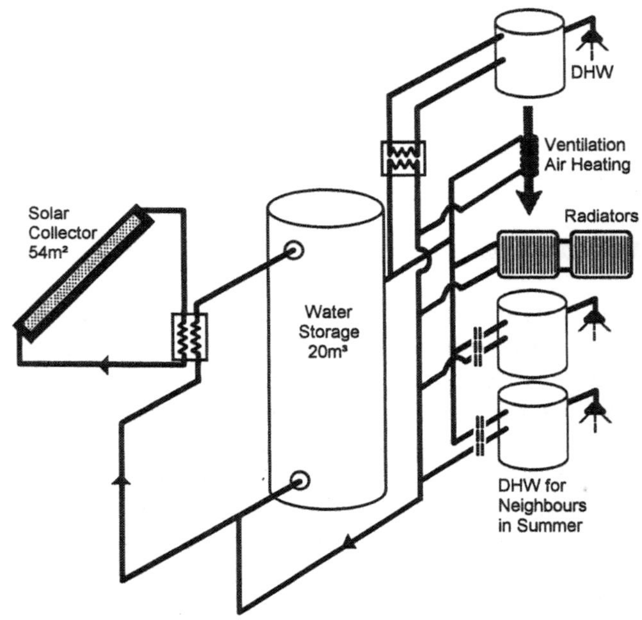

Figure 2.9.2. The active solar water system in the Berlin house (domestic hot-water and space heating)

The top of the domestic hot water tank acts as a buffer for the low-temperature space-heating system

Figure 2.9.3. Location of the solar collectors for domestic hot water and space heating, the Berlin house

Table 2.9.1. Overview of functions and characteristics of the various active solar water system examples

IEA Task 13 House	DHW heating	Space heating	Pre-heating of supply air	Heated area (m^2)	Collector (m^2)	Storage (m^3)	Active water system (kWh/a)	Special features
Waterloo, Canada	X	–	–	230			1580	PV-powered circulation pump – hot fill connection to washing machine and dishwasher
Kolding, Denmark	X	(X)	–	90			1700	Hot-water tank as buffer for space heating system – PV-powered circulation pump
Pietarsaari, Finland	X	X	–	166	10	3	2000	Solar and ground coupled heat pump
Berlin, Germany	X	X	X	170	48	20	6060	Combined seasonal and short-term storage
Amstelveen (one apartment), The Netherlands	X	–	–	100	4.5	0.2	2450	Hot-fill connection to washing machine
Röskär, Sweden	X	X	X	54			990	Heat exchange with floor water-piping system

(floor heating) supplied from a district-heating heat exchanger.

In one of the east–west running rows (house type A, the Task 13 row), a small PV panel supplies and controls the pumps for the water/glycol loop between collectors and storage tanks.

Solar water system with combined seasonal and short-term storage, for space heating and integrated domestic hot-water and supply-air heating – the German Berlin house

In this house, two solar collector fields are mounted as parts of the external cladding. The collectors are integrated into the sunspace envelope with a surface area of 48 m^2 and a slope of 45°. The collector fluid is driven by natural convection and has a flow rate of 10–12 litre/m^{2}h over the collector fields. A small pump is used to compensate for the pressure drop over the counterflow heat exchangers that transfer the solar gains to the storage system. The seasonal storage is a 20 m^3 heavily insulated water tank located in the middle of the house. The system is shown in Figure 2.9.2.

The domestic hot water is stored in a 300 litre tank on the top floor and heat is transferred by a plate heat exchanger. The temperature is maintained at 60°C. The cold supply water is circulated through the heat exchanger before entering the tank. If the storage temperature is too low, the domestic hot water can be heated directly from the solar collectors for both space heating and domestic hot water (Figure 2.9.3).

The hot-water system is also used to heat up the fresh air in the ventilation system after the heat-recovery ventilator. Here the system has back-up support from a thermostat-controlled small radiator placed in the ventilation duct. The combined seasonal and short-term heat storage system can supply heat for domestic hot water to two of the neighbouring houses.

Cross comparisons

Table 2.9.1 shows the various functions provided by the active solar water systems used in the Task 13 houses. The table illustrates that it is possible to combine several additional features without negatively affecting the overall system efficiency.

RECOMMENDED READING

Steinemann P (ed.) (1992). *Technology Summaries for Solar Low Energy Houses*, IEA Task 13 Working Document, Chapters 10 + 11. ETH-Hönggerberg, Solararchitektur; Zürich, Switzerland.

Duff W S (ed.) (1995). *Advanced Solar Domestic Hot Water Systems*. Final Report of the IEA SHC Task 14 DHW Working Group. Colorado State University, USA.

Garg H P (1987) *Advances in Solar Energy Technologies*, Vol. 1. D. Reidel Publishing Company, Dordrecht. (ISBN 90-277-2430-X)

- *Authors*. Rolf Stricker, Michael Beckert and Hans Erhorn (Fraunhofer-Institut für Bauphysik, Nobelstrasse 12, D-70569 Stuttgart, Germany)

2.10 Photovoltaic systems

Introduction

The price of photovoltaic (PV) cells has been and is high, but has been rapidly decreasing for a number of years, and PV systems are expected to meet a considerable part of the future electricity demand. The systems are particularly interesting for advanced solar low-energy buildings, where space-conditioning and hot-water loads are extremely low, and where the total load is often dominated by the electricity needed for lighting, household appliances, and parasitic loads for the technical systems. A PV system thus offers the opportunity of a very large solar contribution to the building.

Concept definition

Photovoltaic cells
A photovoltaic or solar cell consists of two layers of a semiconducting material (most often silicon). When sunlight strikes the junction between these layers, direct current (DC) is generated. The efficiency of the transformation depends on the construction of the cell and the cell material.

The most common types of solar cells are monocrystalline silicon cells, polycrystalline cells, and amorphous silicon cells. Crystalline cells are breakable silicon wafers cut from a large silicon block, while amorphous cells consist of a very thin layer of uncrystallized silicon which is deposited on a substrate (e.g. a glass sheet). The colour and appearance of these cell types are quite different. Polycrystalline cells often look bright blue and 'sparkling', while monocrystalline cells look dark grey, almost black. Amorphous silicon usually looks dark brownish red.

Photovoltaic modules
An assembly of solar cells placed in a sealed unit is called a photovoltaic module. The cells are covered with a layer of glass or plastic to protect them from water and physical damage. Standard rectangular modules (crystalline or amorphous) can be delivered with or without a frame. Frameless modules are called laminates. The minimum thickness of the laminates is about 8–10 mm.

The size of PV modules varies from manufacturer to manufacturer and is mainly determined by the manufacturing process and handling convenience. Typical sizes of crystalline PV modules are 0.5 x 1.0 m^2 or 0.33 x 1.33 m^2; these modules are made up of about 40 cells. Basically, modules of any size could be produced (up to 4 m^2), but in practice modules larger than 1 m^2 are not very common. Standardization of module size is being attempted.

For connecting PV modules to each other (to produce more power) or to the grid, most modules are equipped with a junction box on the rear side. In this box, the plus and minus terminals of the modules are connected to the wiring. Junction boxes are in general 50–70 mm deep.

Some modules have male and female connectors already attached to the terminal wires. These connectors allow interconnecting of individual modules without using junction boxes.

Finally, some modules are equipped with a connector and socket system. In this case the connector of one module can be plugged into the socket fixed on the rear side of the next module.

The efficiency for conversion of sunlight into electricity depends on the cell material. The conversion efficiency of modules of amorphous silicon cells varies from 3 to 8%. This means that at an irradiation of 1000 W/m^2, an amorphous silicon module will produce 30 to 80 W/m^2. Modules of polycrystalline silicon have a conversion efficiency of about 9–12%, while modules of monocrystalline silicon cells have an efficiency of 12–16%.

A number of PV modules connected in series or parallel is called a PV array.

PV systems
A common way to classify PV systems is according to the method of storing and delivering electricity:

- stand-alone systems
- one-way grid-connected systems
- grid-connected systems
- small power systems.

A stand-alone system consists of a PV array and a battery bank wired to the end-use appliances. Appliances that run on DC electricity can be connected directly to the system. To use AC appliances, an inverter is needed, as shown in Figure 2.10.1. Stand-alone systems have been widely used in remote installations where it is very expensive to provide electricity from a power station. Therefore, these systems are often found in lighthouses, mountain lodges and so on. In

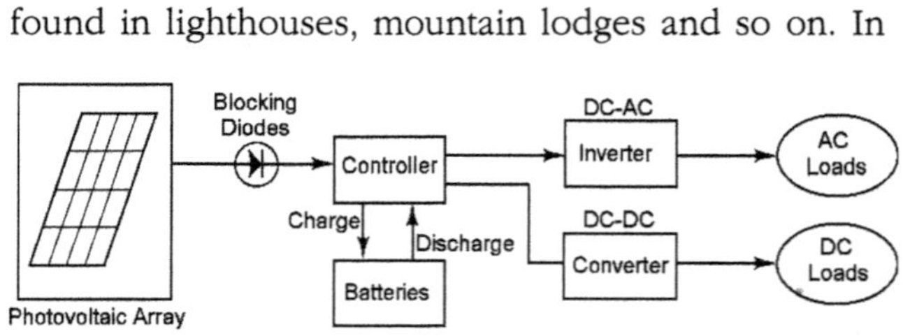

Figure 2.10.1. Stand-alone PV system

urban areas, however, there are also examples of stand-alone systems.

One-way grid-connected systems (Figure 2.10.2) are applied to unitary appliances such as a PV-operated air conditioner. Grid electricity is supplied to supplement PV operation whenever necessary. No feedback is allowed to the grid, even if the PV power is excessive. An example of an application of this type, which is commercially available in Japan, is the solar air conditioner.

In grid-connected systems (Figure 2.10.3) the power generated by the PV system is fed into the grid whenever it is not used on site. Batteries are not required, but a DC/AC inverter and protector must be provided. Surplus PV electricity can be sold to the electricity company. The Task 13 houses in Norway, Finland, and Japan all have this type of system. The power generated during the daytime is fed back to the commercial grid line. In cooling-dominated climates the PV power will relieve a certain amount of demand during the peak hours in the summer time when people use air conditioning.

Small power systems are often used for pumps or fans with a DC motor to circulate water or air to the solar heat collector. They have built-in controllers that operate while the sun shines strongly enough, and batteries are not required. Sometimes small fans are operated to circulate warm air from a sunspace to rooms on the north side. One single PV module is sufficient to operate such small fans. Attached sunspaces may be equipped with PV modules to operate forced ventilation, as these spaces tend to overheat in the summer.

Installation of PV panels

The integration of PV panels into the roof or façade of buildings can be done in many different ways. Innovative solutions include PV panels that replace or assist other necessary functions such as weather skin or solar shading.

There are basically four different photovoltaic module mounting schemes: rack mount, stand-off mount, direct mount, and integral mount.

- Rack mount involves installing the modules on a tilted support frame or rack. There is a wide variety of rack designs made of wood or metal. Most manufacturers offer rack-mounting frames and hardware that specifically fit their modules. Some rack mount designs offer the option of adjusting the tilt. Rack mount provides easy access to the front and rear of the module for maintenance. This mounting scheme also provides good ventilation and 'cooling' of the modules, which increases their efficiency. The cost of rack mount is relatively high compared to other mounting schemes, because of the extra material and labour involved
- Stand-off mount involves a module support structure that stands off from the roof or façade. The frame support usually has to be fixed to the structures under the weather skin which requires careful sealing at the penetration points. This configuration also provides for good ventilation and 'cooling' of the modules. The cost of the stand-off approach is higher than the direct and integral mounts because of the hardware involved and the lack of cost savings in replacing some or all of the roof or façade materials. The stand-off mount is an attractive alternative for retrofit applications where the weather skin is already in place.
- Direct mount involves replacing the outer weather skin with photovoltaic modules. For example, PV modules can be mounted overlapping in a manner similar to conventional roof tiles. This PV skin gives some weather protection to the structure, but is not completely water- or airtight. The cost is fairly low, since there is little mounting hardware involved and savings are obtained as a result of replacement of some weather skin materials.
- Integral mount involves replacing the entire weather skin with PV modules. This requires air- and water-tight junctions between the modules. For example, frameless PV modules could be mounted in a framing system similar to that used in conventional glass roofs or façades. This mounting technique offers substantial cost savings because of the avoided cost of weather skin materials.

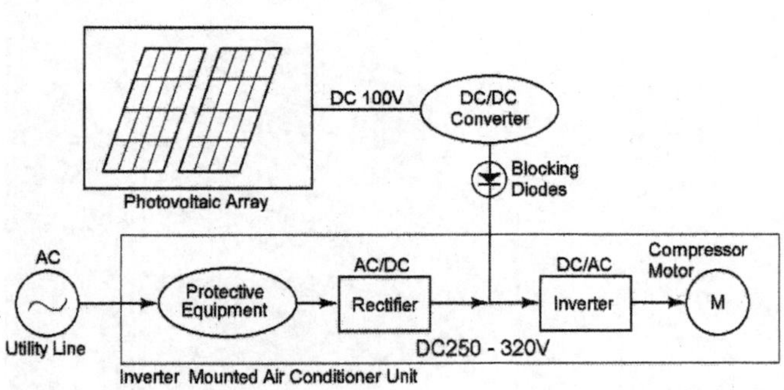

Figure 2.10.2. One-way grid-connected PV system

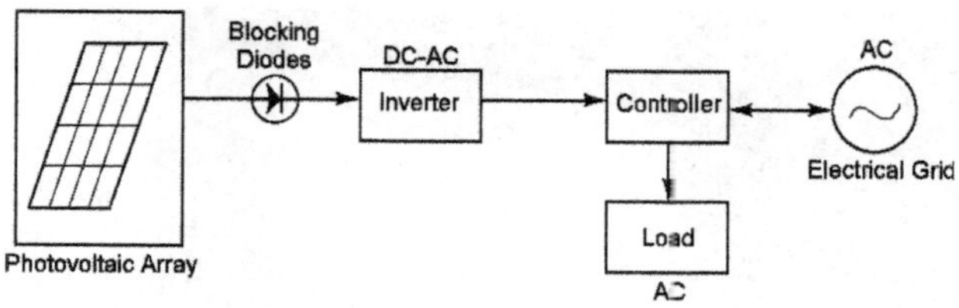

Figure 2.10.3. Grid-connected PV system

Appropriate PV array size

Key factors affecting the system sizing are the load, the operation time (all year, summer only, etc.), and the location of the system (solar availability). Also, the available roof or façade area can restrict the PV array size. Finally, economics may be the most important boundary condition restricting the system size.

The optimum size of a grid-connected system depends mainly on economics: investment cost, interest rate, tax deductions or state subsidies, the energy payback policy of the local utility, and the ratio of PV electricity used directly to that fed into the grid.

Typically, the buyback ratio is less than one, which means that the utility pays less for the PV electricity than the building owner pays for the grid electricity. There-

Figure 2.10.4. Hybrid PV panels on the Japanese WISH house

fore, a high percentage of the PV system production should be used directly in the building. This requires careful matching of the PV production with the house load profile.

The typical size of a grid-connected system for a one-family house (with an annual electricity consumption of 4–5 MWh) is 2–5 kW. This means an annual electricty production of 1.5–3.8 MWh in northern Europe and 1.6–4.0 MWh in central Europe, assuming optimum orientation and design.

For stand-alone systems the sizing must be done more accurately than for grid-connected systems, because the available buffer capacity is quite limited. To compensate for unexpected long cloudy periods some overdimensioning of the battery size is needed. This reduces the average battery depth of discharge and increases the battery life.

In addition to simple sizing tools based on rules of thumb, there exist a number of computer programs that can be used for more accurate simulation of PV systems. The computer programs are based on quite accurate mathematical models for different system components.

VARIATIONS

The Japanese house

A hybrid collector combined with PV panels is used in Japanese WISH House. The PV panels are cooled by the air flowing through the passages in the panels. The heated air is used as a heat source for the heat pump unit in the heating season.

The peak output from the PV panels is designed to be 4.2 kW with the area of 44.8 m². The panels were installed on the flat roof using racks (Figure 2.10.4).

The Finnish house

The PV array in the Finnish Pietarsaari house consists of 45 panels with the size of 800 x 1550 mm (Figure 2.10.5). The array is installed on the south-facing roof. The total array area is 54 m², and the peak output power is 2 kW. Batteries are not used; the output from the PV system is connected directly to the building's electrical system.

While an integrated PV roof system using waterproof PV panels had been designed, the panels were actually installed on the exterior of a traditional roof because of the construction schedule.

The Norwegian house

Monocrystalline PV panels with a total area of 22.6 m² are installed on the south-facing roof (35° tilt) of the Norwegian house at Hamar. The panels were mounted in a stand-off configuration using aluminium profiles, as shown in Figure 2.10.6. The air space at the rear side of the panels provides effective ventilation.

The system is grid-connected with a design peak output of 2.52 kW. The inverter has a rated power of 1.8 kW and an efficiency of 91% (at nominal power).

Energy simulations predicted an annual electricity production of 2550 kWh. Based on an annual electricity consumption of 6000 kWh, about 1380 kWh would be fed into the grid.

The Swiss house

PV panels are also installed on the south-faced roof (45° tilt) of the Swiss house in Gelterkinden. The system is grid-connected with a peak power output of 3 kW. No batteries are used in the system.

Cross comparisons

Table 2.10.1 gives an overview of the large PV systems used in the Task 13 houses. The Canadian Waterloo house and the Danish row houses use

Figure 2.10.5. Installation of PV panels in the Pietarsaari house

Table 2.10.1. Comparison of the large PV systems used in the Task 13 houses

	Finnish house	Japanese house	Norwegian house	Swiss house
System without battery	Grid-connected without battery	Grid-connected without battery Hybrid PV panel for improvement of heat pump system	Grid-connected without battery	Grid-connected
PV array				
– mounting	on roof	with scaffolding	on roof	on roof
– total area	54.0 m²	44.8 m²	22.6 m²	30 m²
– tilt	45°	17°	35°	45°
– peak power	2.3 kW	4.2 kW	2.52 kW	3.0 kW
– module size	800 x 1550 mm	390 x 3967 mm	530 x 1188 mm	445 x 985 mm
– cell type	amorphous silicon	polycrystalline silicon	monocrystalline silicon	polycrystalline silicon
Inverter		5.0 kW	1.8 kW	3.0 kW
Generated power (predicted)	2300 kWh/a	5510 kWh/a	2550 kWh/a	2830 kWh/a

small PV power systems (that are not connected to the grid) for their active solar domestic hot-water systems.

RECOMMENDED READING

Humm O, Toggweiler P (1993). *Photovoltaics in Architecture.* Birkhäuser Verlag, Basel.

Sick F, Erge T (ed.) (1996). *Photovoltaics in Buildings. A Design Handbook for Architects and Engineers.* James & James (Science Publishers) Ltd, London.

Strong S, Scheller W G (1991). *The Solar Electric House.* Sustainability Press, Still River, Massachusetts.

Photovoltaic Systems Design Manual (1991). Energy, Mines and Resources, Quebec, Canada.

- *Authors.* Ken-ichi Kimura, Mitsuhiro Udagawa (Department of Architecture, Waseda University, Shinjuku, Tokyo 169, Japan) and Inger Andresen (SINTEF Architecture and Building Technology, N-7034 Trondheim, Norway)

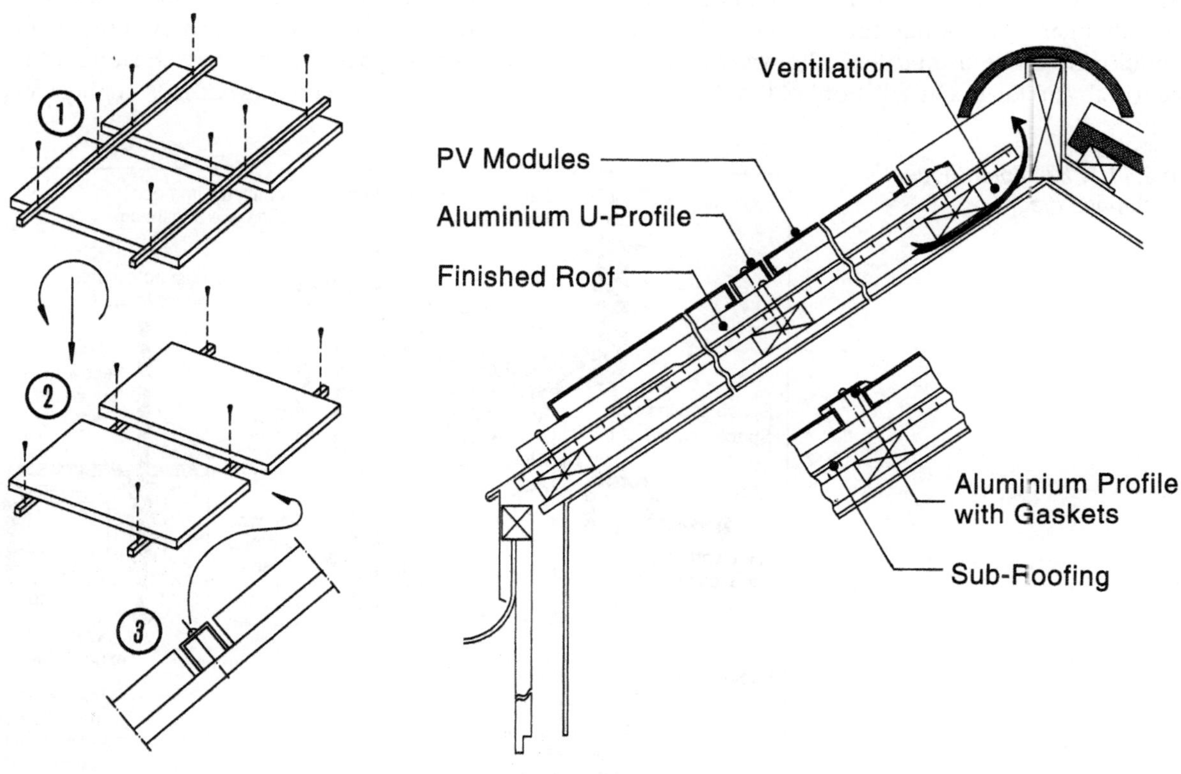

Figure 2.10.6. PV panel install system used in the Hamar House: (1) the PV modules mounted on aluminium profiles; (2) modules mounted on the roof in pairs; (3) aluminium profiles mounted on the roof to prevent the accumulation of snow

2.11 Integrated mechanical systems

Introduction

Integrated mechanical systems (IMS) provide several residential functions, such as ventilation, heat recovery, space heating, and water heating, in one apparatus.

Some systems combine conventional off-the-shelf HVAC components into a package, whereas others are designed from scratch as optimized units, often incorporating heat pumps, to produce a system that operates much more efficiently.

Concept definition

Most dwellings need mechanical equipment to heat and/or cool the space, heat domestic hot water, provide ventilation, and perform heat recovery (from exhaust air, waste water, or flue gases). In most houses, a separate piece of equipment is used to provide each of these functions. An integrated mechanical system is a combination of several of these devices into one system.

Integrated mechanical systems offer several advantages over individual pieces of equipment. First, they can reduce the amount of energy required by the individual systems. For example, by combining a water heater with an air conditioner, the waste heat from the air conditioner can be used to heat the water. Second, the cost (both capital and installation) of the IMS can be less than the sum of the individual pieces of equipment it replaces. Savings can be achieved by having only one cabinet, using a fin coil for both heating and cooling and reduced on-site installation time. Finally, it is also possible to make the equipment more compact and thus save space in the house.

There are three limitations to IMS. First, the building industry (and homeowners) are not used to integrated systems. In some cases the installation and servicing of an IMS is more difficult and crosses trade boundaries. Second, for many IMS to function efficiently, there needs to be a reasonable balance between heating (and/or cooling) loads. For example, in the case of the combined air conditioner/water heater, it may be necessary to dump hot water to keep the air conditioner running if the cooling load is much larger than the water-heating load (either on a peak or seasonal basis). Third, in many cases the IMS uses a heat pump to transfer energy. Heat pumps require electricity, which is, in most countries, an expensive energy source (compared to fossil fuels).

Variations

The Canadian Brampton house

The unit used in this house is probably the most 'integrated' of integrated mechanical systems. It provides space heating, space cooling, water heating, ventilation, grey-water heat recovery, and exhaust-air heat recovery.

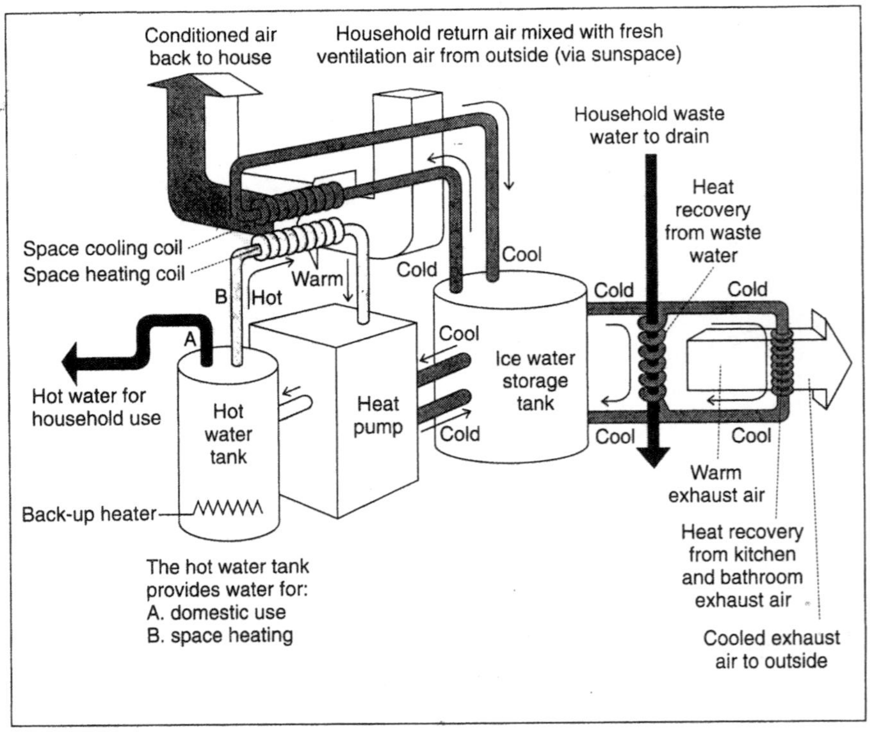

Figure 2.11.1. Brampton house: IMS schematic diagram

The system (Figure 2.11.1) comprises a hot water (or DHW) tank and a cold water (or ice storage) tank thermally connected by a heat pump. The hot water side supplies heat to meet space- and water-heating loads. Hot water demand is met by taking hot water from the top of the DHW tank directly to the taps. Cold mains water enters the bottom of the DHW tank to replace the water used. To provide space heating a pump circulates water from the hot water tank through the heat-pump condenser and a fan coil. Heat is delivered to the space by circulating air over the fan coil.

The air-circulation fan operates continuously to ensure good air circulation and a constant supply of outdoor air to all rooms. The IMS continuously exhausts air from kitchen and bathroom areas. Outdoor make-up air enters the house via an intake grille in the sunspace. The ventilation air is pulled into the return air duct of the IMS and mixes with recirculated building air.

The DHW tank is kept hot by operating the heat pump. The heat pump transfers heat from the ice tank to the DHW tank. Heat is added to the ice tank by circulating a 33% ethylene glycol–water mixture through a grey-water heat exchanger and an exhaust-air coil. If there is no heat in the ice tank (i.e. it is almost all ice), the heat pump shuts off and a 6.0 kW electric element is used to heat the hot-water tank.

Cooling is provided from the ice tank. When the house thermostat indicates that cooling is required, a three-way valve opens to direct the IMS glycol mixture through the house cooling coil. The glycol mixture transfers its heat to the ice storage tank via a heat exchanger in the tank. This system allows heat removed from the house air to be stored in the ice tank for use when water heating is required.

The American Grand Canyon and Yosemite houses

The unit used in these houses (see Figure 2.11.2) combines the functions of space heating, controlled ventilation, heat recovery, water heating, and space cooling in one small and economical unit. The IMS consists of a 300 litre insulated hot water tank with a ventilation heat pump unit on the top. The heat pump unit contains ducting, dampers, a 690 W compressor, two blowers, controls, and two heat exchangers that serve as the heat-pump evaporator and condenser. Exhaust air is drawn from the bathrooms and kitchen causing a slight negative pressure in the house. This causes outdoor air to be drawn into the house through normal cracks in the envelope. The average Coefficient of Performance (COP) of the heat pump is 3.1.

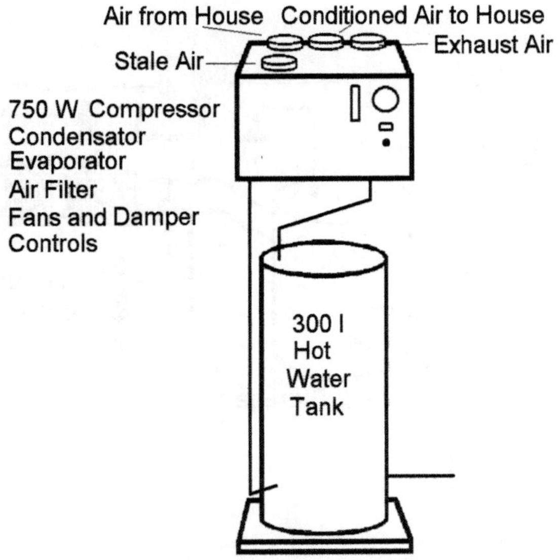

Figure 2.11.2. The IMS unit used in the two American houses

The unit can operate in several modes, as follows:

- *Hot water heating* (priority mode). Heat is recovered from exhaust or cooling air and transferred to the domestic water tank. Hot water heating capacity is 1948 W. Tank recovery time is 8.1 h for a 45 K rise. An optional 4500 W immersion heater is provided for fast recovery. When cooling is not required, heat is transferred from the exhaust air providing forced ventilation of the house. When cooling is needed in summer, the ducting changes to allow heat to be removed from recirculated house air. Net space cooling is 2503 W at an electricity consumption of 960 W. The exhaust air flow is maintained by the other blower
- *Space heating.* Heat is pumped from exhaust air into recirculated house air. Net space heating is 2032 W. Electricity consumption is 805 W
- *Ventilation. 170* m^3/h of air is exhausted from the house. The heat pump is off. Electricity consumption is 70 W
- *Space cooling.* Heat is pumped from recirculated house air to exhaust air. Net space cooling is 2503 W. Electricity consumption is 960 W. Exhaust air flow is 209 m^3/h.

This IMS provides for base-load heating, comprising most of the total load, but it cannot meet peak requirements. The IMS heat is exhausted to the house at a central location, eliminating the need for a distribution system. The remainder of the heat is from electric resistance baseboard (or skirting board) heaters in the various rooms.

The Canadian Waterloo house

Space heating and ventilation for the Waterloo Region Green Home is handled by a combined furnace/heat recovery ventilator (HRV). This system, developed by the Canadian Gas Research Institute, uses a conventional mid-efficiency natural gas furnace connected to a small container of rocks with two compartments (Figure 2.11.3). Stale air from the bathrooms and kitchen mix with the furnace flue gases and pass through the first compartment of rocks. The large surface area of the rocks ensures that almost all of the heat in the exhaust air is removed and stored in the rocks.

Outdoor air is heated by the rocks as it passes through the second compartment of rocks. Approximately every five minutes, a reversing valve switches

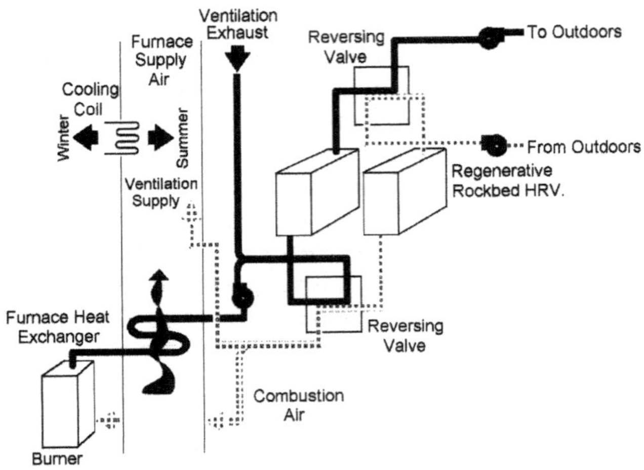

Figure 2.11.3. Furnace/HRV system schematic diagram

the air streams to the two rock compartments, and thus the heated rocks are cooled and cooled rocks are heated. The outdoor air is distributed throughout the house via furnace ductwork. Products of combustion are prevented from re-entering the house by pressurizing the fresh air stream.

This system has two main advantages over conventional furnaces and HRVs. First, it is very efficient. According to previous in-field performance testing, the combined system was more efficient than a condensing natural gas furnace and standard HRV. Second, when commercialized, this system will be less expensive than a condensing natural-gas furnace and standard HRV, because it is integrated into one package and does not require a stainless-steel heat exchanger.

The Norwegian house

The system in the Norwegian house at Hamar provides space heating, domestic hot-water heating, and ventilation heat recovery. A custom-built recuperating ventilation heat-recovery unit (the rotating wheel type) is combined with a heat pump for space heating and domestic hot water, using a ground loop as heat source (Chapter 2.6). Surplus heat from the sunspace is fed into the ground and regained to the heat pump via the ground loop. The heat-recovery efficiency is about 90%. The system is shown schematically in Figure 2.11.4.

Table 2.11.1. IMS function comparison

House	Space heating	Space cooling	Water heating	Ventilation (with heat recovery)	Waste-water heat recovery
Brampton, Canada	X	X	X	X	X
Yosemite, USA	X	X	X	X	
Grand Canyon USA	X	X	X	X	
Waterloo, Canada	X			X	
Hamar, Norway	X		X	X	

Cross comparisons

Table 2.11.1 compares the functions provided by the IMS used in the Task 13 houses. The choice of IMS depends on the loads to be met (and their magnitude) and the availability and cost of energy sources.

RECOMMENDED READING

Morofsky E (1989). *Integrated Mechanical Systems: Recent Developments and Potential Housing Applications, A State-of-the-art Survey prepared by Canada for the First Experts' Meeting of IEA Task 13*, Boulder, Colorado, 1989. Available from Enermodal Engineering Ltd, 650 Riverbend Drive, Kitchener, ON N2K 3S2, Canada.

- *Author.* Stephen Carpenter

Figure 2.11.4. Schematic diagram of the Norwegian IMS

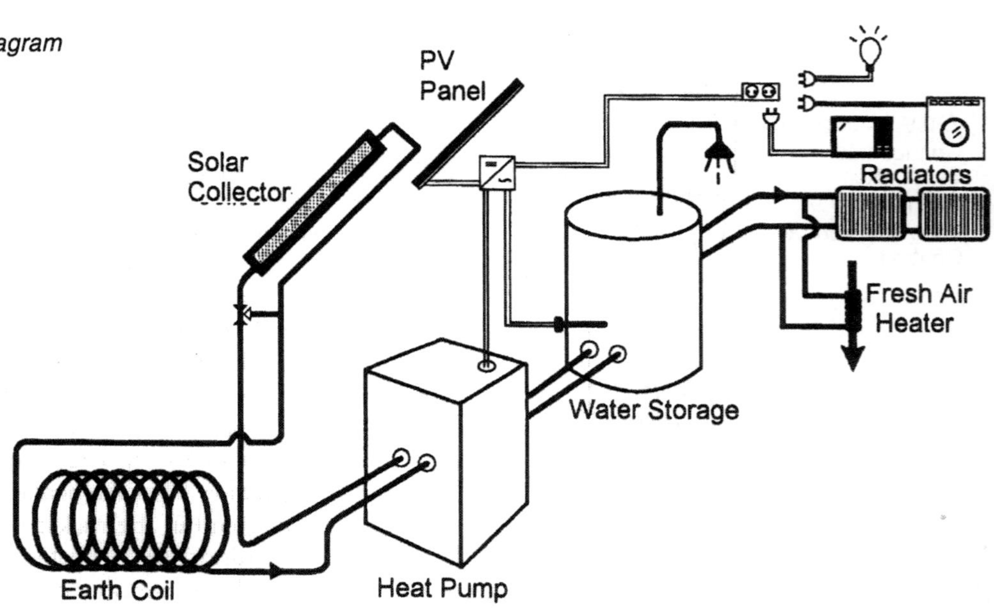

2.12 Home automation systems

Introduction

A home automation system is an electronic multi-task control system with which a number of control functions in the dwelling can be programmed or can be monitored/activated from somewhere else. Used correctly, such a system can decrease energy consumption and improve the indoor thermal comfort. Typically, the system can also be used to enhance the security in the building.

Concept definition

Underlying principles
Home automation systems regulate, for instance, the heating or cooling system, the lighting level of the dwelling, the ventilation system, the communication, the electrical appliances, and the security in accordance with the climate and with the occupants' wishes.

Some applications of the home automation system reduce energy consumption. Thermal losses can be reduced by avoiding unnecessary space conditioning when the occupants are absent (e.g. through use of presence sensors for activating heating/cooling or lighting), or by turning off the heating system if somebody opens a window. The system may also be used to optimize the use of off-peak electricity, or even (based on signals from the utility company) to switch on some of the electrical appliances when the electricity rate is the lowest.

Home automation systems can improve the thermal and visual comfort of the house, e.g. by operating a ventilation system, blinds, other shading devices, or shutters, according to the time of day, solar gains, daylighting, wind, temperature, or other indoor/outdoor climate parameters.

The system can work independently of the occupant or be controlled by the occupant, e.g. the occupant can schedule or override the programmed operation of the heating system by phone.

Advantages/limitations
Home automation systems offer two main advantages: they improve security, and they improve occupants' comfort while reducing energy consumption.

Concerning the home automation system development, two main factors have to be considered.

- The different systems have to be easily accessible to everyone. One major barrier against wide expansion of home automation systems is that the persons who could benefit most (old people, disabled persons) are not at all used to computers. The home automation system designers have to provide good documentation for the user of the system. It is difficult to determine the level of detail in the explanations needed by the occupants, as this depends on their experience with and attitude to computers and electronic equipment
- The user should be able to control the devices managed by the home automation system, i.e. they must be able to override the system.

Overview of variations
There are two different ways to manage the main control functions in a house:

- a stand-alone system, i.e. a system in which the functions are independent of each other. For example, the room temperature can be controlled by room thermostats that are not linked to other parts of a control system
- a home automation system which connects several functions.

System management
A central management unit can control and program all the functions of the home automation system, or the home automation system can be structured around local intelligent units. These two different management systems can also be combined. Generally, the following hierarchy is applied:

Level 1: detectors
Level 2: independent local intelligent units
Level 3: central unit of management which also plays the role of user interface.

Variations

The Japanese house
The components managed by the home automation system in this house (Figure 2.12.1) are: the lighting, the air conditioning unit, the hybrid collector, the PCM thermal storage panel system, and the security. All equipment in the building is controlled from a computer. Simple on/off equipment is controlled through an input/output unit by a programmable logic controller.

The Belgian house
The components managed by the home automation system (Figure 2.12.2) in the Belgian PLEIADE house are the electrical heating system, the ventilation system, the artificial lighting, the movable solar shading, and the security.

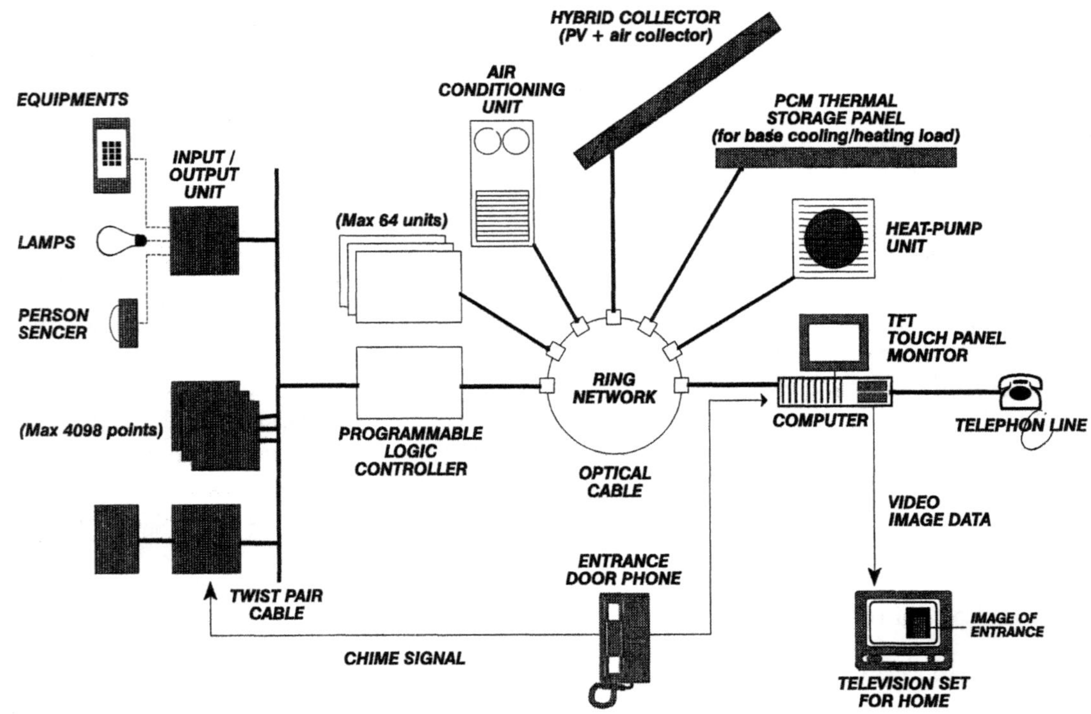

Figure 2.12.1. The home automation system in the WISH house, Japan

The home automation system is structured around local intelligence units (BUS EIB). In Figure 2.12.2, the numbers assigned to each sensor (in the right column) indicate the devices managed as a function of each of these sensors, e.g. in the case of a window being opened (3), the off-peak storage heater and the convection heaters are turned off.

The home automation system reduces energy consumption by an optimal use of solar energy and daylighting, and reduces energy cost by optimal use of the tri-horary rate offered by the local utility company. The rates vary 'randomly' during the day according to the total load on the utility, and information on the current rate or rate change is continuously supplied to the home automation system by the utility.

The Netherlands house

The components managed by the home automation system in the Urban Villa apartment building are the movable solar shading, the opening of the windows, the lighting (atrium and basement), the ventilation (with heat recovery), the hot water system, and the heating system.

The control system of the Urban Villa (Figure 2.12.3) consists of a collective and an individual part. The collective part of the system comprises:

- the sun and wind controller (1) for the outdoor climate and sun shading devices (8)
- the atrium control system (3)
- the four solar control systems, each one providing four apartments with solar energy for domestic hot water (4);

Table 2.12.1. Stand-alone and home automation systems in the Task 13 houses

	B	CDN -B	CDN -W	DK	FIN	D-R	D-B	J	NL	N	S	CH	US-GC	US-Y
Air temperature	H	S	S	S	H	S	S	H	H	S	H	S	S	S
Lights	H	–	–	–	–	–	–	H	–	S	–	–	–	–
Daylighting	H	–	–	–	–	–	–	–	–	–	H	–	–	–
Shading devices	H	–	–	–	–	–	–	–	H	–	H	–	–	–
Ventilation	H	S	S	H	S	S	S	H	H	S	H	S	S	S
Summer ventilation	–	–	S	H	S	–	–	–	H	–	–	–	–	–
Security	H	S	S	–	–	–	–	H	–	–	–	–	–	–
DHW	H	S	S	S	H	S	S	S	H	S	H	–	–	–
Humidity	–	–	–	–	–	–	–	H	–	–	–	–	–	–
Solar system	–	S	–	S	H	–	S	–	H	S	H	–	–	–
Fire protection	–	–	–	–	–	–	–	–	H	–	S	–	–	–
Cooling	–	S	S	–	H	–	–	H	H	–	–	–	–	S

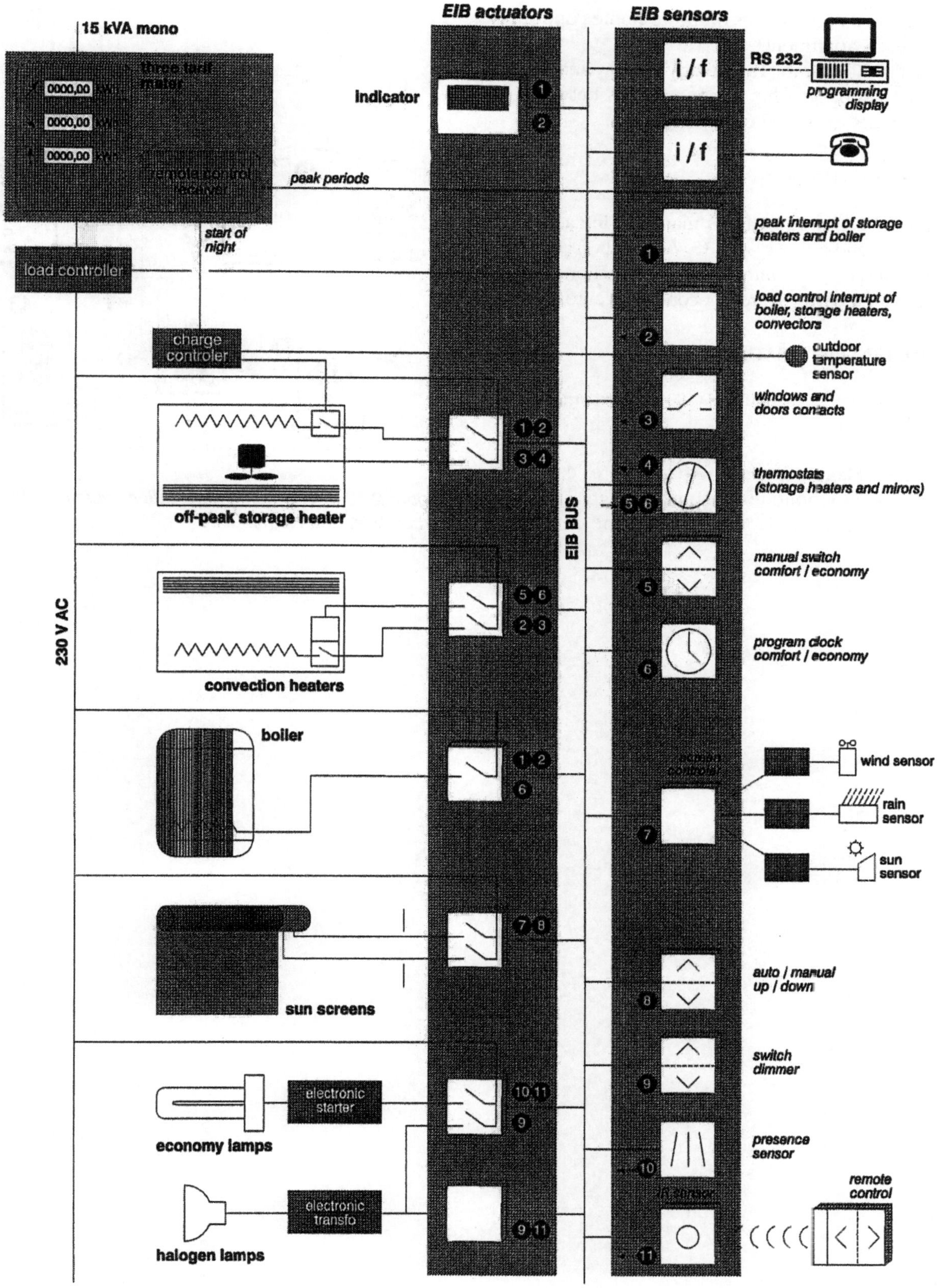

Figure 2.12.2. The home automation system in the PLEIADE house, Belgium

The individual part consists of the 16 home controllers, one for each apartment.

A local CTR (Combi Talk Ready) bus system (based on RS 485) supports the communication between the different parts of whole control system.

Cross comparisons

In Table 2.12.1, the functions managed in each house either by a stand-alone system (S) or by a home automation system (H), are summarized. The functions that are marked with a dash (–) are not controlled automatically.

RECOMMENDED READING

Colomès J, Mérieux P, Schouker J (1991). *Domotique*, Edition Eyrolles, Paris.

• *Author.* Magali Bodart (Architecture et Climat, Université Catholique de Louvain, Place de Levant 1, B-1348 Louvain-la-Neuve, Belgium)

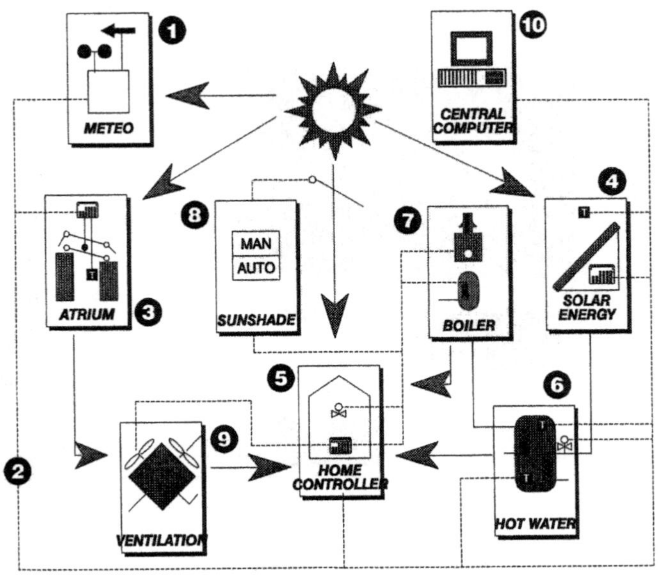

Figure 2.12.3. The home automation system in the Urban Villa, the Netherlands

2.13 Energy efficient lights and appliances

Introduction

Heat generated in a building by people, electric lights, and appliances – and in some cases waste heat from heating and storage systems – help reduce the demand for auxiliary space heating. However, normally only a part of these internal gains is usable. For certain installations, some of the converted energy is lost, e.g. in the heated water from washing machines and dishwashers flowing down the drains, and in water evaporation by food processing. The energy source is usually electricity, but may also be gas for some appliances, e.g. vented gas stoves.

It would be appropriate to limit the internal heat generation so there would never be a surplus. This is of course impossible, but a possibility is to reduce the energy consumption for lights and appliances by influencing both the habits of the residents and by use of more efficient equipment. Both aspects are important. Presently the on-going development of efficient appliances is counteracted by use of more electrical equipment in homes and the trend to have an increasing number of devices on stand-by, increasing parasitic energy use.

Saving electricity also has major environmental aspects. Electricity is a high-exergy form of energy, and may (depending on means of production) be highly polluting, rated per end-use kWh. The efficiency of electricity production in fossil-fuel power plants without co-generation is about 40%, so a reduction of about 60% of the primary energy consumption can be obtained by substituting a kWh of electricity by a kWh of heat. When using coal, oil, or natural gas in power plants, a number of chemical compounds are released that seriously pollute the environment, chiefly carbon dioxide (CO_2), nitric oxides (NO_2) and sulphur dioxide (SO_2). Any reduction of the electrical energy consumption in buildings will thus be an advantage for the environment, even if the reduced internal gains to some extent have to be compensated by a higher output from the heating systems.

This chapter focuses on the efficiency of domestic appliances today and gives an indication of how much energy could be saved per year for an average household, if the appliances were replaced by the best ones on the market. The use of automatic controllers for electricity savings has been discussed in Chapter 2.12.

Concept definition

The electricity consumption for lights and appliances in a dwelling is primarily dependent on the following factors:

- the number of members in the household
- the user habits in connection with the use of electricity in the household
- the size of the dwelling
- the number of electric appliances
- the efficiency/power consumption of each appliance
- how often and for how long the appliances are used.

The average consumption of electricity may vary strongly from one household to another. In Table 2.13.1 average annual electricity consumption is shown for Europe and for Japan.

Figure 2.13.1 shows typical electricity consumption by appliances in normal one-family houses.

An analysis carried out by the Physics Department at the Technical University of Denmark of the electricity consumption for an average North European household showed savings of approximately 46% if the lights and appliances used today were replaced by the best ones on the market (Table 2.13.2).

Three strategies for saving electricity and gas savings are:

- information campaigns to change user habits
- promotion at the market place of selection of efficient devices (labelling based on standardized testing of lights and appliances)
- use of automatic controllers (e.g. presence sensors to save electricity by switching off the light);

and, additionally, to save primary energy:

- substitution of electricity with gas.

Variations

Most of the Task 13 houses have to some degree included use of energy-efficient lights and appliances to obtain a low total energy consumption. In the following chapter, information on the most efficient appliances available on the market today is given. When a range of electricity consumption is indicated, the term 'normal' designates the typical consumption

Table 2.13.1. Average electricity consumption in kWh/year for a family with two children

	One-family house	Apartment
Europe	4000–5000	2000–2500
Japan	5000–6000	4000–4500

Table 2.13.2. Electricity consumption per year for a typical North European household before and after changing lights and appliances to the best on the market

kWh/year	Consumption with average lights and appliances			Consumption after replacement by the best on the market		
	Max.	Mean	Min.	Max.	Mean	Min.
Refrigerator	–	–	–	100	100	100
Freezer	600	–	–	180	180	180
Refrigerator/freezer combined	650	650	600	–	–	–
Washing-machine	400	300	200	120	90	60
Drum drier	400	–	–	300	–	–
Dishwasher	300	–	–	60	–	–
Electric cooker/ oven combined	900	900	550	600	600	370
Lights	900	900	550	360	360	220
Plug loads	450	450	300	400	400	270
Total	4600	3200	2200	2120	1730	1200
Savings (%)				54	46	45

for the best appliances on the market. Typically, these are products used in the Task 13 houses.

Ovens and range tops

The energy consumption of the most efficient oven available (a convective oven) is 0.4 kWh when heating the oven to 200°C. The consumption is the same (0.4 kWh/h) for maintaining the temperature. The electricity consumption is approximately the same for many different makes (0.4–0.6 kWh). The use of microwave ovens (0.6–1.2 kW) will also be energy efficient because of their short operation periods. Ceramic range tops (called hobs in the UK), some of them with halogen heating units, are the most efficient ones available, but the actual electricity consumption depends strongly on the user (e.g. using the correct size pots and pans, heating very little water, etc.).

An example of an innovative product used is the halogen light stove in the Canadian Brampton house. The Canadian Waterloo house has an innovative combined vented gas oven and ceramic range top. This is a sealed unit that allows intake of combustion air and venting of combustion products directly to the outside. A small fan is the only electricity consumer in the unit.

In the German Berlin house, an efficient gas stove (combined oven and range top) is installed. To avoid problems with the open combustion system in an airtight mechanically ventilated house, a separate ventilation system for the kitchen was installed (cf. Chapter 2.5).

Figure 2.13.1. Pie charts showing a typical division of electricity consumption by appliances in North America (Canada – top) and Europe (bottom)

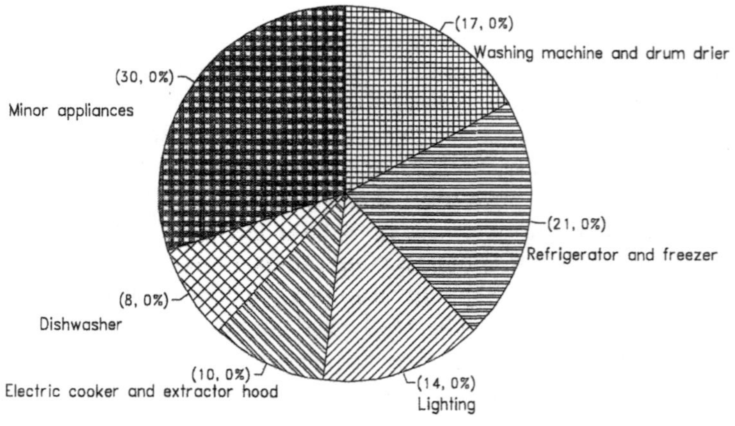

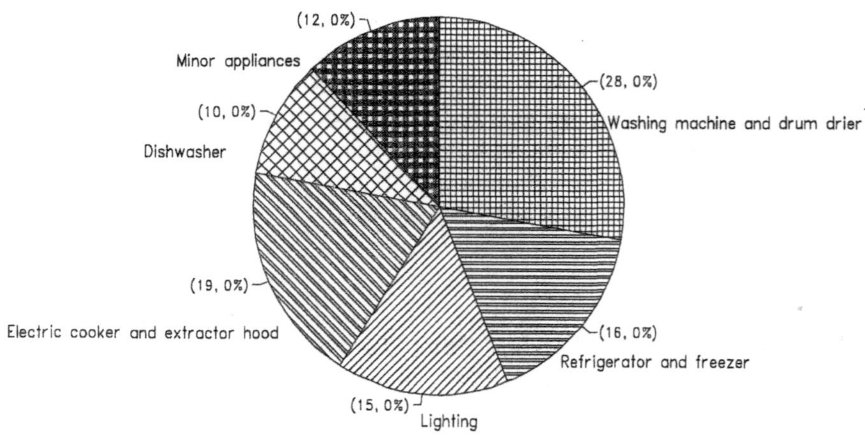

Refrigerators and freezers

Separate refrigerators. The most efficient refrigerator available uses 0.28 kWh per day. This corresponds to 89 kWh/year. The size is 200 litres. 'Normal' consumption lies between 0.6 and 0.8 kWh/day for sizes between 90 and 274 litres. A refrigerator of this type is used in the Danish row houses at Kolding.

Combined refrigerators/freezers. There are two types of combined refrigerator/freezers:

- *Refrigerator with freezer compartment.* A 200 litre refrigerator with a 20 litre freezer compartment has an electricity consumption of approximately 1.9 kWh per day. The consumption varies very little with varying sizes of the freezer or the freezing capacity (the storage temperature).
- *Combined refrigerator/cabinet freezer.* The most energy efficient combined refrigerator/freezer available in Europe is reported to be using 1.1 kWh/ day, which corresponds to 390 kWh/year. This includes a 182 litre refrigerator and a 48 litre cabinet freezer at –18°C with the ability to freeze down 4.5 kg/100 litres within 24 hours. However, the unit installed in the Norwegian house uses only 0.85 kWh/day. 'Normal' energy consumption lies between 1.1 and 1.6 kWh/day, where 1.6 kWh corresponds to a 197 litre refrigerator with a 153 litre freezer.

A large combined refrigerator/cabinet freezer with the extremely low energy consumption of 0.6 kWh per day (a 290 litre refrigerator with a 110 litre freezer) is available in North America. This model was used in the Canadian Brampton House.

In Northern Europe, it is common to have refrigerator and freezer (–18°C) as two separate units. This makes good sense from an electricity conservation point of view. It is thermodynamically inefficient to create the +5°C temperature through a cycle reaching –20°C or even lower. There are other reasons why a combined refrigerator/freezer system is problematic energywise. If one evaporator serves both freezer and refrigerator compartment, as in typical American models, the evaporator will be subjected to moisture from both. Since the refrigerator door is opened more frequently than the door in the freezer, the frost build-up will typically be more than ten times higher than for the freezer alone. A European chest freezer requires defrosting (manually) only every one or two years, whereas a combined American model typically is set to defrost automatically more than once a day. Defrosting a refrigerator evaporator requires no electricity, since the ice melts off between compressor runs.

Another advantage of separating refrigerator and freezer is better temperature control in both, which also results in electricity savings. Finally, the units can be separately switched off, repaired, and replaced. The disadvantage of separate refrigerator and freezer units is mainly the extra cost for another compressor. A solution could therefore be to run two refrigeration circuits alternately with one compressor.

Separate freezers. There are two types of freezers: chest freezers and cabinet freezers. The energy consumption of a chest freezer depends (of course) on the size of the freezer. The best 120 litre freezer on the market today has an energy demand of 0.5 kWh/day and a 300 litre freezer a demand of approximately 0.9 kWh/day. A model with a predicted consumption of 0.6 kWh/day is used in the Danish row houses at Kolding.

The energy demand of the best cabinet freezer of a normal size on the market is 0.9 kWh/day (222 litres). 'Normal' consumption lies between 0.85–1.35 kWh/ day, which corresponds to the energy consumption of 50–258 litre freezers.

Washing machines

The washing temperature and capacity are two factors which must to be considered when the aim is to find the best washing machines. If the washing temperature is 90°C (normal programme without pre-wash), the lowest electricity consumption will be 1.9 kWh corresponding to 1.0 kWh when the washing temperature is 60°C. There is only a small difference in the capacity (40–48 litres, mainly 48 litres), and 'normal' consumption lies between 1.9 and 2.5 kWh (90°C), and between 1.0 and 1.3 kWh (60°C). The development of new detergents allows most clothes to be washed in cold or luke-warm water. However, the use of these products may have a negative environmental impact.

Washing machines and dishwashers are sometimes connected to the domestic hot-water supply (ideally solar heated) to save electricity for water heating. Units with a low water consumption should be chosen, for water and electricity saving. To operate efficiently with a fairly small water consumption, the required temperature must be immediately available, thus requiring circulation of the hot water. The machines should never be connected to hot water only, but should have a two-pipe system to allow for cold- and hot-water connection.

An example of a two-pipe connection to a solar water heating system is found in the Canadian Waterloo house. In this case, the washing machine spin-dries clothes at high revolutions.

In the Danish row houses at Kolding, washing machines are not installed in the individual dwellings. A small communal building for the 33 dwellings also houses efficient washing machines for the occupants.

Clothes drying

The most energy efficient way of drying clothes is the old-fashioned way, taking them outside to dry in the wind and sun, if available. The second best way is to make use of waste heat (or solar heat) in the house, e.g. to dry clothes in a room benefiting from boiler or heat-storage heat loss, or in a drying closet supplied from solar heating.

Table 2.13.3. The importance of the spin-drying process for the time of drying. The spin-drying of a household washing machine is 300–1400 rpm (rotations per minute). A separate spin-drier goes up to 3000 rpm

Spin-drying (rpm)	Water content of laundry %	Water content of laundry litre/kg	Time of drying (minutes)	Energy consumption (kWh)
305	150	5.0	155	5.5
460	110	3.6	95	4.1
800	80	2.6	70	3.0
2800	45	1.5	40	1.7

In the Finnish house an electrically operated drying cabinet has been installed, and in the Canadian Waterloo house a natural gas clothes drier, controlled by a moisture sensor, is used.

If electric driers (drum driers) are used, the preliminary conditions for high efficiency are that the laundry has been spin-dried well (max. 50–60% remaining moisture), that the drier is not overloaded, and that the laundry is not dried for too long.

An example of the importance of a good spin-drying process for the drying of the laundry is given in Table 2.13.3. The energy consumption of the best appliances available today is about 2.3–3.0 kWh per 5 kg laundry (according to the size of the spin-drier).

Of electrical driers, a condenser drier is often considered the most suitable for low-energy dwellings, because it runs without disturbing the balance of the ventilation system. However, heat is transferred from the clothes to the water and thus drained away. On the other hand, it is also difficult to connect a conventional electric drier to a ventilation heat-recovery system, because many of the fine textile fibres will pass the system filters and may lower the recovery efficiency or even clog the heat exchanger.

Dishwasher

Again, the most energy-efficient way of washing dishes etc. is to do it manually, using cold and efficiently heated (preferably solar-heated) hot water. If dishwashers are chosen, the dishwashers should be carefully sized according to the average wash size, i.e. typically half the standard size is desirable. The washing temperature, the capacity of the machine, and the drying process (which uses a large heating element) are the primary parameters for electricity consumption. If a division of the washing and the drying process can be achieved (and the electric drying eliminated), it is an advantage. An example of this is found in the Canadian Waterloo house.

The following electricity consumption per wash can be given for a dishwasher designed for 12–14 persons:

- a normal programme without pre-washing, washing at 50/50°C: 0.9–1.4 kWh
- cold pre-washing and washing at 50/55°C: 1.0–1.7 kWh
- cold pre-washing and washing at 70/65°C: 1.4–2.0 kWh.

Lighting

As far as possible, all rooms should receive natural light. In dwellings, this is usually accomplished with normal windows (in some of the partition walls as well as in the exterior walls), with daylight windows, and with skylights. More advanced daylighting techniques are too expensive and sometimes (often) not suitable for dwellings. The Norwegian and Belgian houses offer the most focused daylight design with, respectively, transparently insulated daylight windows, and a multistorey hollow core accessing light from a large skylight to all floors through a 'translucent' staircase design.

Daylighting should be supplemented with appropriate task lighting. Halogen lamps are often chosen for task lighting and give excellent light, but are not especially energy-efficient.

Over the last decade the development of efficient, long-lasting Compact Fluorescent Lamps (CFL) have brought new possibilities for illuminating buildings. The technique is based on small fluorescent light tubes folded to almost normal bulb size.

In most cases these bulbs can replace incandescent bulbs directly and give considerable energy saving, especially in places where many lamps are used or where light is required for long periods. The average lifetime of the fluorescent bulbs is claimed to be 8000 h (eight times better than the incandescent). The power consumption for the two types relates as shown in Table 2.13.4. Compact Fluorescent Lamps are used in several of the Task 13 houses. They are probably most extensively used in the Norwegian house, where they are combined with an advanced control system.

Use of automatic controls (e.g. presence sensors) can further reduce the electricity use for lighting, as discussed in Chapter 2.12. The Norwegian house and the Belgian PLEIADE house are good examples of use of this technology.

Cross comparisons

Most of the Task 13 houses include energy-efficient electricity use to a certain degree in their designs. Table 2.13.5 shows the electricity consumption of the Task 13 houses in Canada, Germany, Denmark, Finland, and Norway, where the selection of appliances was given

Table 2.13.4. Electricity consumption (W) for incandescent light bulbs and fluorescent bulbs with equal light performance

Incandescent light bulbs	Compact fluorescent bulbs
25	5
40	7–9
60	11–14
75	15–18
100 and more	20–25

Table 2.13.5 Overview of electricity use in kWh/day in selected Task 13 houses

Electricity use (kWh/day)	CDN-W	D-R	DK	FIN	N
Refrigerator	1.15	0.34		0.35	
Freezer	1.15	0.70		0.89	
Refrigerator/ freezer	0.84	0.80		0.85	
Washing-machine	0.33[a]	1.8[b]	1.30[c]	1.8[f]	0.80[k]
Drum drier	1.68	3.1[b]			1.90
Drying cabinet				3.60[g]	
Dishwasher	0.54[a]	1.5[b]	(1.20)[d]	0.30[h]	1.40[k]
Electric cooker/ oven	0.95		0.55[e]	0.40/0.60[j]	
Lights	2.15		1.75		2.00
Plug loads	3.00		2.20		
Total	**10.95**	–	**6.60**	–	–

[a] Not including energy use for hot water
[b] Per wash
[c] 114 washes at 95°C, 130 washes at 60°C and 172 washes at 30°C
[d] 290 washes per year (unit not installed at beginning of monitoring period)
[e] 3–4 hours of use every week
[f] Measured value, full load (3.5 kg), 90°C
[g] Capacity 5 kg
[h] Capacity 14 IEC-norm units, 55°C (1.0 kWh/day at 65°C)
[j] Heating-up to 200°C/continuous use at 200°C
[k] Connected to 55°C DHW, produced by a heat pump

specific attention, covering different products and sizes of appliances. It gives an indication of the emphasis placed on electricity-saving lights and appliances (the gas consumption for the stove in the Canadian Waterloo house is converted to kWh, 99 m^3 gas equivalent to 1040 kWh).

Conclusions

Today it is possible to obtain large electricity savings by influencing both the habits of the residents and by gradually introducing more efficient appliances.

Over the last ten years much work has been done to develop more energy-efficient appliances. The electricity consumption for new appliances is today about half that of ten-year-old equipment, but the ultimate efficiency has by no means been reached yet, and compared to the best appliances available today it is still possible to make significant improvements. It is to be expected that the energy consumption of the best appliances will be reduced by a further 50% over the next ten years. Also, efficient control of the use of daylight and artificial lighting can further reduce the electricity consumption.

The Task 13 experience can be briefly summarized as this:

- Avoid unnecessary use of equipment and appliances – and watch for stand-by parasitic electricity consumption
- Select all appliances carefully – a wide range of very efficient equipment is available
- Use natural light first and then Compact Fluorescent Lights (CFLs) wherever possible
- In food preparation, the use of pots etc. with built-in heating elements, e.g. the familiar electric kettle, should be encouraged, as well as the use of ceramic range tops (hobs), convector ovens, and microwave ovens. A prerequisite for energy-efficient food preparation is to heat very little water
- In selecting refrigerators and freezers, the appliances should not be oversized. Generally, separate refrigerator and freezer units are the most efficient, but some combined units have two compressors or run two cooling circuits alternately on one compressor and thus achieve high efficiencies
- When selecting washing machines, look for high-revolution spin-drying, and investigate combined hot- and cold-water supply
- Dry clothes outdoors or in drying cabinets using solar or waste heat – if using electric driers, look for high-revolution units, and select condenser driers if the building is equipped with a balanced ventilation system
- When using dishwashers, downsize the units, and investigate combined hot- and cold-water supply.
- When electricity is produced from non-renewable sources, look for gas substitution equipment, such as gas ovens.

RECOMMENDED READING

Johansson T B (ed.) (1989). *Electricity: Efficient end-use and new generation technologies, and their planning implications.* Lund University Press, Lund, Sweden.

Nørgård J S, Gydesen A (1993). *Energy efficient domestic appliances – analyses and field tests.* Energy Group, Physics Department, Technical University of Denmark. Invited presentation at NATO workshop, Espinho, Portugal, June 1993. Off-print available from Department of Buildings and Energy, Technical University of Denmark.

Thomsen K E (1994). *Analysis of Energy Efficient Lights and Appliances in Well Insulated Buildings.* Paper for IEA Future Building Forum 'Efficient use of electricity', Sophia Antipolis, France, 1994. Off-print available from Department of Buildings and Energy, Technical University of Denmark.

- *Author.* Kirsten Engelund Thomsen (Department of Buildings and Energy, Technical University of Denmark, Building 118, DK-2800 Lyngby, Denmark)

3

Examples

Robert Hastings (editor)

Introduction

The 15 houses presented here represent an interesting range of approaches to achieving very low energy consumption. Some include highly innovative technologies and complex systems, like the Norwegian and Finnish houses with their roof-integrated photovoltaic system and heat pumps; or the German Berlin house, which achieves zero auxiliary heating by means of an active collector sloping up the height of the south façade, supplying heat to a large storage tank that is a central feature of the house. These houses, still far from being economical, given present energy prices, are contrasted by the less dramatic, but builder-constructed, houses that also achieve very low energy consumption. Three examples: the Belgian and Danish row houses and the Swiss duplex rely on very good insulation and interior construction mass to enhance the usability of solar gains through generous south-facing window areas with high-quality fenestration. Because the shells of all the houses had been so thermally optimized, minimizing the heating required for tempering fresh air became essential. With such tight construction, mechanical ventilation is necessary to ensure adequate room air quality. Recovery of heat from the exhaust air is then only logical.

Because 100% solar coverage is still too costly, all but the Berlin house require back-up heating. The most sophisticated back-up systems can be found in the American, Canadian, and Netherlands houses, which all use integrated mechanical systems (combining heat recovery, heating, cooling, and ventilation). The most complex control system can be found in the Belgian house, which has a decentralized binary unit system interacting with the heating, ventilation, lighting, shading, and security systems.

Cooling was a special design issue in three of the projects. In the Yosemite house night flushing was the chosen cooling strategy, with cross-ventilation enhanced through windows strategically placed relative to the prevailing evening breezes. The Japanese house makes use of phase-change storage materials to shift electricity demand to off peak. The Italian apartment building concept (the only project not built) explored the use of solar-driven absorption cooling.

Innovative construction techniques were also tried out for some of the houses. The Swedish prototype house makes use of precut insulation panels splined together. The American Grand Canyon house uses prefabricated sandwich panel construction. The Swiss house mixes conventional masonry construction for the north-facing walls and a light, wooden-post construction carrying concrete slabs for the south-facing zone.

Table 3.1 overleaf provides an overview of the strategies and technologies used in each of the 15 houses.

Table 3.1. Strategies and technologies used in each of the Task 13 houses

	US-GC	US-Y	B	CDN-B	CDN-W	DK	NL	FIN	D-B	D-R	I	J	N	S	CH
A Space heating reduction															
1 *Reduce transmission losses*															
compactness	x	x	x		x	x	x	x	x	x			x	x	x
super-insulation	x		x	x	x	x	x	x	x	x			x	x	x
high-performance windows				x	x	x	x	x	x	x			x	x	x
2 *Reduce infiltration losses*	x	x	x	x	x	x	x	x	x	x		x	x	x	x
3 *Heat recovery (heat pump/heat exhanger)*															
exhaust air	x	x	x	x	x	x	x	x	x	x		x	x	x	x
waste water				x											
ground								x	x				x		
4 *Passive solar systems*															
Trombe wall	x														
transparent insulation													x		
sunspace				x			x	x	x	x			x		
direct-gain systems	x	x	x	x	x	x	x		x	x		x	x	x	x
5 *Active solar systems*															
with short-term storage systems						x		x	x		x		x		
with seasonal storage systems								x	x						
6 *Energy-efficient heating systems*															
integrated mechanical systems	x	x		x	x						x		x		
floor/ceiling heating systems						x									
					x		x		x	x		x			
B Cooling avoidance															
1 *Reduce solar gain (shading)*		x	x	x	x		x		x	x	x	x		x	
2 *Natural ventilation*	x	x	x		x	x	x		x	x	x	x	x	x	
3 *Thermal mass*															
phase-change materials												x			
building mass	x	x	x			x	x		x	x	x			x	x
4 *Energy-efficient cooling systems*															
heat pumps															
heat exchangers															
C Water heating reduction															
1 *Reduce demand (saving devices)*				x	x	x		x						x	
2 *Active solar systems (DHW)*					x	x	x	x	x		x			x	
3 *Heat recovery*	x	x		x							x			x	
4 *Energy-efficient heating systems*	x	x				x		x					x	x	
D Electricity reduction															
1 *Efficient lights and appliances*															
special daylighting features					x									x	
lights and appliances	x	x		x	x	x		x	x		x		x	x	
2 *PV systems*															
grid-coupled systems								x				x	x		x
small power systems					x	x									
3 *Fans and pumps*															
energy-efficient devices					x	x			x	x				x	
energy-saving control						x					x	x		x	
E Home automation systems															
heating			x				x	x				x	x	x	
cooling							x					x			
DHW															
electricity			x									x		x	

3.2 The American house at Grand Canyon

SUMMARY

This house demonstrates that the total energy use can be reduced to about 25% of the energy used by contemporary houses in the same region. Normal electrical use is reduced to one-half of typical levels by use of compact fluorescent lights and efficient appliances.

The annual heating load is reduced by 91%, compared with typical houses being constructed in the area. The second-largest load, water heating, is reduced by 65% by the use of the exhaust-air heat pump, part of the integrated mechanical system.

KEY FEATURES

- structural insulated panels, providing good insulation and airtightness
- direct gain passive solar heating
- trombe wall passive solar heating
- integrated mechanical system, providing exhaust-air heat recovery
- efficient lights and appliances.

LOCATION

Arizona, USA. Latitude 36°N, longitude 110°W. The house is located near the South Rim of the Grand Canyon in a Ponderosa pine tree forest at an altitude of 2135 m.

CLIMATE

The winter is cold and the summer is mild with no requirement for mechanical cooling (Figure 3.2.1). The

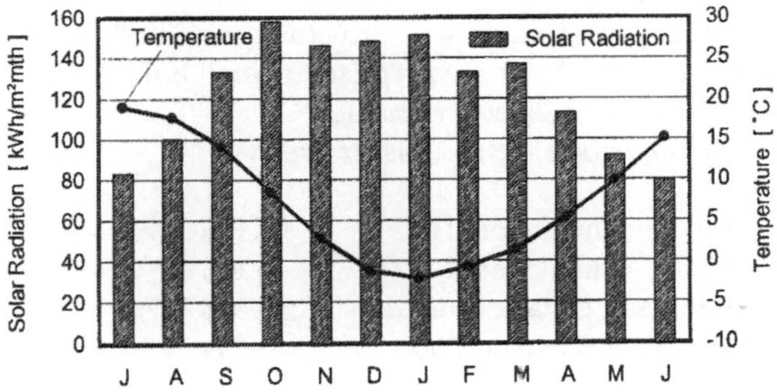

Figure 3.2.1. Grand Canyon climate diagram

combination of an extremely cold climate and abundant winter sunshine provides an ideal solar design opportunity.

CONCEPT

The house (Figures 3.2.2 and 3.2.3), a 125 m^2 (net area), two-storey, three-bedroom unit, is one of the first 16 houses to be built in a 59-unit extension of the employee housing area on the south rim of the Grand Canyon.

Heating loads have been minimized by combining a compact design and an airtight envelope constructed with structural insulated panels. These prefabricated units simplify erection of the house in a remote location where minimum site disturbance is a priority. The panels consist of two oriented strandboard external skins glued to a foam core providing very good insulation with a minimum thickness. No additional wall structure is required. Heat losses to the ground are minimized by insulating both under the concrete floor slab and outside the foundation perimeter. The effective leakage area is specified to be within a range of 190 cm^2 to 320 cm^2, as measured in a blower-door test.

High-performance windows supply direct solar gain, and a Trombe wall supplies delayed heat. Each provide about the same net annual heating energy benefit. The solar savings fraction is 65% of the net heat load. The unvented Trombe wall is covered with double water-white glazing. The mass wall consists of a 20-cm-thick concrete half-height wall that extends across the entire south side of the house beneath the windows. A selective-surface foil is adhered to the outside of the wall to reduce losses by thermal radiation transfer to the glazing. This combination yields a seasonal efficiency of 56%, which is nearly double that of a typical Trombe wall.

CONSTRUCTION

The pre-cut structural insulated panels are an engineered system. Splines, pinned and glued, join the panels without the need for wood studs. Wood thermal bridges occur only at the top and bottom plates, at each corner, and around openings. Electrical service is run through pre-cut holes. The foam insulation is an expanded polystyrene made with fire retardants but without CFCs. The construction is inherently tight and rigid because all joints are glued. Component tests of small structures made with structural insulated panels have confirmed that the predicted overall building loss coefficient is realized in practice.

The envelope characteristics are:

roof U-value, overall	0.126 W/m^2K
wall U-value, overall	0.166 W/m^2K
perimeter U-value, added	0.568 W/m^2K
under-slab U-value, added	0.568 W/m^2K
door U-value	0.963 W/m^2K
window U-value, overall	2.00 W/m^2K
window solar transmittance	76%
controlled ventilation	0.50 ac/h

SOLAR SYSTEMS

The good performance of the Trombe wall (Figure 3.2.4) is achieved by using a simple design without the complication of thermocirculation vents (which would be counterproductive in this application). The wall is an upward extension of the high-conductivity concrete foundation wall, finished on the inside with sheetrock (to match the other walls) and on the outside with a selective surface (a metal foil glued-to-the-wall surface that has a solar absorptivity of 0.93 and an infrared emittance of 0.07). The double glazing is a low-iron glass to maximize solar transmittance. This combination minimizes thermal back-losses to the outside, provides for adequate heat storage, and encourages conduction through the wall. Heat transfer to the room is by radiation and convection. The floor slab is heated somewhat because the wall extends slightly below it.

Direct-gain solar heating is achieved simply by locating most of the windows on the south side. The U-value of these windows is 2.00 W/m^2K. There would be little advantage in using windows with a lower U-value in this sunny climate because reduced solar transmittance almost cancels the benefit of reduced heat losses.

HVAC

The house uses an integrated mechanical system (IMS) that combines the functions of auxiliary space heating, controlled ventilation, heat recovery, hot water heating, and auxiliary cooling in one small and economical unit. It consists of a 300 litre insulated hot water tank, two blowers, controls, and two heat exchangers that serve as the heat-pump evaporator and condenser. Exhaust air is drawn from the bathrooms and the kitchen causing a slight negative pressure in the house. This causes fresh air to be drawn into the house through normal cracks in the exterior. The average coefficient of performance (COP) of the heat pump is 3.1.

Indoor air quality is maintained by continuous outside-air ventilation of 170 m^3/h and a particulate-arresting filter on the recirculated indoor air that removes 98% of airborne particles 6 microns or larger.

The IMS provides for base-load heating, comprising about half the total load but cannot meet peak requirements. The IMS heat is exhausted in a central location, eliminating the need for a distribution system. The remainder of the required backup heat is from electric-resistance base-board (skirting board) convectors in the various rooms, providing good zone control in response to room-by-room needs. Backup heat is largely in the late night hours, out of phase with the peak loads of the electric utility system.

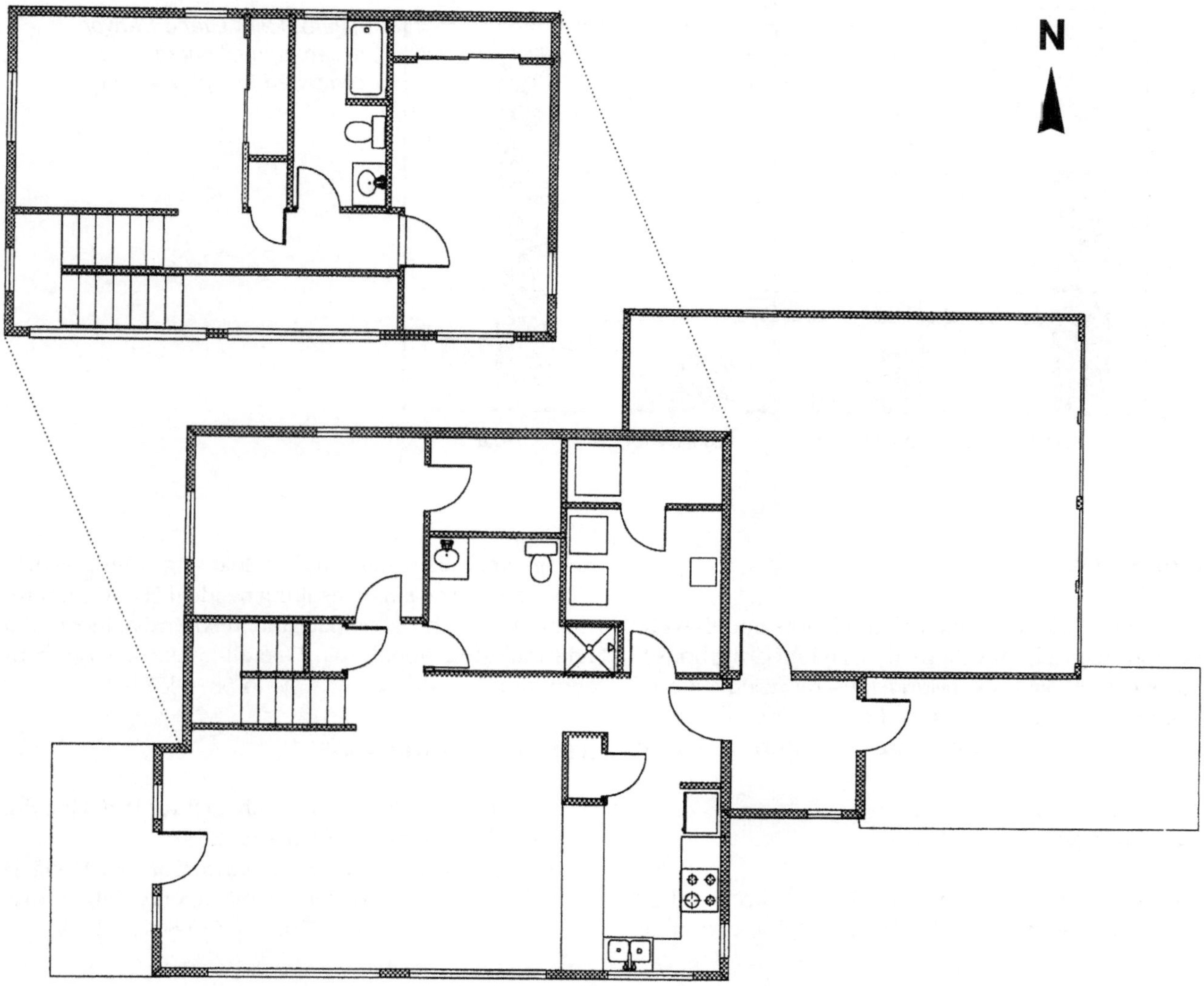

Figure 3.2.2. Floor plan (Grand Canyon house)

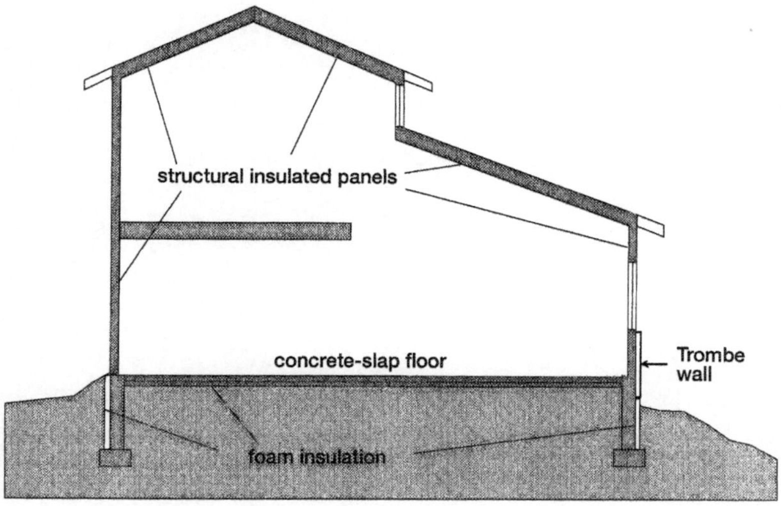

Figure 3.2.3. North–south Section (Grand Canyon house)

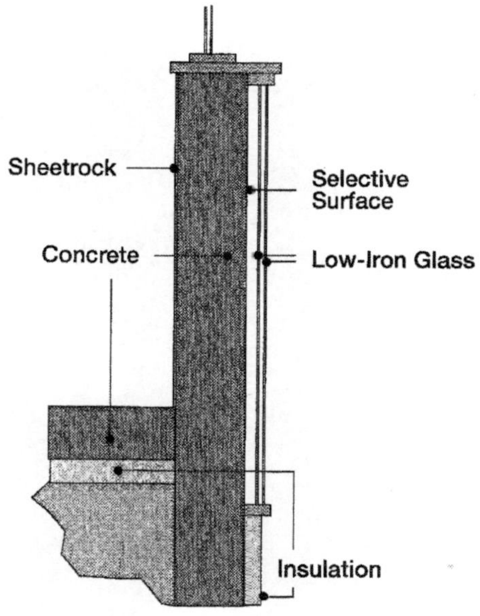

Figure 3.2.4. Grand Canyon house: Trombe wall detail

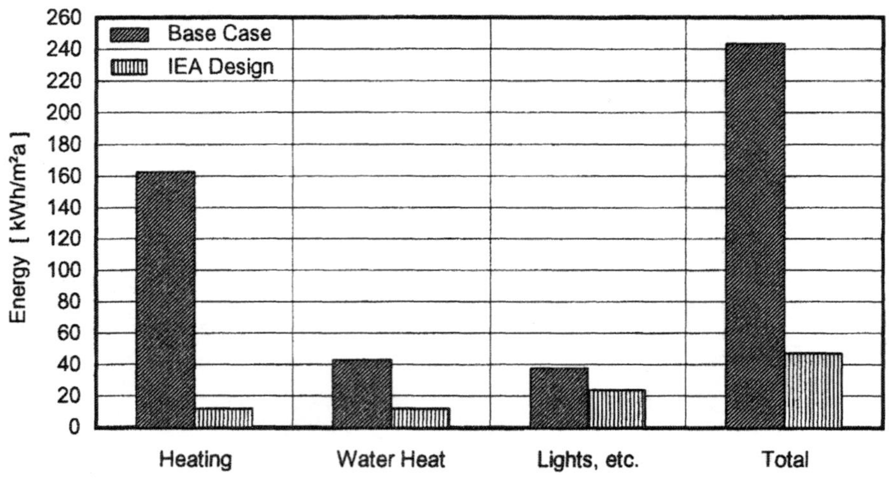

Figure 3.2.5. Grand Canyon house: annual energy use, compared to typical houses

ENERGY USE

Thermal network mathematical simulation models were developed to predict the thermal behaviour of the house. Input weather data are hourly files representative of long-term average patterns at the site. Annual energy use, compared to typical houses, is shown in Figure 3.2.5.

CONCLUSIONS

Good performance at a low added cost is achieved by keeping the design simple, by paying attention to energy analysis results, and by learning from practical experience. The small resulting residual HVAC requirements are met by a compact and inexpensive integrated and multi-functional unit. Overall annual energy consumption is reduced by 76%, to 5876 kWh.

ACKNOWLEDGEMENTS

- *Architectural design.* OZ Architecture, 1580 Lincoln, Suite 200, Denver, CO 80203, USA.
- *Energy design and analysis and author.* Dr J. Douglas Balcomb, National Renewable Energy Laboratory, 1617 Cole Boulevard, Golden, CO 80401, USA.

3.3 The American house for Yosemite at El Portal

SUMMARY

In this house summer cooling loads are nearly eliminated; mechanical cooling will be required only toward the end of a string of hot summer days. Most of this benefit is achieved by night ventilation working in conjunction with added thermal mass for removing heat during the day. The ceiling fans play an important, but secondary, role. Direct gain passive solar heating reduces annual heating loads substantially. The next largest load, water heating, is reduced by 65% by means of the exhaust-air heat pump, which is part of the integrated mechanical system.

KEY FEATURES

- high-mass foam-core walls, providing thermal storage
- night-vent cooling strategy
- direct-gain passive solar heating
- integrated mechanical system, providing exhaust-air heat recovery
- efficient lights and appliances.

LOCATION

El Portal, California, USA. Latitude 38°N, longitude 119°W. The house is situated 16 km to the west of Yosemite National Park at an altitude of 610 m on the north side of the steep and narrow Merced River Canyon.

CLIMATE

The climate (Figure 3.3.1) is continental with mild winters and hot summers. The average daily temperature swing in July is from 12°C at night to 38°C in the afternoon with extreme peak conditions 7 K higher. Cool down-canyon breezes provide a reliable natural cooling source during summer nights.

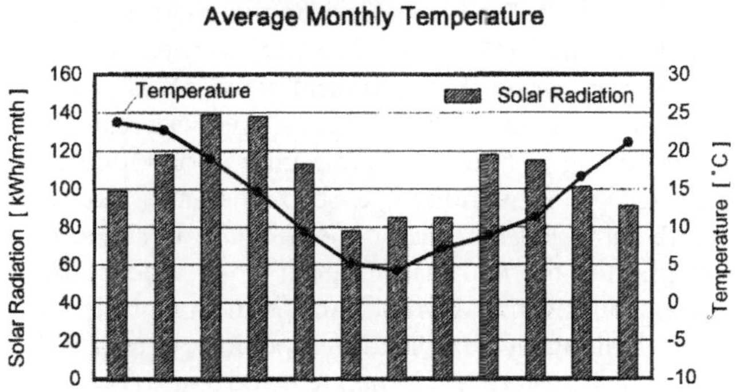

Figure 3.3.1. El Portal climate diagram

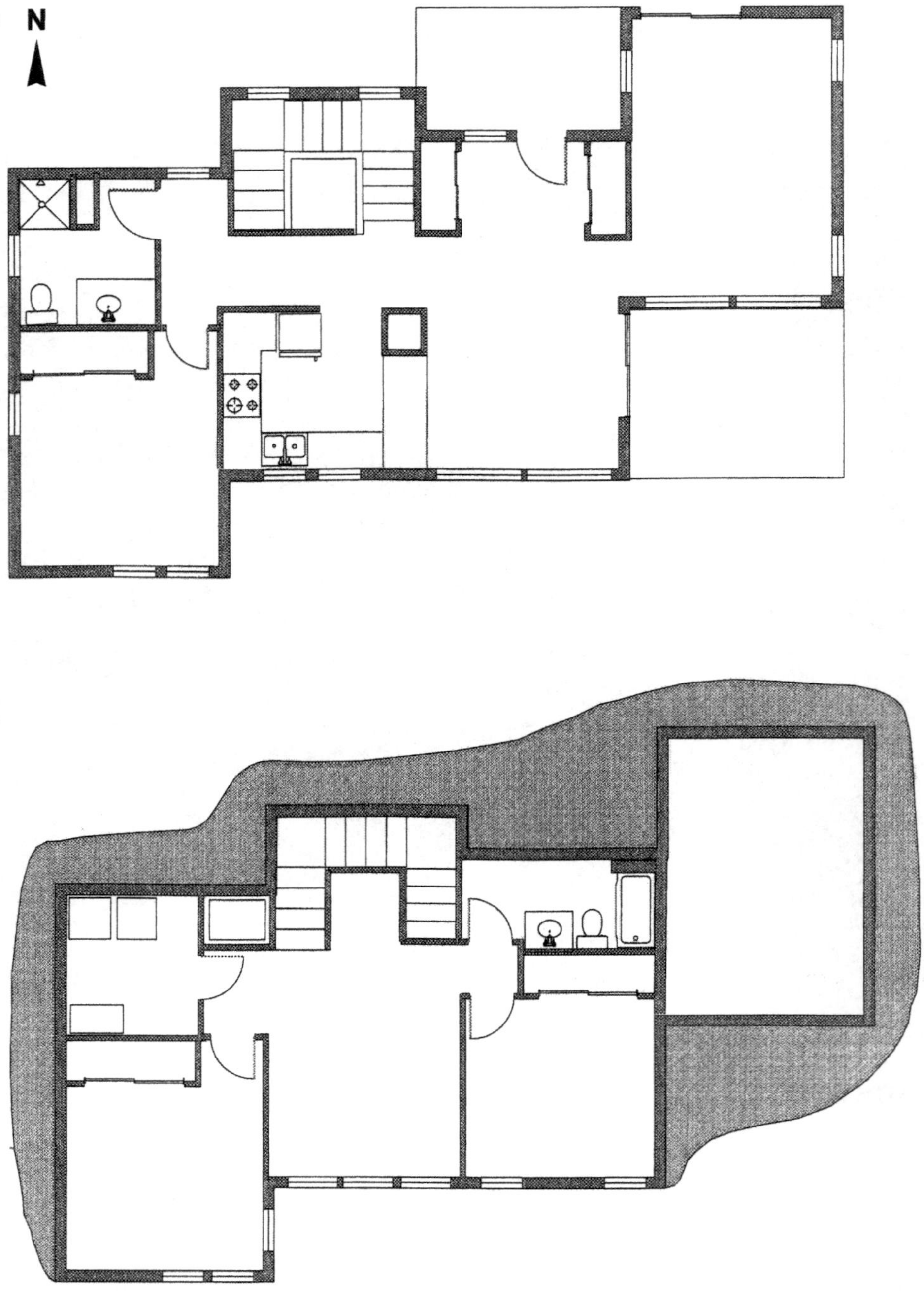

Figure 3.3.2. Floor plan (Yosemite house)

CONCEPT

The house is a 130 m², two-storey unit with one bedroom on the upper level and two bedrooms in a half-basement that is exposed on the south side only. The house plan (Figure 3.3.2) is not rectangular but has two offsets along the north and south sides, forming staggered façades. These jogs in the wall catch the easterly night breeze encouraging cross-ventilation through open windows east-to-west and north-to-south through each major room of the upstairs, thus cooling the thermal mass (Figure 3.3.3). The house is closed during the daytime, and heat is stored in the mass, limiting the peak inside temperature to reasonable levels. This stored heat is then rejected to the cool ventilation air on the following night, and the cycle repeats. Direct gain passive solar heating is achieved by locating a majority of the windows on the south side.

The lower level is fully earth-bermed on the east, north, and west sides, effectively sheltering it from summer overheating. The south side is exposed for passive solar heating.

Large-bladed paddle fans are hung from the ceilings in the main upstairs rooms. These force a large volume of downward-flowing air, creating circular air flow patterns in the room with typical air velocities of 25 to 75 cm/s. This is sufficient to make a significant improvement in thermal comfort on summer afternoons but not enough to disturb loose papers. The effect on comfort is equivalent to a reduction in still-air temperature of 2.2 K. Thus the cooling set point can be raised by this amount, resulting in a major savings in

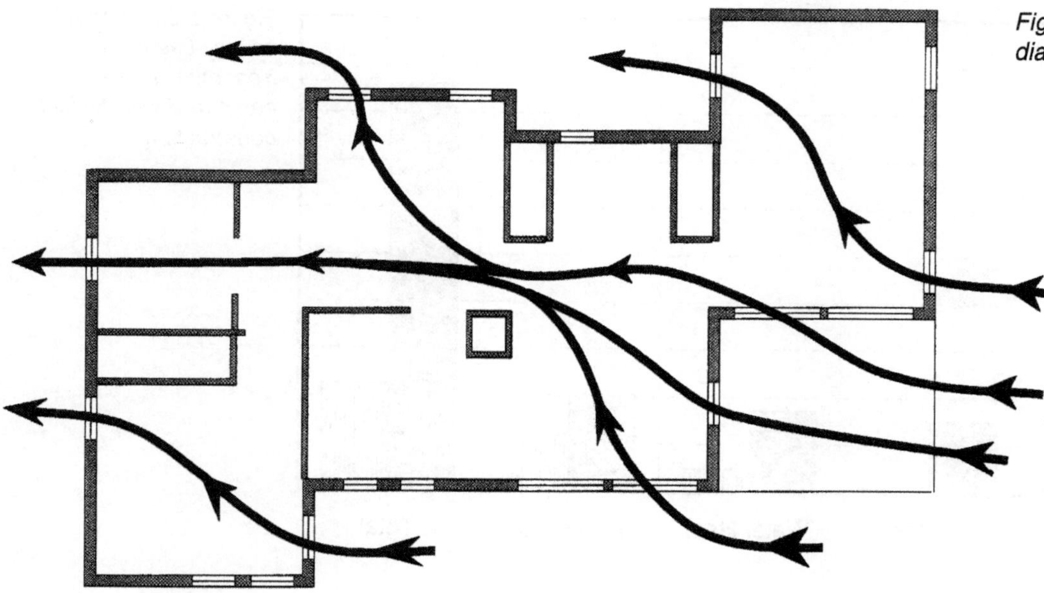

Figure 3.3.3. Air-flow diagram (Yosemite house)

auxiliary cooling energy with no sacrifice in comfort. Fan-blade rotation is reversible to provide half-speed up-flow in winter, thus minimizing the cooling effect.

CONSTRUCTION

To increase mass, a construction system is used different to the typical lightweight wood-frame-and-fibreglass residential construction prevalent in the USA. The selected system combines thermal mass, good insulation, and high structural integrity into a low-cost

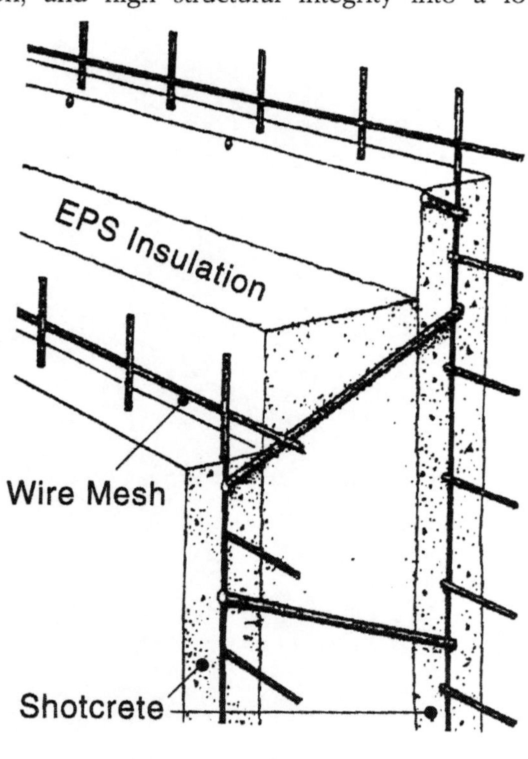

Figure 3.3.4. Yosemite house wall construction

exterior wall. It consists of a wire-frame structure bracketing an EPS foam, as shown in Figure 3.3.4. The units, which are lightweight and easily handled, are erected on site, wired together to form a reinforcing network, and then sprayed with concrete on both sides (a process called shotcrete). The end result is an insulating sandwich consisting of two 5 cm thick reinforced concrete panels enclosing a 10 cm foam core. The U-value of the wall through the foam is 0.29 W/m²K. The unit includes a few thin reinforcing wires that tie the inside and outside wire meshes together, resulting in thermal bridging of about 30%. The resulting structure is strong, meeting strict seismic building codes.

The envelope characteristics are:

roof U-value, overall	0.13 W/m²K
wall U-value, overall	0.21 W/m²K
perimeter U-value, added	1.13 W/m²K
window U-value, overall	1.83 W/m²K
door U-value	0.96 W/m²K
window solar transmittance	58%
ventilation	0.45 ac/h

HVAC

The house has the same type of integrated mechanical system as the Grand Canyon house. The cooling feature of the unit is used at this site, pumping heat either into hot water or into exhaust air.

ENERGY USE

The annual energy use compared with typical construction is shown in Figure 3.3.5.

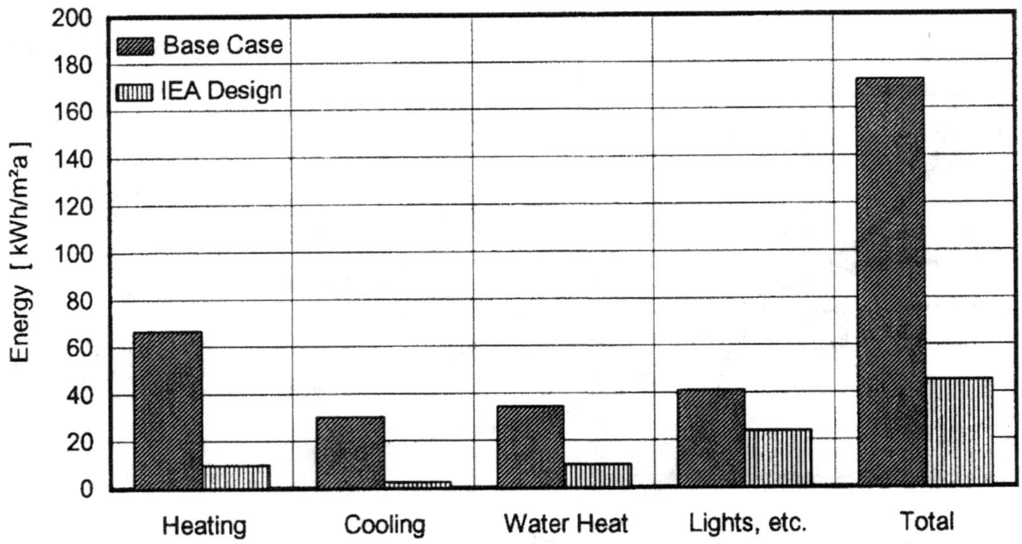

Figure 3.3.5. Annual energy use in the Yosemite house compared with typical construction

CONCLUSIONS

Attention paid to the micro-climate of this canyon site resulted in a well-suited design solution. The search for an economical wall construction method led to a novel product made in the unlikely state of Florida. Annual mechanical cooling is reduced by 98% compared with a base-case house in the same climate. The primary strategies used are shading, night ventilation, added building mass, and ceiling fans. The night-vent strategy requires the active participation of the occupants and therefore may not be suitable for all situations. Overall annual energy consumption is reduced by 73%, to 8220 kWh. Perhaps too much attention was paid to the summer cooling issue, leaving the winter heating as the largest energy use in this mild, sunny climate.

ACKNOWLEDGEMENTS

- *Architectural design*. OZ Architecture, 1580 Lincoln, Suite 200, Denver, CO 80203, USA.
- *Energy design and analysis and author*. Dr J. Douglas Balcomb, National Renewable Energy Laboratory, 1617 Cole Boulevard, Golden, CO 80401, USA.

3.4 The Belgian PLEIADE row house

SUMMARY

During the design of this two-storey row house, with a 180 m² heated floor area, special attention was given to thermal comfort, good indoor air quality, and daylight. Both a gas air and an electrical heating system were installed for purposes of comparison. The house also contains a home automation system, which ensures optimal energy use and thermal comfort.

KEY FEATURES

- super insulation
- thermal mass
- direct gain
- solar control
- natural lighting
- heat-recovery system
- home automation system
- airtightness and air quality.

LOCATION

The house is located in Louvain-la-Neuve, 30 km south-east of Brussels, at latitude 50°N.

CLIMATE

The climate is cold temperate (Figure 3.4.1).

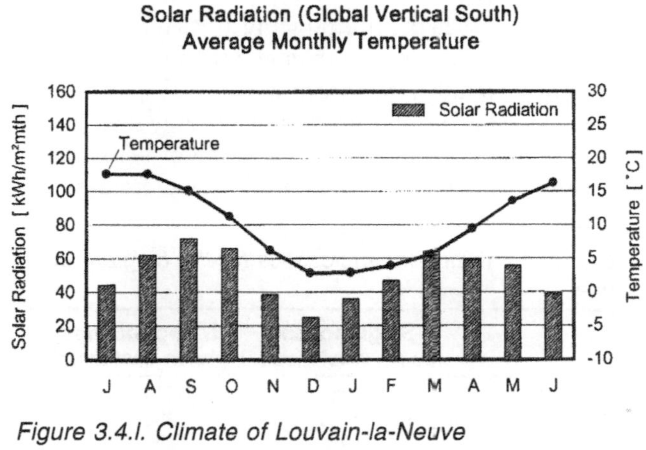

Figure 3.4.l. Climate of Louvain-la-Neuve

CONCEPT

The ground floor is arranged according to the temperature requirements of the different spaces. The living room is placed along the sun-lit façade, the dining area is set back, and the kitchen, entrance, and garage face north east. The central hall is daylit from a roof skylight. The first floor is planned symmetrically around the lightwell. The office is under the large southern-end skylight, and the bedrooms in the four corners are by the blind gables. Plans and section are shown in Figure 3.4.2.

The walls are an insulated, prefabricated wood construction with internal insulated facing that incorporates wiring and plumbing. The slate-covered roof is very well insulated, and incorporates an insulated false ceiling. The glazed openings are small on the north-east-facing street side and large on the south-west-facing garden side. All the sun-lit bay windows, and the skylight, are equipped with automatically controlled mobile blinds. Initial simulations indicated that thermal mass would reduce the energy consumption by about 10%, and help combat overheating. Thermal mass is therefore provided in the form of concrete floors, a heavy limestone interior, and masonry party walls.

The envelope characteristics are:

outside walls (25 cm of insulation)	0.14 W/m^2K
floor at ground level (12 cm of insulation)	0.19 W/m^2K
roof (33 cm of insulation):	0.12 W/m^2K
outside doors:	0.70 W/m^2K

The south-west façade is equipped with double glazing with a vacuum-deposited low-emissivity coating, while the north-east façade and the glass roof have double glazing with two low-emissivity coatings. The 15 mm gap between the glass panes is filled with argon gas.

The glazing characteristics are given in Table 3.4.1.

Simulation studies carried out before the project was started have indicated that:

- thermal mass in the south-facing areas lessens heat requirements and overheating appreciably
- thermal mass in the north-facing areas reduces overheating, particularly if movements of air between the north and south zones are induced
- a sunspace achieves only minimal energy savings
- in a super-insulated house with night setback the heating power determined by static analyses is underestimated.

Table 3.4.1. Glazing characteristics for the Belgian row house

	Single coating (south-west facade)	Two coatings (north-east facade)
Solar transmission	65%	55%
Visible light transmission	75%	72%
Light reflection	13%	11%
U-value	1.32	1.14

HVAC

Preliminary studies have shown that in the case of a super-insulated house with considerable thermal mass, the heating power calculated according to the norm NBN B62-003 must be increased to allow for non-stationary situations, particularly if the heating has to be restarted after a long absence of the occupants. The heating power installed in the house is 8.6 kW.

The heating system of houses with low energy consumption must be able to respond adequately to fluctuations in solar and internal gains. In this house, two heating systems were tested: an electric system, partially based on heat storage, and a gas system with pulsed air.

Each of these systems is independent of the other and was tested independently.

In the initial phase, a fired boiler provides domestic hot water. Later, solar collectors will be installed on the south-west-facing roof.

Electric heating

Production and regulation of electric heating is decentralized and consists of:

- dynamic off-peak storage heaters, in areas with high heat demands
- direct convectors, in areas with low heat demands
- radiating mirrors in the bathrooms.

During peak hours the electricity network sends a signal to the home automation system, which then cuts out the storage-heater resistors, thereby minimizing consumption at the peak hourly rate. The regulation of the electric heating by the home automation system includes both the programming and manual control of temperatures, the shutting off of the convectors or storage heater fans in the case of windows being opened, and the cutting of power to certain appliances during peak hours and during periods of large electricity consumption. This is done in order to limit the total connection power of the house to 15 kW.

Gas heating system and ventilation

A hot-air system was chosen, both because there was an interest in testing it and because it could be combined with the mechanical ventilation system. The ventilation system provides the required air quality and prevents overheating during hot periods (night-time flushing).

A four-zone regulation system regulates the heat requirements in each zone and prevents overheating, especially in the south-west part of the house. The system is of the 'mechanical pulse and extraction with static recuperator' type. Figure 3.4.3 shows the combined heating and ventilation system. An air-to-air exchanger is included in the system. The four zones,

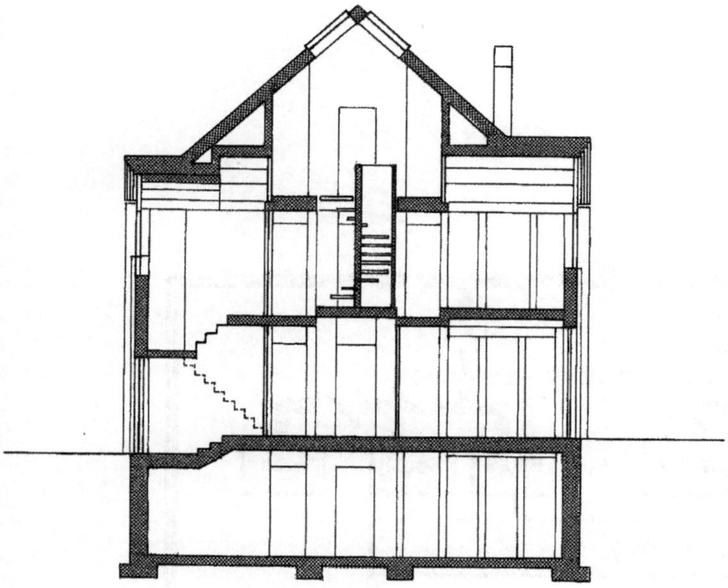

Section

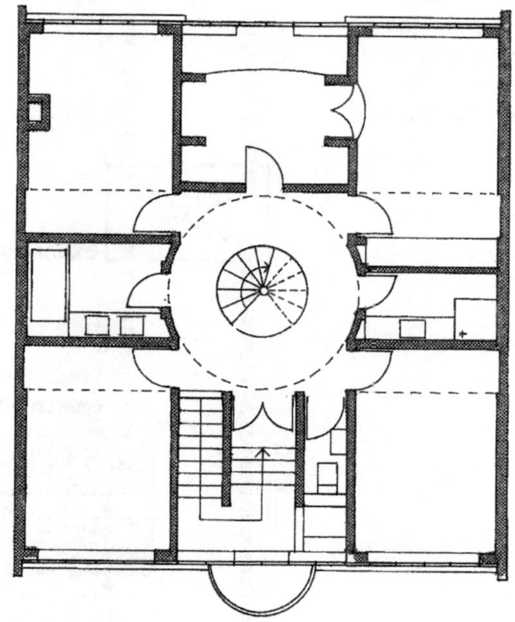

First floor

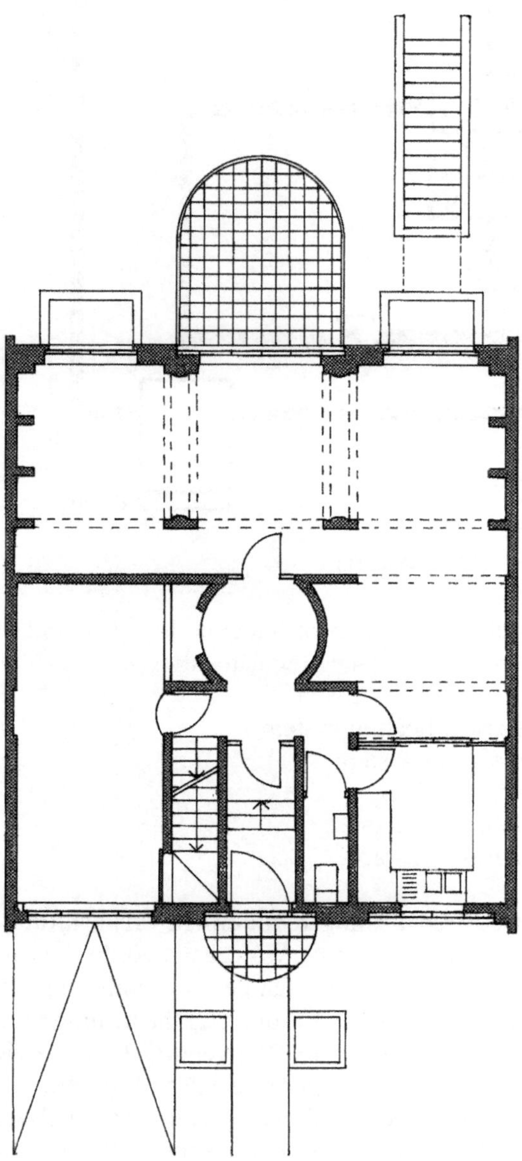

Ground floor

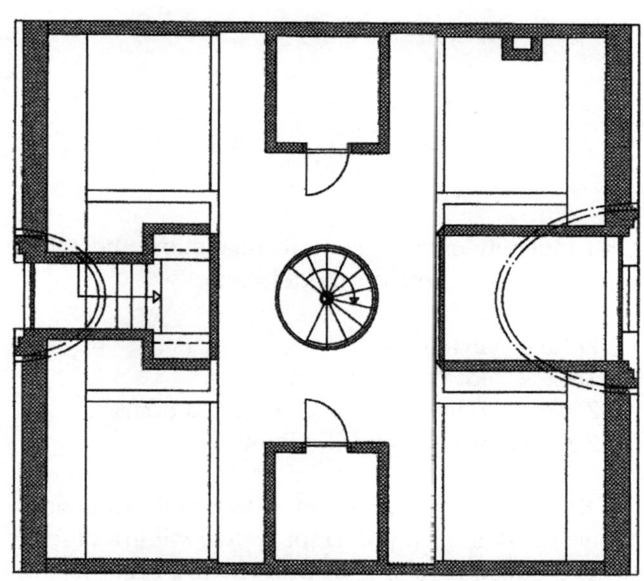

Second floor

Figure 3.4.2. Plans and section of the Belgian row house

Figure 3.4.3. Gas heating system and ventilation in the Belgian row house

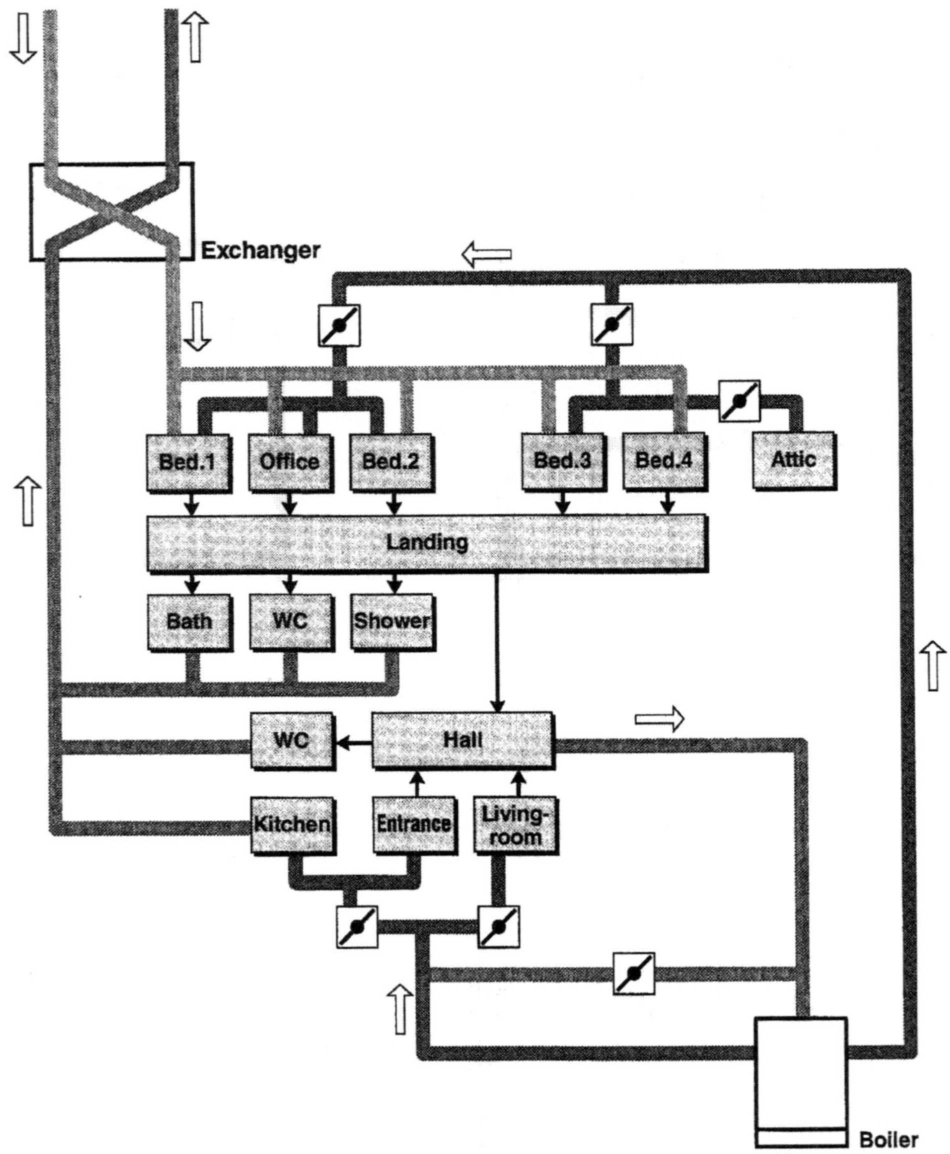

with independent temperature regulation and therefore separate thermostat modules, are:

Zone 1: living room
Zone 2: kitchen and entrance hall
Zone 3: office and south-facing bedrooms
Zone 4: north-facing bedrooms.

An EXCEL IRC system provides decentralized control. It consists of a microprocessor system with local intelligence linked by a special BUS (Binary Unit System). It is particularly well adapted to the management of current loads and integrates well into the electrical installation. All the appliances are connected in parallel to the BUS, which covers the entire building. Certain modules use infrared signals for communication. This simplifies installation and wiring. The user–system interface consists of a wall control panel with a screen and is operated with the help of menu-driven software.

HOME AUTOMATION

The automatic system of integrated control and management (Figure 3.4.4) regulates the:

- electrical heating system
- ventilation system
- electric lighting
- shading devices
- security sensors.

The predicted heating requirement (disregarding the efficiency of the system) was calculated. In the case of a house with a heat exchanger and with intermittent day/night heating it was found that the heating requirement was 15.3 kWh/m^2. Conventional new Belgian houses have a consumption of 135 to 170 kWh/m^2 per year. Strategies to avoid overheating were analyzed in greater detail because in many super-insulated dwellings, overheating occurs during hot weather. Extreme

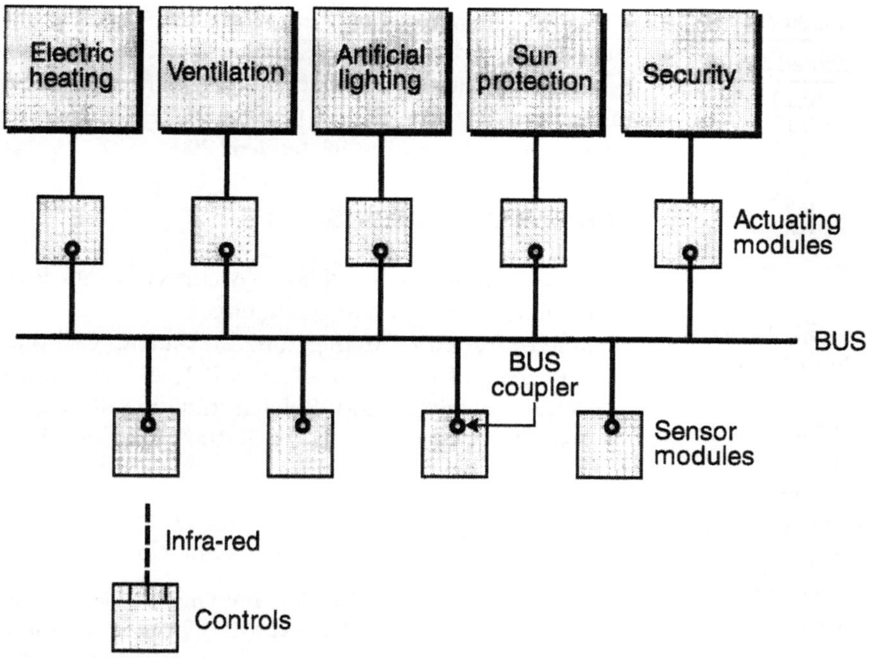

Figure 3.4.4. The home automation system in the Belgian row house

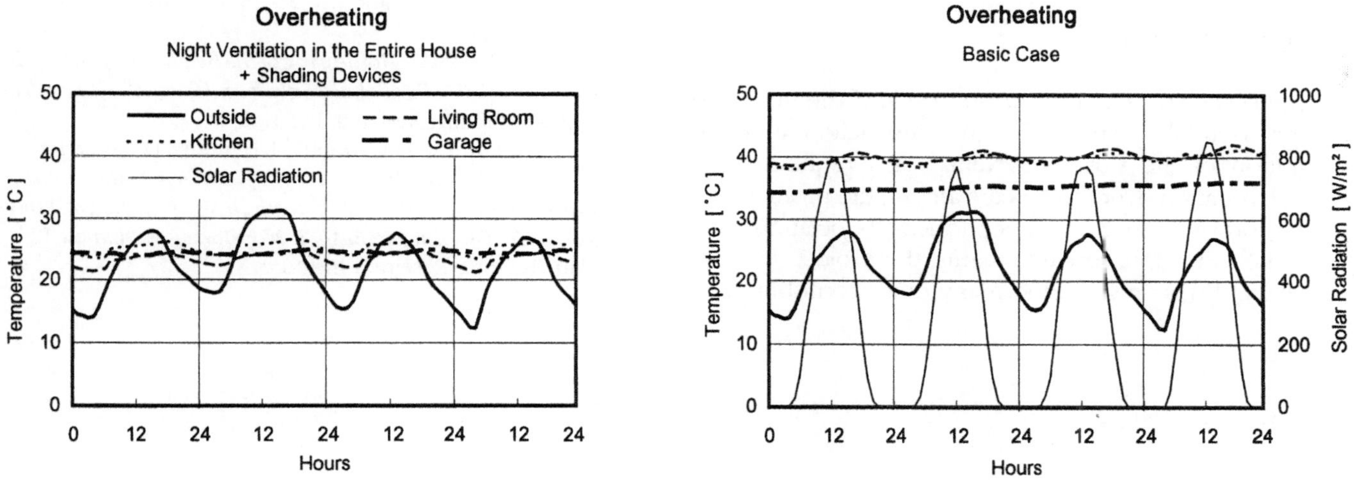

Figure 3.4.5. Simulation results for the Belgian row house

weather conditions were considered (high outdoor temperature and intense solar radiation). In the base case studied, there was no artificial or natural ventilation and in each room an infiltration of only 0.05 ac/h was assumed. It was found that the temperature in the rooms was very high, around 40°C, and it was also very stable. This is due to the inertia of the building.

In the case of natural night ventilation and with shading, all indoor temperatures were 15 K lower than the base case values. The temperature of the living room fluctuated between 21°C and 25°C, the kitchen between 23°C and 27°C (due to internal gains), and the bedrooms between 22°C and 25°C. Sun screens were placed in front of the southwest bay windows and over the glass roof. In the simulations (Figure 3.4.5), it was assumed that the bay windows were shaded when the global solar radiation passing through the glass exceeded 150 W/m².

MONITORING

Monitoring was conducted in order to analyse the end-uses of energy, thermal comfort in winter and summer, advantages of natural lighting, and the responses of the heating and ventilation systems to different weather conditions.

The first airtightness measurements, done when the dwelling was completed yielded an n_{50} value of 5 ac/h, which was clearly unacceptable. This leakiness comes from the lack of experience of the Belgian building contractors with regard to construction details. A series of pressurization measurements identified numerous

Table 3.4.2. Theoretical and measured U-values (W/m²K)

Envelope element	Theoretical values	Measured values
North façade	0.12	0.13
South façade	0.16	0.17
North roof	0.10	0.11
South roof	0.10	0.11
Garage floor	0.40	0.49
Living room floor	0.20	0.21
Common walls	3.20	3.60

leakage paths. After improvements by the building contractors and BBRI, an overall n_{50} value of 2.0 ac/h was obtained.

Monitoring has allowed the UA value of the components of the building envelope to be quantified, based on a global energy balance approach and direct measurements of the specific elements of the envelope. Table 3.4.2 compares the theoretical and measured U-values of the envelope elements. Differences between measurements and theoretical values of the common walls and the garage floor come from measurement inaccuracy.

CONCLUSIONS

The success of this project is a result of the integrated approach taken during design, construction, and integration of the various systems. This required close collaboration over three years among the promoters of the project, different laboratories, associations, the Walloon Region, research teams, the project designer, several firms, and a good many other contributors.

Not all the phases of the project have been completed. The proof of the effectiveness of the concepts by monitoring during actual occupancy still remains to be carried out. The ultimate aim is to get the results applied on a wide scale starting in the year 2000.

ACKNOWLEDGEMENTS

- *Design Team.* SPRL Atelier d'Architecture, 45 Rue Richier, B 5500 Dinant, Belgium.
- *Research Team.* Architecture et Climat, Catholic University of Louvain-la-Neuve, B-1348 Louvain-la-Neuve, Belgium, and Belgian Building Research Institute, Violetstraat 21–23, B 1000 Brussels, Belgium.

PUBLICATIONS

Wouters P, L'Heureux D, De Herde A, Gratia E (1993). *The Pleiade Dwelling : an IEA Task XIII Low Energy Dwelling with Emphasis on IAQ and Thermal Comfort, Energy Impact of Ventilation and Air Infiltration*, 14th AIVC Conference, Copenhagen, Denmark, pp. 21–23. Belgian Building Research Institute.

Wouters P et al. (1995). *Monitoring of the Belgian IEA Task XIII Dwelling PLEIADE: Global Context and Some Basic Results.* Belgian Building Research Institute.

Wouters P, Martin S, Ducarme D, L'Heureux D, Somogyi Z, Bossicard R, Voordecker P (1995). *Determination of the Basic Thermal and Solar Performances of the Belgian IEA Task XIII Dwelling PLEIADE: Monitoring Activities.* Belgian Building Research Institute.

3.5 The Canadian Advanced House in Brampton

SUMMARY

The Brampton house, although conventional in appearance, contains many innovative energy conservation and solar technologies that reduce total energy use by 70% compared to standard new Canadian housing. Energy conservation features include a well insulated and airtight lightweight building shell and energy-efficient appliances and lighting. Solar features include high performance windows and a two-storey sunspace which preheats ventilation air through ducts in the sunspace wall and floor slab ducts.

An integrated mechanical system provides the heating, cooling, ventilation and water heating for the house. The system comprises a hot water tank and an ice storage tank thermally connected by a heat pump. Heat, recovered from exhaust air, grey water, and excess passive solar gains is fed to the ice storage tank. A heat pump upgrades this heat for the hot-water tank, which supplies space and water heating. Fluid from the ice tank is circulated through a fan coil to provide summer air conditioning.

KEY FEATURES

- super-insulation
- high-performance windows
- integrated mechanical system
- sunspace preheat of ventilation air
- energy-efficient lights and appliances

LOCATION

Brampton is located just north-west of Toronto at latitude 43° N.

CLIMATE

The climate (Figure 3.5.1) is continental with relatively long cold winters and short but warm and humid

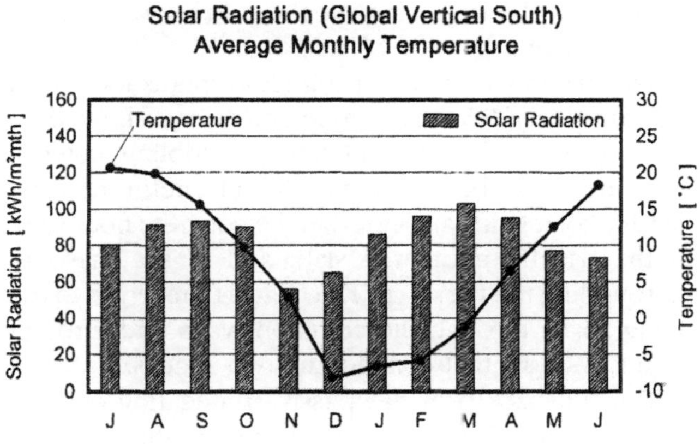

Figure 3.5.1. Brampton climate conditions

summers. Solar availability is moderate for most of the year, although the fall tends to be quite overcast.

CONCEPT

The Brampton Advanced House is a two-storey home with a full basement and two-storey (unconditioned) sunspace. The house design is typical of upscale suburban housing in Canada. Building floor plans are given in Figure 3.5.2. Although the house appearance is conventional, the house contains many novel energy-efficient features.

CONSTRUCTION

The walls are of double-frame construction with 240 mm of wet blown cellulose in the cavity and 25 mm of rigid fibreglass wall sheathing added to increase the thermal resistance and reduce thermal bridging (Figure 3.5.3). The ceiling consists of 13 mm of drywall and 300 mm of blown cellulose. The basement walls are comprised of 13 mm of drywall, 175 mm of blown cellulose, poured concrete wall, and 50 mm of rigid fibreglass sheathing foundation (Baseclad). The basement floor is 100 mm of poured concrete with 50 mm of rigid fibreglass foundation sheathing (Baseclad) underneath. The sunspace floor is an uninsulated slab-on-grade. The wall between the living area and the sunspace has 100 mm of cellulose insulation.

The primary air/vapour barrier is 0.15 mm polyethylene located directly behind the drywall. All seams and headers are caulked. The house was designed to have an airtightness of 0.75 ac/h at 50 Pa pressure difference; this corresponds to a natural air leakage rate of approximately 0.05 ac/h.

SOLAR FEATURES

A large passive solar contribution was achieved by using high-performance windows and locating the majority of windows on the south side. With the exception of the sunspace skylight, all the windows in the house are wood-framed, triple-glazed, with two low-emissivity coatings, two argon-gas fills and butyl-rubber edge spacers. Including frame and edge-of-glazing effects, the living area windows were calculated to have a U-value of 1.10 W/m^2K.

The majority of south-facing glass area is located in the sunspace. Because temperatures are allowed to float (the sunspace has no heating or cooling system), glass areas can be increased without energy penalty. Passive solar gains are stored in the concrete floor slab, in the backfill under the slab, and in the masonry surrounding the fireplace. The heat is transferred to the living space through the common walls and through the glass doors that connect the two spaces.

The sunspace provides passive preheating of fresh ventilation air. Outdoor air is ducted into the sunspace through ceiling diffusers. A return-air duct mounted high on the wall pulls the fresh, preheated air down to

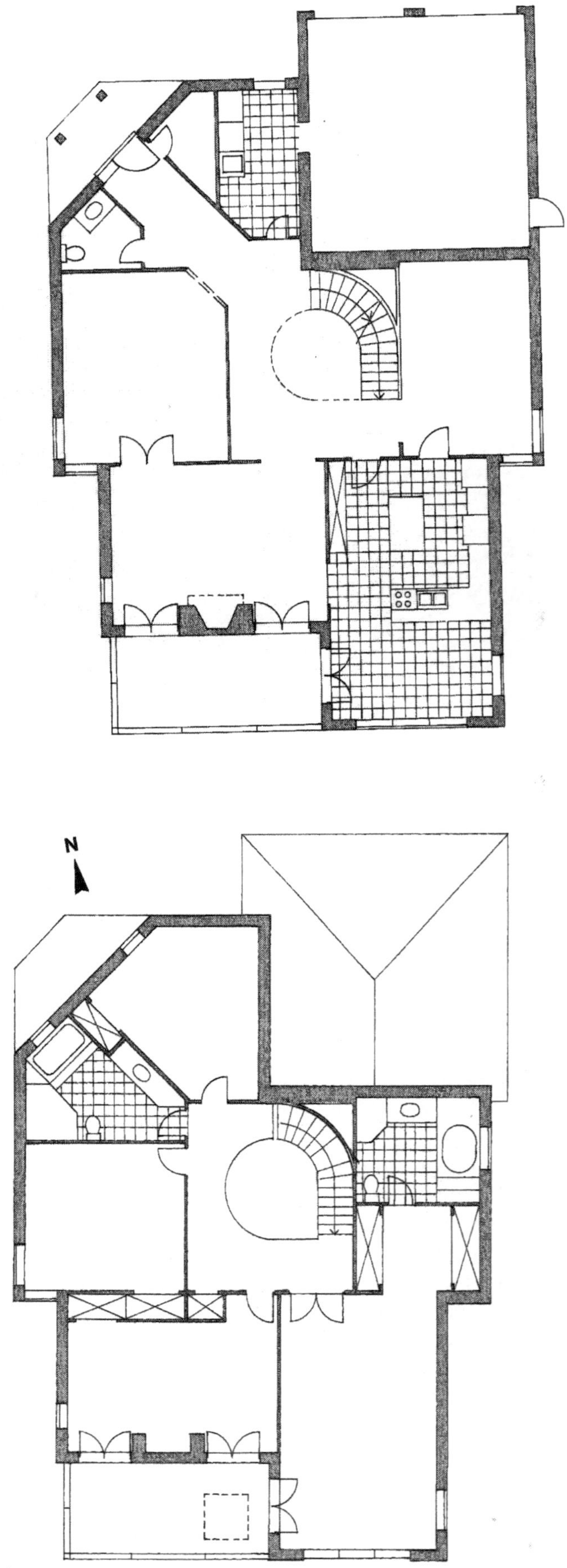

Figure 3.5.2. Brampton Advanced House floor plans

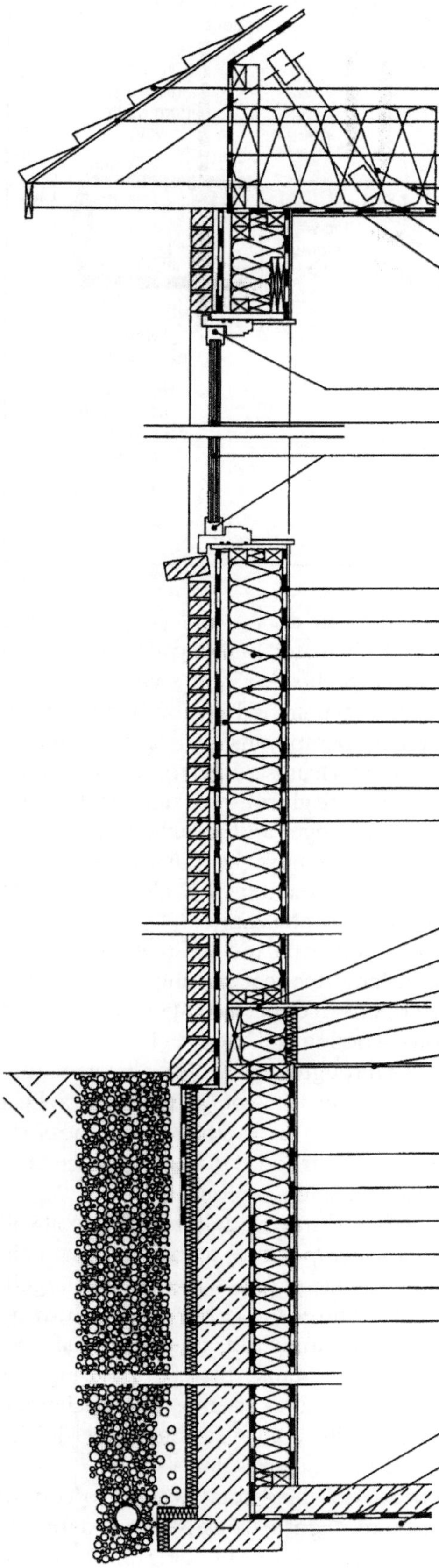

Layer	Description	Width (cm)	Thermal Conduct. (W/mK)	U-Value (W/m²K)
Roof				
1	Asphalt Shingles	-	-	
2	Plywood Sheathing	1.50	-	
3	Cardboard Insulation Retainer	-	-	
4	Wood Training on 600 Centres	14.00	0.140	0.09
5	Cellulose Insulation	46.00	0.040	
6	Vapour Retarder	-	-	
7	Gypsum Board	1.25	0.210	
Window				
8	Wood Frame			1.58
9	Triple Glazing			0.83
10	Glazing and Frame			11.10
Exterior Wall				
11	Gypsum Board	1.25	0.210	
12	Vapur Retarder	-	-	
13	Double Stud Wallon 400 Centres	8.90	0.140	
14	Cellulose	24.00	0.040	0.15
15	Rigid Fibreglass	3.80	0.040	
16	Building Paper	-	-	
17	Air space	2.50	-	
18	Brick Veneer	10.00	-	
Ceiling				
19	Wood Subfloor	1.90	0.140	
20	Wood Joists On 400 Centres	25.00	0.140	
21	Cellulose Insulation	18.00	0.040	-
22	Extruded Polystyrene Blocking	5.00	0.025	
23	Gypsum Board	1.25	0.210	
Below Grade Wall				
24	Gypsum Board	1.25	0.210	
25	Vapour Retarder	-	-	
26	Wood Frame On 400 Centres	3.80	0.140	0.18
27	Cellulose Insulation	17.80	0.040	
28	Concrete	25.40	2.00	
29	Rigid Fibreglass	5.00	0.040	
Below Grade Floor				
30	Concrete Slab	12.70	2.10	
31	Poly Daniproof Membrane	-	-	0.63
32	Rigid Fibreglass	5.80	0.040	

Figure 3.5.3. Above- and below-grade wall section of the Brampton house

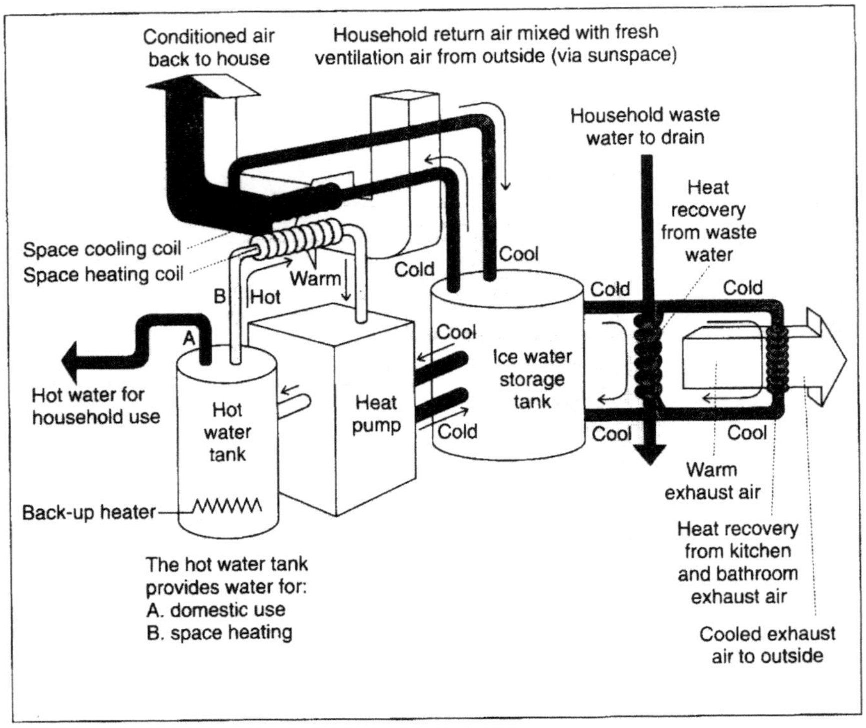

Figure 3.5.4. Integrated mechanical system (IMS) schematic (Brampton house)

the floor slab, where it is circulated through a series of parallel-flow ducts and then to the integrated mechanical system (IMS) for distribution to each room. When the sunspace temperature exceeds a set point, the skylight opens automatically to provide passive cooling. Reflective blinds and an exterior trellis provide summer shading for the sunspace.

HVAC

One of the most innovative features of the Advanced House is the IMS (Figure 3.5.4 – the 'Solmate', developed by Allen Associates) which provides heating, ventilation, cooling, and water heating for the house. The mechanical IMS was an engineering prototype specifically built for this project. Although similar systems were installed in a few earlier homes, the Advanced House was the first detailed assessment of this technology.

The system comprises a hot water (or DHW) tank and a cold water (or ice storage) tank thermally connected by a heat pump. The hot water side supplies heat to meet space- and water-heating loads. Hot water demand is met by taking hot water from the top of the DHW tank directly to the taps. Cold mains water enters the bottom of the DHW tank to replace the water used. To provide space heating, a pump circulates water from the hot water tank through the heat-pump condenser and a fan coil. Heat is delivered to the space by circulating air over the fan coil.

The air circulation fan operates continuously (at 285 litre/s) to ensure good air circulation and a constant supply of outdoor air to all rooms. The IMS maintains a continuous exhaust of 64 litre/s to provide ventilation. Outdoor make-up air enters the house via an intake grille in the sunspace. The ventilation air is pulled into the return air duct of the IMS, and mixes with 221 litre/s of recirculated building air. Figure 3.5.5 shows the sunspace preheating of the ventilation air.

The DHW tank is maintained hot by operating the heat pump. The heat pump transfers heat from the ice tank to the DHW tank. Heat is added to the ice tank by circulating a 33% ethylene glycol–water mixture through a grey-water heat exchanger and an exhaust air coil. If there is no heat in the ice tank (i.e. it is almost all ice), the heat pump shuts off and a 6.0 kW electric element is used to heat the hot-water tank.

Cooling is provided from the ice tank. When the house thermostat indicates that cooling is required, a three-way valve opens to direct the IMS glycol mixture through the house cooling coil. House circulation air is cooled as it passes through the coil. The glycol mixture transfers its heat to the ice storage tank via a heat exchanger in the tank. This system allows heat removed from the house air to be stored in the ice tank for use when water heating is required.

In the summer, heat removed by the cooling coil will cause the cold-tank temperature to rise. To maintain adequate cooling capacity, hot water is discharged. Dumping hot water causes the compressor to turn on to supply hot water. Running the compressor removes heat from the cold-side tank, thereby allowing the cooling system to continue to operate. The discharged hot water is mixed with cold water and used in the underground yard irrigation system.

A standard heat/cool thermostat is used to control room air temperature. Night setback is not used for two reasons. First, because of the long building time constant, the building temperature would not drop significantly overnight. Second, returning the house to normal temperatures in the morning would require expensive resistance electric heating. Because the higher-efficiency

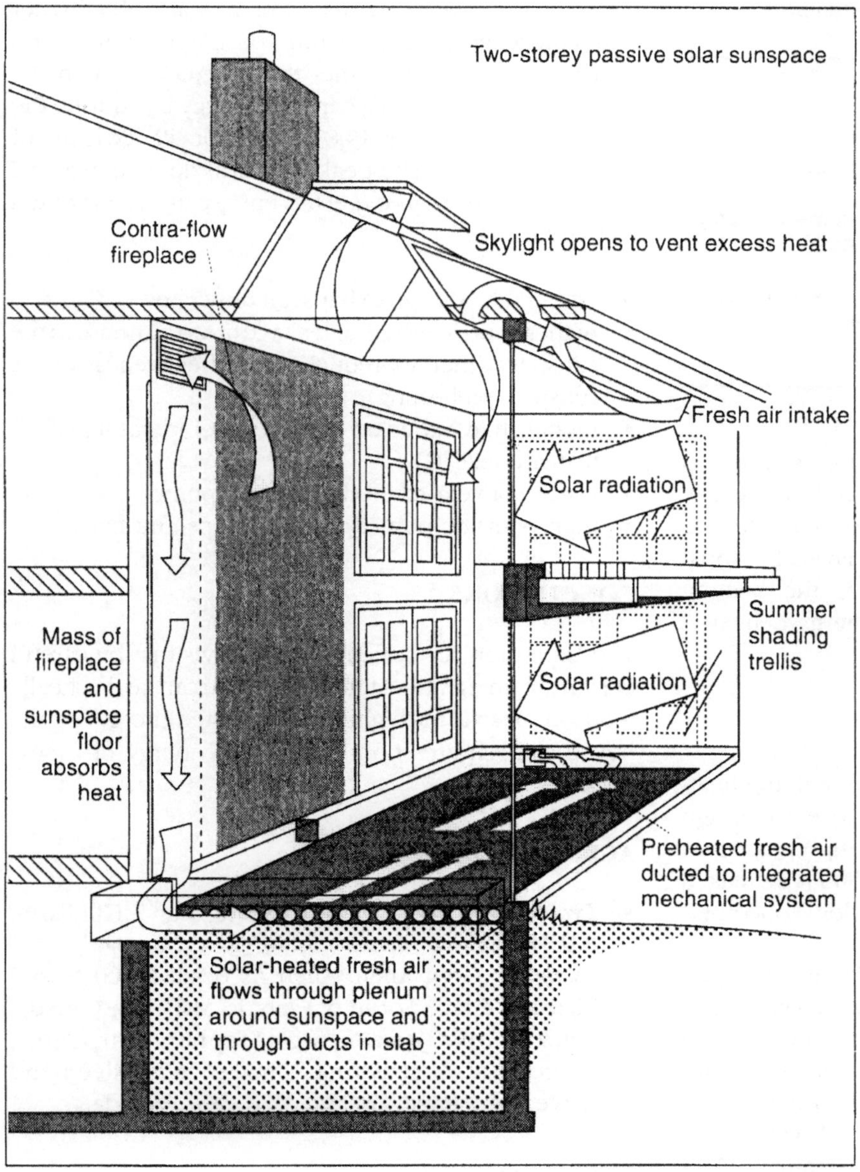

Figure 3.5.5. Sunspace preheating of ventilation air in the Brampton house

heat pump maintains the temperature overnight, less electricity is used.

The IMS was designed to provide 2.3 kW of cooling and 6 kW of heating. Although this cooling capacity might be sufficient for small homes, it was not intended to meet the full cooling load of the Advanced House. The IMS heating output is also slightly low for the expected maximum house heating load of approximately 8 kW. It was expected, however, that the long building time constant would allow the building to coast through extremely cold periods and the energy-efficient fireplace would provide sufficient auxiliary heating. Nevertheless, in-duct electric heating coils were installed in the fall of 1991 to ensure that the house temperature could be maintained at the thermostat setting.

The IMS was predicted to have an instantaneous coefficient of performance of between 2.5 and 3.0 depending on the temperature of the ice tank. Energy to operate the exhaust-air and recirculation fans was estimated to be 1400 kWh annually (based on a continuous draw of 160 W).

ELECTRICAL SYSTEMS

The house has all the major appliances found in most Canadian homes. In this case, however, energy-efficient appliances are used instead of standard products. These appliances are expected to require only half the energy of conventional products. They include a front-loading clothes washer, a halogen light stove, and one of the most efficient refrigerators in the world (20 kWh/month).

No incandescent lighting was used in the house. Most lighting was tube and compact fluorescent, except for the Halogen lights in the kitchen and front hall. Lighting and receptacle loads are expected to consume approximately 1500 kWh annually.

ENERGY USE

The predicted energy consumption over a typical year for the Advanced House is shown in Table 3.5.1.

Because the IMS links various energy loads in the building, the monitored distribution of compressor energy may differ from the numbers in Table 3.5.1, but

Table 3.5.1. Energy consumption of the Brampton Advanced House

Component	Annual Consumption (kWh)	
Space heating	4822	Compressor power, pumps and back-up
Hot water	2016	Compressor power to supply a 5130 kWh load
Lights/Appliances	4042	
Air conditioning	225	Compressor power to heat-dumped water
Fans	1402	
Total	12507	

the total should be correct. Thus, the house is expected to consume 30.7 kWh/m^2 of conditioned floor area, of which 11.8 kWh/m^2 is for space heating. According to the simulations, the IMS is expected to have a Seasonal Performance Factor (SPF) of 2.0 over the heating season. The SPF is the ratio of useful energy delivered divided by all energy input.

MONITORING RESULTS

Tests of the building shell show that the building heat-loss coefficient and airtightness are close to design targets. Some decrease in building airtightness over time was noted: the air change rate at 50 Pa increased from 0.9 to 1.3. Formaldehyde and radon concentrations are well within accepted guidelines.

Hot water use for the three-person family averaged 164 litres per day at 45°C. The water-conserving appliances are credited with reducing the water-heating requirement to only 60% of the typical residential load. An average of 260 litres per day was used for normal cold water demands; in addition 140 m^3 of water were dumped during the cooling season to maintain IMS cooling capacity.

Monthly lighting and receptacle loads averaged 400 kWh; of this, 125 kWh per month was consumed by the major appliances. Fan energy use averaged 134 kWh per month. These loads are close to design values.

Monitored energy consumption was 28% higher than computer-predicted values during the demonstration period and 60% higher than predicted when the house was occupied. The annual energy use during the occupied period was 19,834 kWh or 49 kWh/m^2 of heated floor area. The peak electrical demand was 8.7 kW. The higher-than-expected energy use is attributed to five factors:

- higher than design exhaust air flow rates
- higher than expected air leakage from the house to the sunspace (thereby reducing the effectiveness of the sunspace preheating)
- higher than expected indoor air temperature (23°C instead of 21°C)
- lower than expected COP of heat pump
- high parasitic power of IMS pumps and fans.

CONCLUSIONS

The Brampton Advanced House shows that by careful design and integration, total energy use can be drastically reduced. Despite the higher than expected energy use, the house still requires only 40% of the energy of a new standard house.

ACKNOWLEDGEMENTS

- *Project Manager.* Elizabeth White, Marsh Hill Farm, R.R. 4, Stirling, Ontario KOK 3EO, Canada.
- *Designer.* Greg Allen, Allen Associates, 400 Mount Pleasant Road, Toronto, Ontario M4S 2L6, Canada.
- *Author and Monitoring.* Stephen Carpenter, Enermodal Engineering Limited, 650 Riverbend Drive, Kitchener, Ontario N2K 3SR, Canada.

PUBLICATIONS

Enermodal Engineering Limited (1992). *Performance of the Brampton Advanced House*, CANMET, Natural Resources Canada, Ottawa, Ontario K1A 0E4.

3.6 The Canadian Green Home in Waterloo

SUMMARY

The Green Home is a 230 m^2 raised bungalow in Waterloo, Ontario. The goal of the project was to demonstrate that houses can be built and operated in a manner that is environmentally responsible. The house contains the latest developments in energy efficiency, water conservation, waste management, CFC reduction, and environmentally appropriate material use. The house was designed to consume less than 30% of the energy and water used in a conventional new home. Monitoring results show that the house is performing largely as expected and is achieving superior indoor air quality.

KEY FEATURES

- super-insulation
- high-performance windows
- integrated furnace/ventilation recovery system
- cistern/ground cooling
- PV-pumped solar domestic hot water
- energy-efficient and CFC-free appliances
- re-used and recycled materials

LOCATION

Waterloo is located 100 km west of Toronto at latitude 43° N.

CLIMATE

The climate (Figure 3.6.1) is continental with relatively long cold winters and short but warm and humid summers. Solar availability is moderate for most of the year, although the fall tends to be quite overcast.

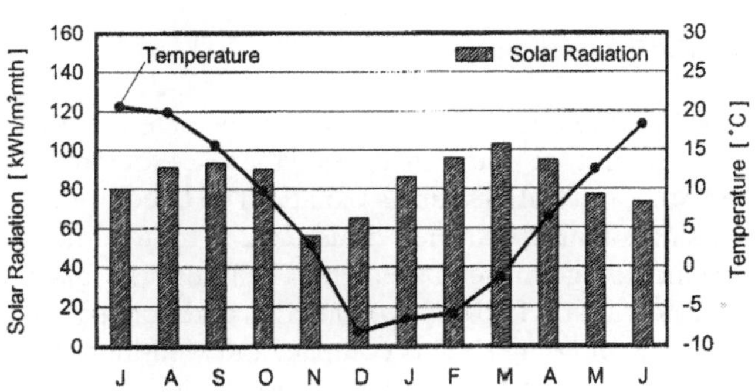

Figure 3.6.1. Waterloo climate conditions

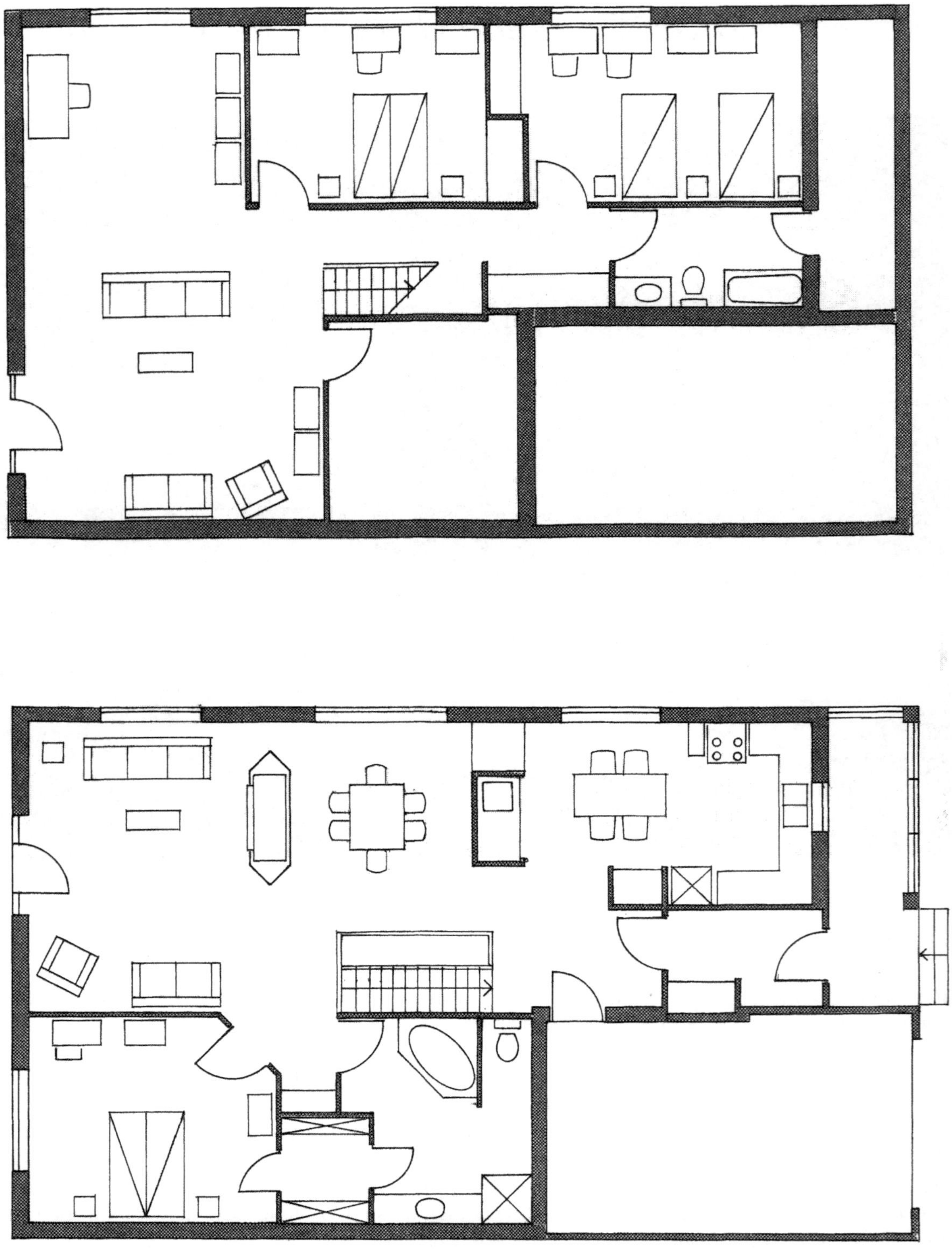

Figure 3.6.2. Waterloo Region Green Home floor plans

CONCEPT

The Waterloo Region Green Home is one of ten houses built as part of the Advanced Houses Program of CANMET, Natural Resources Canada. The Green Home is a single-family detached house of 230 m² total floor area including finished basement. The house is a raised bungalow that is fairly conventional in exterior appearance. The building shell is compact and well-insulated to minimize heating and cooling loads and required equipment size. Exposed surface area is kept to a minimum by using a fairly square house footprint and avoiding irregular façades. A boxy house form is avoided by the garage, front and back porches, and south-side overhang. A dormer on the south side allows natural light to penetrate to the north side. The living room, master bedroom, and dining room have cathedral ceilings to create the feeling of larger rooms and achieve a moderately steep roof slope for proper orientation of solar collectors.

Because of the regional importance of a basement regarding marketability, the house is raised so that the

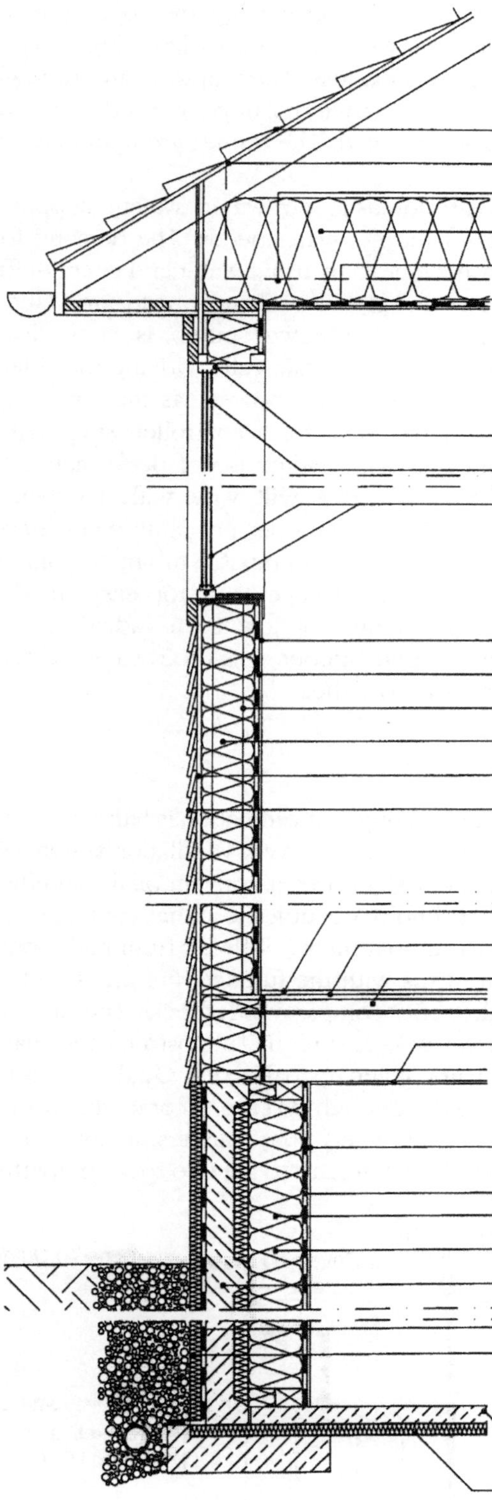

Layer	Description	Width cm	Thermal Conduct. W/mK	U-Value W/m²K
Roof				
1	Steel Roofing	–	–	0.094
2	Wood Waferboard	1.60	0.14	
3a	Wood I-Beam On 600 Centres	8.90	0.14	
3b	Cellulose Insulation	46.00	0.04	
4	Vapour Retarder	–	–	
5	Gypsum Board	1.20	0.21	
Window				
6	Fibreglass Frame			1.07
7	(Triple) Centre- Glazing			0.95
8	Glazing and Frame			1.02
Exterior Wall				
9	Gypsum Board	1.20	0.21	0.16
10	Vapur Retarder	–	–	
11a	Cellulose Insulation	24 00	0.04	
11b	Wood I-Beam On 600 Centres	24 00	0.14	
12	Insulated Sheating	1.30	0.06	
13	Wood Siding	1.30	0.14	
Ceiling				
14	Hardwood Flooring	1.30	0.14	–
15	Waferboard Subfloor	2.20	0.14	
16	Wood I-Beam On 600 Centres	36.60	0.14	
17	Polyurethane Foam	15.00	0.025	
18	Gypsum Board	1.20	0.21	
Below Grade Wall				
19	Gypsum Board	1.20	0.21	0.20
20	Vapour Retarder	–	–	
21a	Cellulose Insulation	24.00	0.04	
21b	Wood Studs On 600 Centres	6.40	0.14	
22	Reinforced Preacst Concrete	5–20	2.00	
23	Moisture Barrier	–	–	
24	Semi- Rigid Fibreglass	5 00	0.04	
Below Grade Floor				
25	Concrete	7 60	2.00	0.61
26	Polystrene Insulation	5.00	0.035	

Figure 3.6.3. Above-and-below-grade wall section of the Waterloo house

basement floor is only 1.2 m below grade on the south side instead of the more conventional 2.0 m. This allows for larger basement windows, which create a more desirable living space. The basement includes two bedrooms and a full bathroom so that it can be used as a separate apartment if required. Floor plans are shown in Figure 3.6.2.

CONSTRUCTION

The building shell is well insulated and airtight. The wall section taken above and below grade is shown in Figure 3.6.3. The basement wall is made up of five meter long precast panels with steel mesh reinforcing. The panels are flat on the outside and resemble a waffle on the inside. The panels are 200 mm thick at the edges but are only 50 mm thick in the middle. Thus only half the concrete of a conventional poured wall with voids is used and it can be easily insulated. The system does not require continuous footings, but only pads at each end. This makes it easier to insulate fully under the floor slab. The floor slab is insulated with 50 mm of steam-blown styrofoam insulation. Including any thermal bridging effects, the below-grade wall and floor have U-values of 0.18 and 0.6 W/m²K respectively.

The walls and above-grade floor were constructed with engineered wood products. For the floor, 350 mm deep engineered wood I-beams were spaced at 600 mm instead of the more conventional 300 to 400 mm, without any loss in floor strength. A slightly thicker sub-floor was used to eliminate any sagging between the joists. Nevertheless, the total wood requirement for the floor was reduced by one third with the largest piece of dimensional wood being a '2 x 4' (38 x 90 mm).

The walls are framed with engineered wooden I-beams to create a 235 mm deep insulation cavity. The added strength of the I-beam studs allows them to be spaced at 600 mm on centre instead of 400 mm. A thin web of oriented strandboard connecting the two chords not only reduced wood requirements but also reduced thermal bridging. The wall cavity is filled with wet-blown cellulose insulation made from recycled newspapers to achieve a U-value of 0.14 W/m²K. A 0.15 mm polyethylene vapor retarder is used on the interior. All seams are sealed with non-drying two-sided tape. Drywall, containing a minimum of 25% recycled wastes, is used for all interior wall finishes. Siding made from wood wastes is used as the exterior wall covering.

SOLAR SYSTEMS

Passive solar heating is a major part of the house design. The majority of the windows are located on the south side.There are no windows on the north side, in order to reduce further heating and cooling demands. The windows were selected to have high solar heat gain and low heat loss. The windows are triple-glazed with two low-emissivity coatings, foam edge seal spacer, and argon-gas filled. The frames are foam-filled fibreglass.

A solar domestic hot-water system supplies the majority of the hot-water needs. The demand for hot water was reduced by using water and energy-efficient appliances and low-flow shower heads and faucets. A high-efficiency low-power pump is controlled and powered by photovoltaic cells. Driving the solar heat collection pump by PV power has the advantages of eliminating the need for a controller, simplifying the installation, and providing better performance (since the pump flow rate will vary with incident solar radiation). The system does not require any electricity to operate and uses thin flexible tubing to connect the collectors to the storage tank for easy installation. During the months of low solar radiation, a high-efficiency, instantaneous, gas-fired water heater supplements the solar heat.

HVAC

The space heating and ventilation is handled by a combined furnace/heat recovery ventilation system (Figure 3.6.4). The system uses a conventional mid-efficiency furnace technology coupled to a small container of rocks with two compartments. Stale air from bathrooms and kitchen mixes with the furnace flue gases and passes through the first compartment of rocks. The large surface area of the rocks ensures that almost all of the heat in the exhaust air is given up to the rock. Outdoor air is heated up as it passes through the second compartment of rocks. Approximately every four minutes, a reversing valve switches the air streams to the two rock compartments,

Figure 3.6.4. Integrated furnace/heat recovery ventilator in the Waterloo house

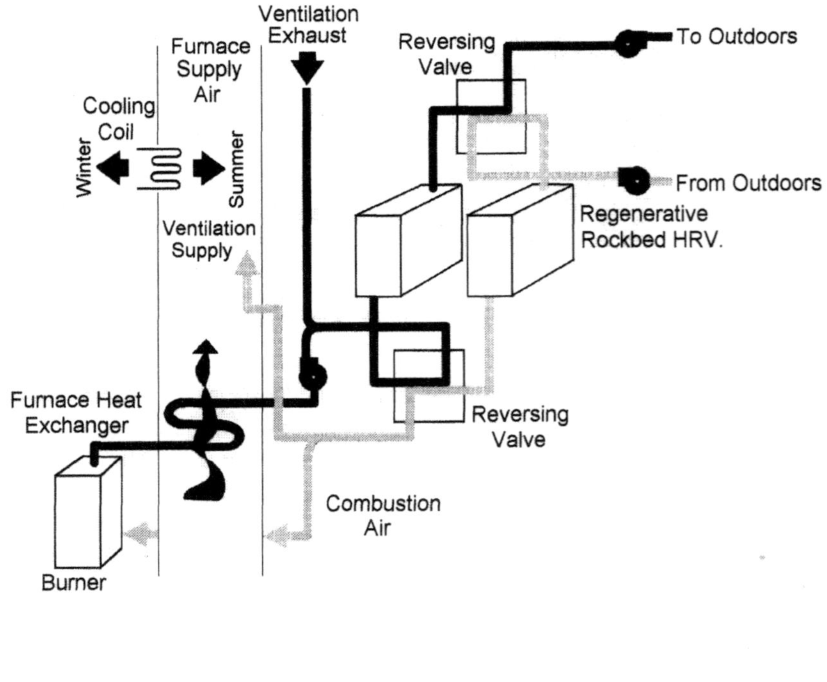

allowing the heated rocks to be cooled and cooled rocks to be heated. The outdoor air is distributed throughout the house via the furnace ductwork. Products of combustion are prevented from re-entering the house by pressurizing the outdoor air stream and shutting off the furnace just before every reversing valve operation.

The house and landscaping were designed, and appliances selected, to minimize the need for mechanical cooling. Nevertheless, a small cooling system was installed. To avoid the use of freon-based cooling systems, it was decided to make use of the cistern water and ground as the cooling source. When cooling is required, cistern water is circulated through plastic tubing buried below the basement floor slab and through a coil in the air-handling system.

ELECTRICAL SYSTEMS

A wide range of energy-efficient appliances are used in the house. An energy-efficient, CFC-free refrigerator and freezer are used for storage of foods. The washing machine is front-loading with a high-speed spin dry. The dishwasher, produced by the same manufacturer, uses only 20 litres of water per standard wash cycle, half the amount of conventional dishwashers.

The clothes dryer and stove operate on natural gas because of its environmental benefits over electricity. The dryer operation is controlled by a moisture sensor to take advantage of the high-speed spin washer. The gas stove uses a ceramic counter top to improve burner efficiency and to permit venting of combustion products to the outside.

The house was designed to make maximum use of daylighting, including an interior glass wall to allow natural light into the north-side bathroom. Most of the electric lighting systems are full-spectrum T8 fluorescent lamps with electronic ballast. Solar-powered lighting fixtures are used for outdoor lighting.

ENERGY USE

The predicted annual energy use is summarized in Table 3.6.1. Energy use is reduced by over 70% (gas consumption is converted to an equivalent kWh).

MONITORING RESULTS

The building airtightness was measured at 0.8 air changes per hour (ac/h) at 50 Pa; well below the R2000 requirement of 1.5 ac/h. The formaldehyde levels were measured at 0.03 ppm, well below the Canadian guideline of 0.1 ppm. The radon concentration was measured at 0.003 WLU, again well below the conservative US guideline 0.02 WLU. VOC concentration was 0.18 mg/m^3, half the European guideline of 0.3 mg/m^3.

Initial results show that the electricity use is 12.9 kWh/day (excluding power required for the monitoring system). This is slightly higher than the predicted value of 11.2 kWh/day but a significant reduction from the 24 to 30 kWh/day used in conventional housing.

The 15% higher electricity use is attributed to the receptacle loads, sump pump, and, surprisingly, the gas stove. The difference in receptacle load is due to the transformer that converts the 120 V supply to 240 V for the European refrigerator (0.5 kWh/day) and an underestimation of miscellaneous energy use. The sump pump ran almost continuously in the spring; the average annual energy use is expected to be lower. The stove's electricity consumption, however, was surprisingly high, at 1.0 kWh/day, because the stove ignition runs continuously when the oven is on, and the purge fan runs for over ten minutes when the stove is shut off.

For the four-month period of occupancy the total requirement for municipal water was 288 litres per day, of which 40% was for hot water. This is a 60% reduction from conventional housing. Also within this period the two-panel solar system provided 74% of the water heating load. The annual solar fraction will be lower with the winter months included.

The Green Home seldom overheated during the 1994 summer. Passive cooling features, including high insulation levels, high-performance windows, wide overhangs, and clerestory windows for venting, appeared to be successful in reducing the need for mechanical cooling. According to monitored results, the cooling system delivered 1.2 kW of cooling to the space. This represents a 2.9 K drop in air temperature across the cooling coil at a flow of 350 litres. The system, while clearly insufficient for conventional houses, was able to maintain the room temperature on hot days.

Table 3.6.1. Predicted annual energy use in the Waterloo house

	Electricity (kWh)	Gas (m^3)	Gas e(kWh)
Space heat	0	503	5282
Water heating	0	73	767
Cooling	213	0	0
Gas appliances			
Dryer	30	52	546
Stove	13	99	1040
Electrical appliances			
Freezer	420	0	0
Refrigerator	420	0	0
Dishwasher	198	0	0
Washing machine	119	0	0
TV/small appliances	546	0	0
Lighting			
Indoor	479	0	0
Outdoor	20	0	0
Fans	1632	0	0
Total	*4090*	*727*	*7634*
Total equivalent kWh			*11724*

CONCLUSIONS

The Green Home successfully demonstrates that through careful design the energy use of all systems can be dramatically reduced. Furthermore, there is an interaction between the systems. For example, improving the building shell allowed a much smaller non-compressor-based cooling system to be used. The house also demonstrates approaches to more environmentally appropriate construction practices, such as recycled and re-used materials, water conservation, improved indoor air quality and elimination of ozone-depleting chemicals.

ACKNOWLEDGEMENTS

- *Project Manager and author.* Stephen Carpenter, Enermodal Engineering Limited, 650 Riverbend Drive, Kitchener, Ontario N2K 3S2, Canada.
- *Architect.* Richard Reichard, Snider Reichard March Architects, 145 Columbia Street West, Waterloo, Ontario N2L 3L2, Canada.

PUBLICATIONS

Grady, W. (1993). *Green Home,* Camden House Publishing, Camden East, Ontario.

3.7 Two Danish low-energy row houses

SUMMARY

Two different super-insulated row-house types were designed, one for east/west-running rows and one for north/south-running rows. The latter was designed as a deliberate attempt to meet the challenge of an unfavourable building site, lacking the potential to orient the buildings to give large south-facing façades. A further objective was to express energy consciousness in the architectural idiom of the buildings, as well as to obtain an attractive, well daylighted living environment in the limited space available.

KEY FEATURES

- super-insulation (quantity and quality)
- passive solar
- active solar DHW
- efficient low-temperature heating system
- heat-recovery ventilation
- water-saving devices
- efficient use of electricity.

LOCATION

The houses are located in Vonsild, south of Kolding, Jutland, at latitude 55°N.

CLIMATE

Climate details are summarized in Figure 3.7.1.

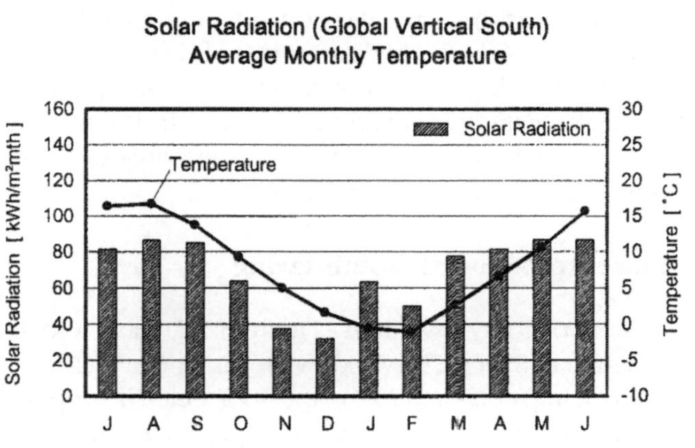

Figure 3.7.1. Climate diagram for the Danish houses

CONCEPT

In the original Danish Task 13 design, described in *Solar Low Energy Houses of IEA Task 13* (1995), there were two types of dwellings, one in the house type for the east–west axis having 106 m^2 floor area in two storeys, the other in the type for the north–south axis having 111 m^2 floor area partly in one, partly in two storeys. In both cases, the main rooms were organized around a two-storey-high family room. All rooms receive daylight, through skylights or normal windows – or a combination of the two. The same thermal-envelope construction, ventilation and heating system, combined active solar system (mainly for domestic hot water), water and electricity saving devices, lights, and appliances were used in the two house types.

As it turned out, the only opportunity to realize the project was to build the houses with 33 dwellings as part of an experimental quota under the Subsidized Dwellings' Act, which meant that the buildings had to be redesigned to bring them within the rather strict economic and average size limits for government-subsidized dwellings. By introducing two small one-storey dwelling types, the designers managed to have a number of fairly large two-storey dwellings built and to keep all the key energy features of the original project, although at a more moderate level. The major change in appearance is that the characteristic curved roofs were replaced with sloping conventional roofs, but, in addition, there is a larger difference in the south-facing window area between the two types in the built version. The insulation level was reduced compared to the original design, especially for windows (to a U-value of 1.5 W/m^2K) and walls, the latter for the combined reasons of cost and the effect on usable interior floor area. For economic reasons, ventilation heat recovery systems were only installed in one of the row house types, type B with the north–south axis. Light interior curtains were used to shade of the windows in the two-storey rooms.

However, two of the large dwellings (one in each row house type) were designated as Task 13 houses and better equipped than the others in the row, e.g. with high-performance windows having a centre U-value of 0.7 W/m^2K, giving an overall U-value of 0.8 W/m^2K. Also, the Task 13 house of type A was equipped with a ventilation heat recovery system. In the row of the Task 13 house, the solar domestic hot-water systems are operated by photovoltaics.

House type A – north/south-facing

These two-storey dwellings (Figure 3.7.2) have a total floor area of 85 m^2. The window area is 24.5 m^2 (21.8 m^2 south-facing) equalling 29% of the floor area. The sloping ceiling creates a special cave-like quality in the inner rooms. All primary rooms are south-facing – the master bedroom on the top floor has a small window to the north, and faces the two-storey family room, separated by single-pane glazing. The other bedroom on the top floor has a south-facing window, but also has a single-pane glass wall towards the double-height room. In total, four blocks were built, three of them with 15 two-storey dwellings and one with eight one-storey dwellings.

House type B – east/west-facing

The two-storeyed dwellings in this type (Figure 3.7.3) have a total floor area of 90 m^2. The window area is 17 m^2 (6.4 m^2 south-facing, and the rest facing east and west, split about fifty-fifty) equalling 19% of the floor area. The one-storey dwellings in between the two-storey dwellings, and the sloping roof giving a high south wall, make it possible to have a large south-facing window high up on the wall to provide direct sunlight for the mezzanine on the top floor and the two-storey family room and kitchen on the ground floor. Substantial passive solar gains as well as unusual room shapes and light effects are thus achieved. The living room on the top floor (the mezzanine) is separated from the two-storey family room by a railing only. In total, two blocks were built, one with four and one with six dwellings, alternately one-storeyed and two-storeyed.

CONSTRUCTION

For insulation of walls and roof, a high density mineral wool with a design thermal conductivity of 0.039 W/mK was used, giving the constructions a U-value of 0.11–0.18 W/m^2K. Most of the structural concrete walls are partition walls or party walls between dwellings, so the façades are not load-bearing.

Large factory-produced building panels were used for façades as well as for roofs. The typical room element is 2.4 m wide and spans the room or dwelling – the wall elements are one or two storeys high and at least 2.4 m wide; some of the one-storey elements are up to 5.1 m wide.

The large panels were constructed indoors from thin wooden studs, plywood, and mineral wool insulation under strict quality control. The wall elements were delivered complete with polyethylene vapour barrier and interior gypsum boards, windows, and doors already in place. The joints between the wall panels were sealed with mastic sealants.

Similarly, the roof elements were delivered with roofing material so that the building could be quickly closed, and all air tightness operations performed under dry indoor conditions. The polyethylene sheets were put in on site with lap joints squeezed between laths, and also squeezed against the walls, again with wooden laths. An additional 50 mm layer of mineral wool insulation was installed under the vapour barrier, and this space was also used for running electrical wires for lamp outlets etc.

The insulation thickness in the built version is 245 mm in the roof and 200 mm in the walls, a reduction from the

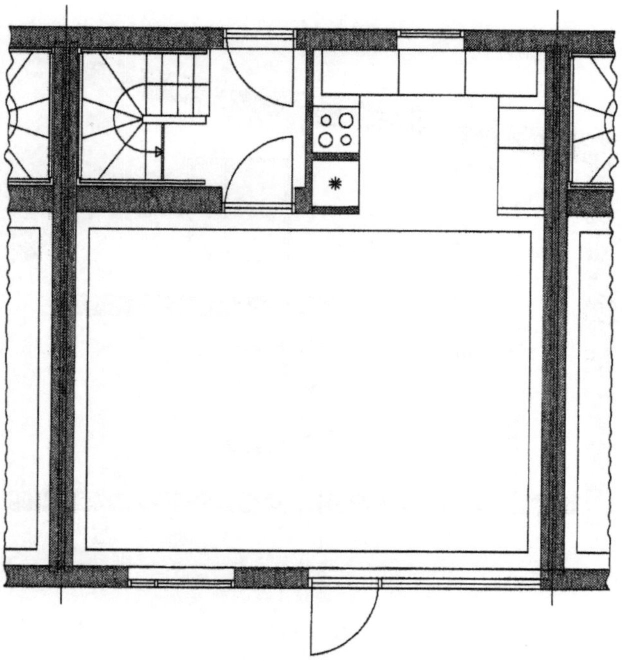

Ground floor

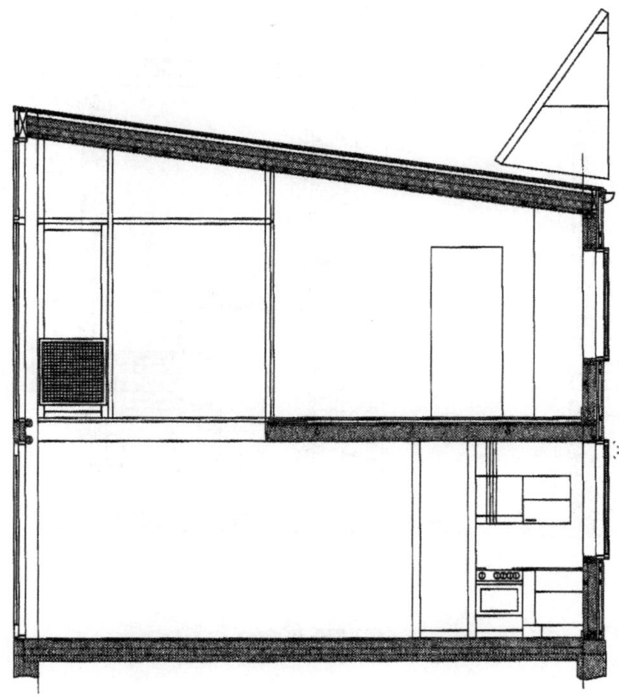

Section

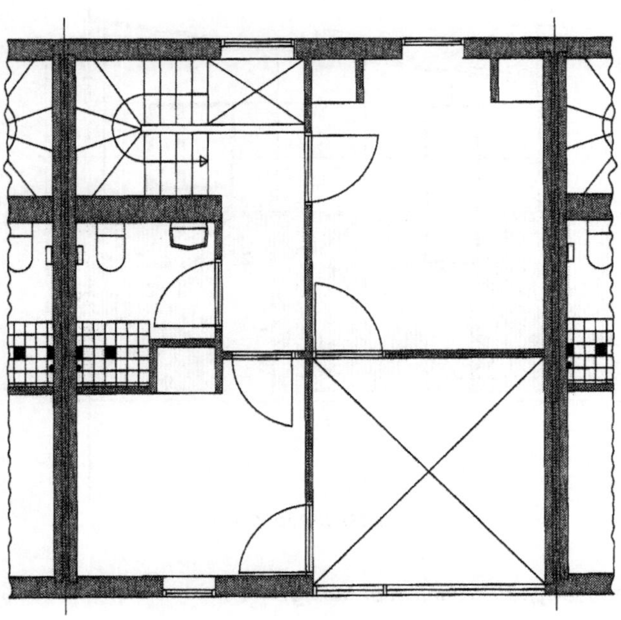

Top floor

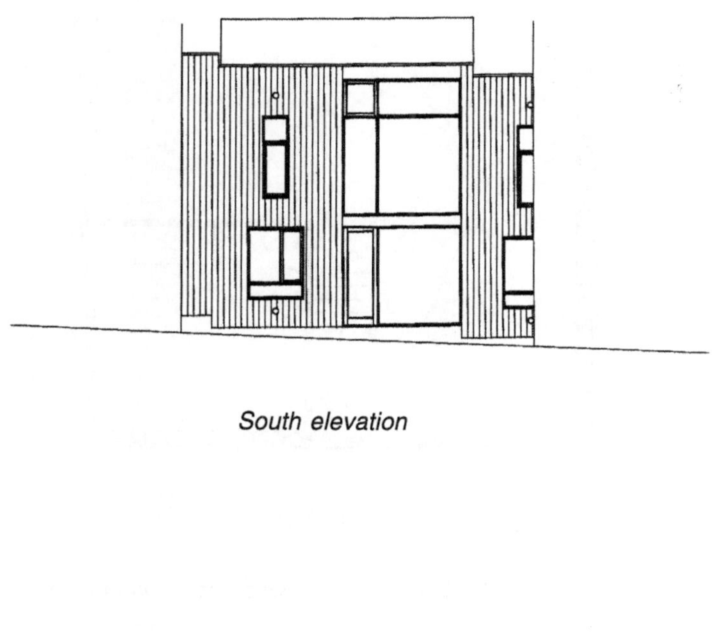

South elevation

Figure 3.7.2. Plan, section, and south elevation of Danish row house type A (north/south-facing)

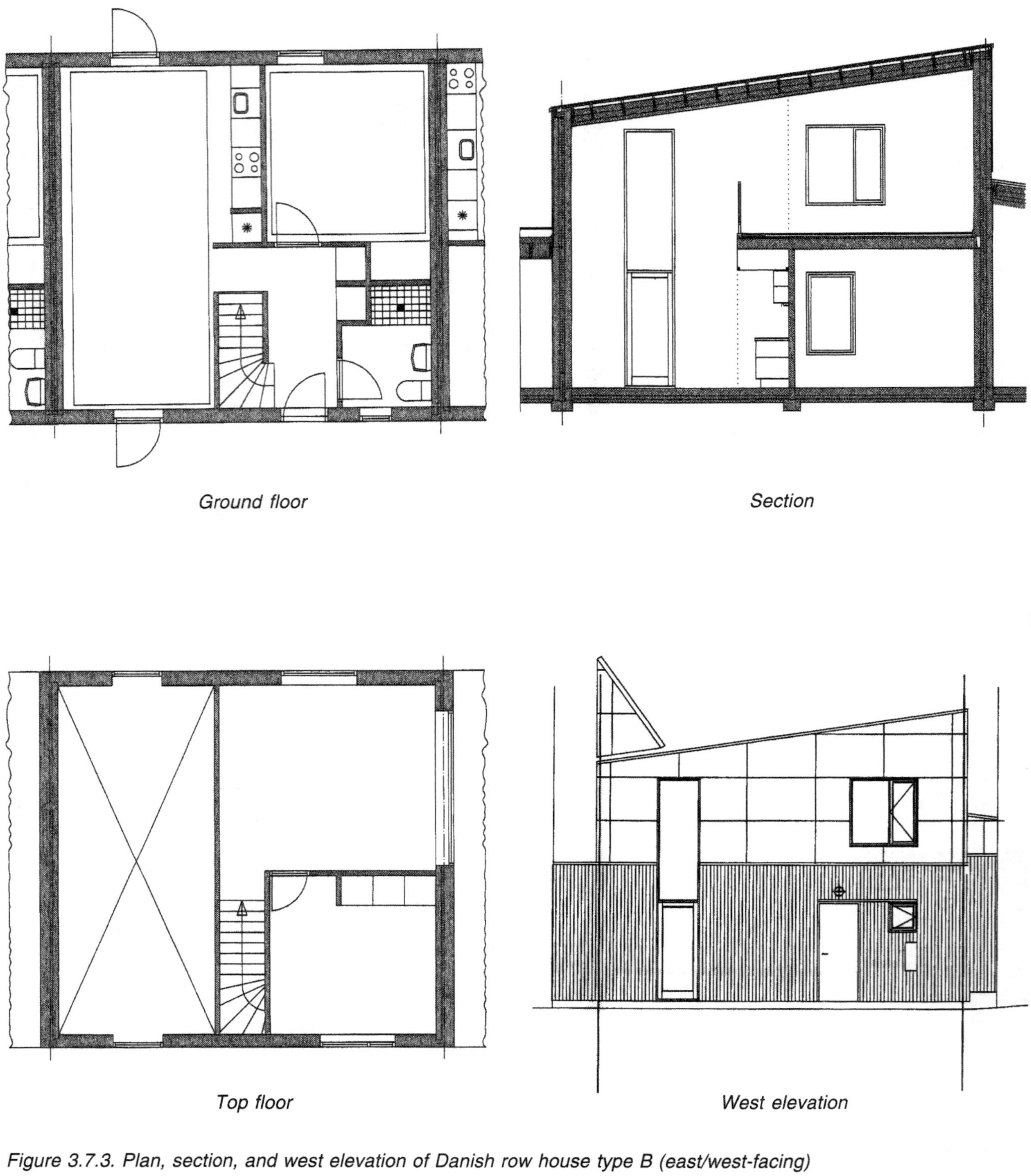

Figure 3.7.3. Plan, section, and west elevation of Danish row house type B (east/west-facing)

original 300 mm wall insulation, due to the redesign of the original Task 13 project into low-cost housing. The flat roofs of the one-storey dwellings in the north/south-running rows (house type B) have an additional layer of rigid 50–100 mm insulation on top. This layer is wedge-shaped to create the slight gradient required to drain the roof surface.

Under the slab-on-grade floor, 100 mm of rigid expanded polystyrene with the same thermal conductivity is used on top of draining and insulating expanded clay clinkers. The usual thermal bridge at the foundation is partially broken through use of lightweight concrete foundation blocks and polystyrene parameter insulation, separating the foundation and the slab. The construction details are shown in Figure 3.7.4.

HVAC

Auxiliary heating is provided, through a heat exchanger, by a local low-temperature district heating system, operating with a 200 litre domestic hot-water (DHW) tank as a buffer (see Figure 3.7.5). Heat is distributed in a

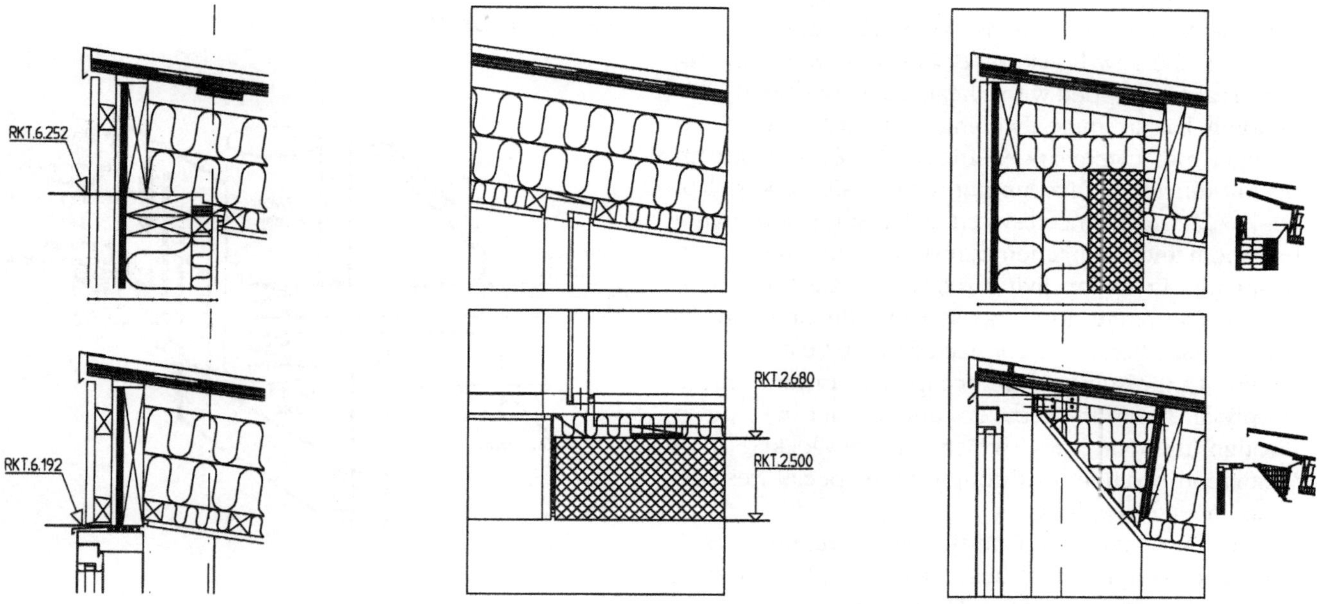

Rowhouse type A
Roof/south wall or window

Rowhouse type B
(top: normal)
(bottom: at south-facing window)

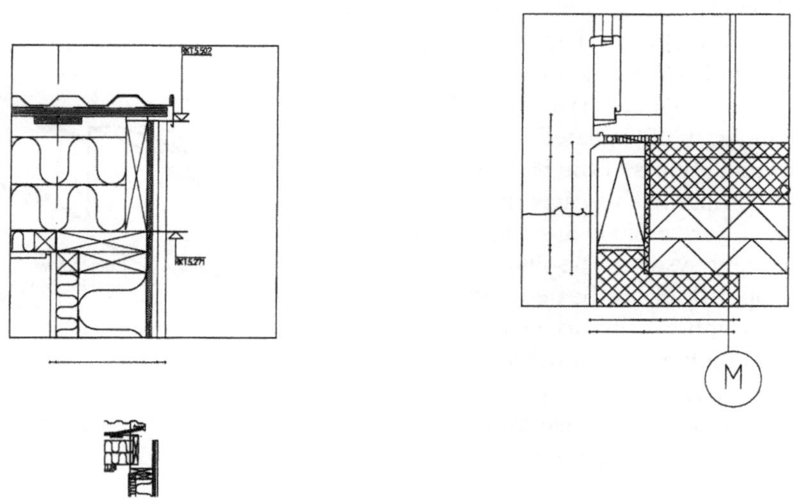

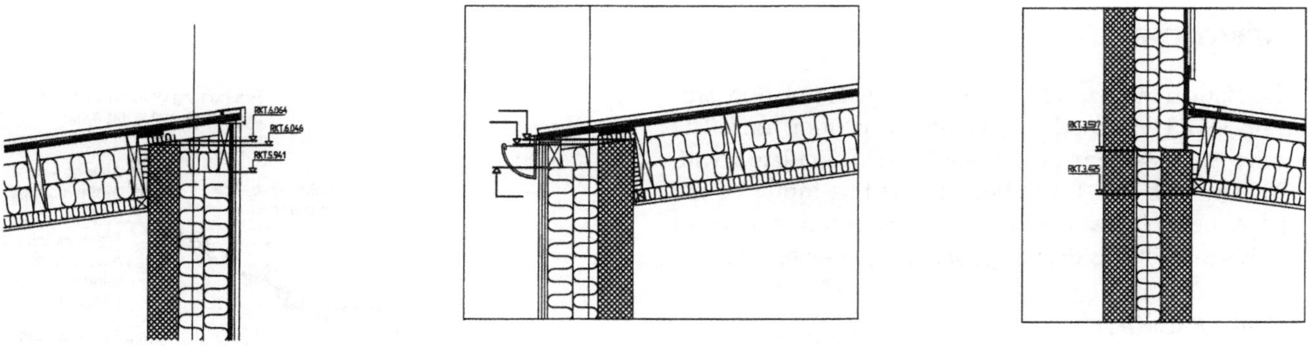

Load-bearing wall/roof

Figure 3.7.4. Envelope construction in the Danish row houses

hydronic system with small radiators or floor heating, controlled by room thermostats. The low-temperature system operates with a small volume of water, so as to be able to respond more quickly to the demand situation. The active solar system is a low-flow DHW system with collectors inclined at 45°, mounted on racks on the roof. A collector array serves up to four dwellings, each with a 200 litre tank. In one of the east–west running rows (house type A, the Task 13 row), a small PV panel supplies and controls the pumps for the water/glycol loop between collectors and storage tanks.

For economic reasons (based on first cost only, not on future trade off with running cost), only the rowhouses with north/south-running rows are equipped with heat-recovery systems. However, the Task 13 house in the row-house type with east/west-running rows is also equipped with a heat-recovery system. This building has a single-dwelling counterflow heat exchanger with an exchanger efficiency of 80–85% and an overall efficiency (accounting for parasitic energy) of 70–75%. It is a balanced system exhausting air from the bathroom and the kitchen part of the living room, and supplying air to the living room and the three bedrooms. The range hood has separate ducting with its own grease filter, but is connected to the heat-recovery system. In the summer the occupants can turn off the supply air fan in the recovery unit and let fresh air in through the windows, which can be locked in a ventilation position, or through three special fresh-air vents in each dwelling.

The house type with north/south-running rows is equipped with a central heat-recovery system serving from four to eight dwellings. The advantage is a slightly higher exchanger efficiency, and lower parasitic energy use per dwelling, set off against the inconvenience of no individual control. The central system is linked to a computerized remote control system from where, for example, the supply air fan can be turned off during warm weather and the conditions of the filters can be monitored. This is also a balanced system with a counterflow heat exchanger, with insulated ducts running under the slab-on-grade floors of the houses. The insulation thickness is 100 mm. In this case, the range hoods (each with its own grease filter) are connected to the central heat recovery system. The living room and bedroom on the upper floor are heated with the supply ventilation air through a water-to-air heat exchanger supplied from a combination of the solar and (chiefly) the district heating system.

In the two-storey rooms of both house types, the temperature build-up under the ceiling is counteracted with a propeller fan.

ENERGY USE

The Danish multizone building energy simulation program TSBI3 was used to calculate the heating load. The active solar system performance was calculated with EMGP3, with the TSBI3 heating load as input for the space heating calculation. Some selected parameter analyses from the design phase are presented below.

BASE ASSUMPTION

A middle-unit house is divided into three zones. All thermostats are considered ideal and set at 20°C. The heating system is turned off from 22 to 05 h. The controlled air change is constant, at 0.5 ac/h with a heat recovery efficiency of 60% (allowing for a slight infiltration). Venting at 3 ac/h occurs if the zone temperature exceeds 23°C. The internal gains are 2849 kWh/a from people (a family of four) and 1270 kWh/a from electricity. The heating season runs from October through April.

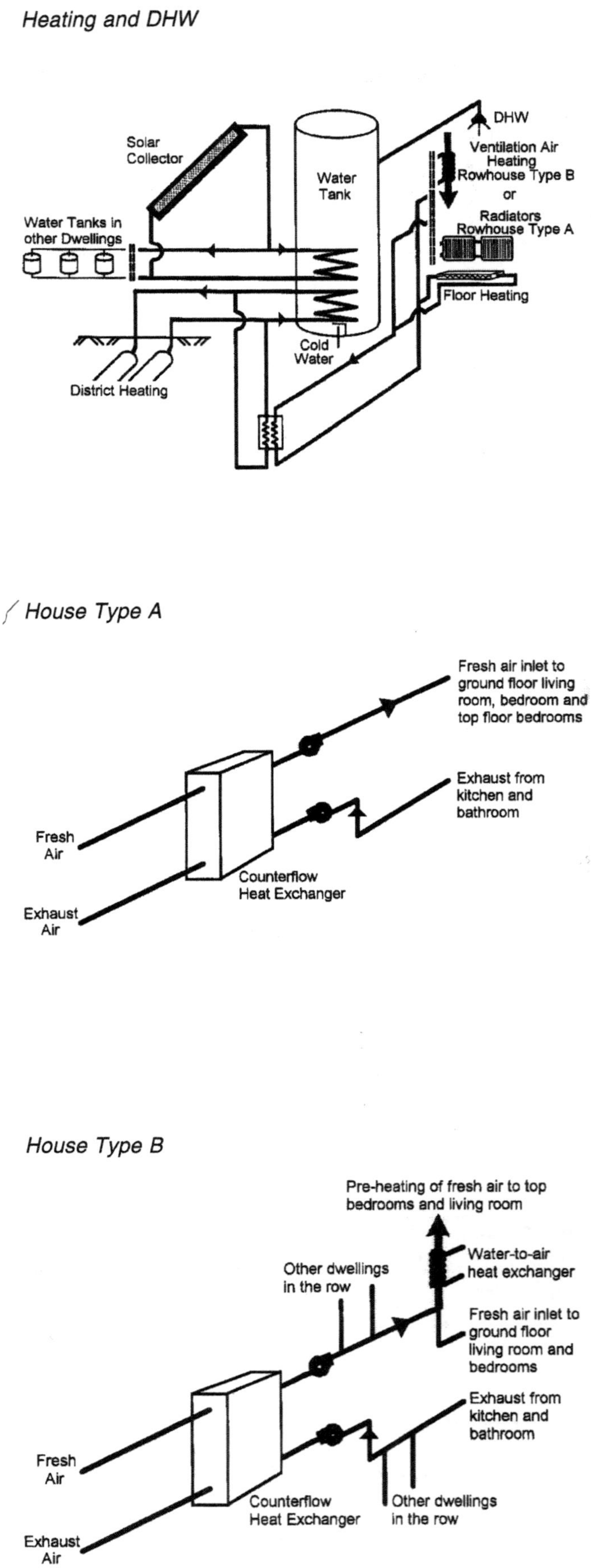

Figure 3.7.5. Simplified diagram of ventilation and heating systems in the Danish row houses

The final low-energy cases (A1 and B1) were super-insulated with 300 mm walls and roof insulation, aerogel skylights and elevated ribbon windows, triple-glazed windows with superspacers, two low-emissivity coatings, and argon filling. Insulating frames were used. The partially evacuated (~80–100 mb) aerogel windows (double-glazed) have a centre U-value of 0.50 W/m^2K and a total solar transmittance of 0.75 (space between panes filled with 20 mm monolithic silica aerogel [Airglass], special stainless steel spacer, and seal). For cost and liability reasons, the aerogel windows and insulated frames were replaced with triple low-E coated gas-filled glazings in wooden frames in the built Task 13 house version. Also, in the built version there are no skylights in the two-storey dwellings.

The cases and their descriptions are summarized in Table 3.7.1.

SIMULATION RESULTS

The annual net heat demands (in kWh) for the four cases in Table 3.7.1, with corrections for two-dimensional heat flows, were as shown in Table 3.7.2.

The total annual energy consumption end uses in kWh/m^2 for the final design were:

net heating load	12.0
auxiliary DHW heating	5.0
lights and appliances	13.0
fans & pumps	5.0
total energy	35.0

A rough calculation indicated that it would be possible to lower the heating load per year by 500–600 kWh through the use of south-facing ventilated solar walls (walls externally covered with transparent insulation, with air gaps ventilated to the indoor air), but that it may prove difficult to control overheating.

ACTIVE SOLAR SYSTEM PERFORMANCE

To determine how much of the DHW and space-heating demand it is reasonable to cover by an active solar heating system a parametric study was performed. In a super-insulated house, it is difficult to justify the capital cost of a space-heating solar system. Although a large part of the heating load can be covered, the absolute savings are small. The heating system must therefore be very inexpensive. One solution is to provide an over-sized DHW system with a separate convector for the largest room, connected to the mantle of the storage. A submerged heat exchanger in the top of the DHW tank allows for back-up to the solar hot-water system, as well as solar contribution to the heating system, thus eliminating the need for an extra tank in the system. In particular, if a boiler system is used for auxiliary heating, a buffer is needed in any case to decrease the frequency of on/off cycling and thereby increase efficiency. A low-flow solar system was analysed, with a flow rate of 0.15 litre/m^2 min versus the normal 0.5–1 litre/m^2 min. The low-flow concept increases the solar system performance by more than 10 %.

The system performance was analyzed using the PC-based simulation program EMGP3. As EMGP3 is not an integrated part of the building simulation program which was used, the hourly space-heating demand from TSBI3 was input. The storage was a 400 litre cylindrical tank with a mantle covering the lower 75% of the storage. The DHW demand was 130 litres per day, heated from 10 to 50°C (2200 kWh/a). An annual space-heating demand of approximately 2000 kWh/a was assumed.

The collector area and tilt were varied. A 45° tilt performed better for small collector areas, where the contribution to space heating is low. For larger collector areas, with a larger space-heating contribution, a 90° collector tilt performed better. In the latter case the collector was oversized for DHW supply and better oriented during the winter. An important advantage of a 90° tilt is that the maximum temperature of the collector is lower than for a 45° tilt – 100 and 163°C, respectively, for a system with a collector area of 8 m^2.

To quantify the contribution, the reference system was compared with two systems for DHW only. The only differences from the reference system were that there is no space heating load and that the storage is 300 litres. The absolute contribution to space heating was almost insignificant. The space-heating solar contribution is taken from the contribution to DHW. The oversized solar system seems pointless, given that the performance was almost identical to that of a less expensive small DHW system. The combined active solar system covered about 75% of the DHW load and typically about 10% of the heating load for the base-case house and a 45° tilt gave the same result for this collector and storage size. To cover a large part of the

Table 3.7.1. The four cases for Danish row house types A and B

Cases	Description
A1 & B1	Houses insulated with 300 mm mineral wool, triple-glazed low-E-coated windows with Superspacers and insulated frames, and with aerogel windows in skylights and elevated ribbon windows (B1)
A2 & B2	Houses insulated with 300 mm mineral wool, triple-glazed low-E-coated windows with *aluminum spacers* and insulated frames
A3 & B3	Houses insulated with 300 mm mineral wool, *double-glazed* low-E-coated windows, aluminum spacers and insulated frames
A4 & B4	Houses insulated according to the *current building code*

Table 3.7.2. Annual net heat demands (kWh) for the four cases in Table 3.7.1

	House A	House B
1 Final designs	1180	1400
2 Option 1	1550	2100
3 Option 2	1750	2440
4 Present standards	7240	9000

space heating demand, a district solar heating system with a seasonal storage would make more sense.

CONCLUSIONS

The analyses of this project clearly demonstrate that it is possible to design row houses with a predicted space-heating consumption of less than 1500 kWh/a in a climate with approximately 1500 sunshine hours and 3000 degree days (base 17°C). While the main orientation of the windows is important, it is possible to design row houses with east- and west-facing façades which require minimal auxiliary heating and which are not prone to overheating. The low energy consumption of this project is achieved through very good insulation and airtightness, minimization of thermal bridges, high-performance windows and frames, and active and passive solar use.

Simulations results from EMGP3 show that solar energy can cover about 75% of the hot water heating demand. It does not appear economical to dimension an active solar system to cover the space-heating demand, unless seasonal storage is introduced. Here, it was decided to augment the conventional low-temperature heating system with the solar DHW heating system.

ACKNOWLEDGEMENTS

- *Energy design and analysis.* Bjarne Saxhof, Jørgen M. Schultz & Kirsten Engelund Thomsen, Department of Buildings and Energy (IBE), Technical University of Denmark (DTU), Building 118, DK-2800 Lyngby, Denmark.
- *Architectural design.* Boje Lundgaard, Boje Lundgaard & Lene Tranberg, Architects MAA/PAR, Pilestrede 10, DK-1112 Copenhagen, Denmark.

PUBLICATIONS

Engelund Thomsen K, Jensen S Ø, Saxhof B (1992). Optimization of a Danish 3rd Generation Low-Energy House Concept. *Proceedings of the 5th International Conference North Sun '92, Solar Energy at High Latitudes*, Trondheim, 65–71.

Engelund Thomsen K, Wittchen K B, Saxhof B (1992). Advanced Solar Low-Energy Buildings – Danish Work within IEA Task 13. *Proceedings of the ASHRAE/DOE/BTECC Thermal Performance of the Exterior Envelope of Buildings Conference*, Clearwater, Florida, Vol. V, 614–620.

Saxhof B, Schultz J M, Engelund Thomsen K (1993). Two Danish Task 13 Low-Energy Houses – Designs and Parametric Studies. In Erhorn H, Reiss J, Szerman M (ed.), *Proceedings of the International Symposium Energy Efficient Buildings*, Leinfelden-Echterdingen, Germany, 211–225

Saxhof B, Schultz J M, Engelund Thomsen K (1993). 2- and 3-Dimensional Heat Losses in Superinsulated Buildings. In Saxhof B (ed.), *Proceedings of the 3rd Symposium on Building Physics in the Nordic Countries: Building Physics '93*, Copenhagen, Vol.1, 109–116.

Saxhof B, Schultz J M, Engelund Thomsen K (1993). *Thermal Analyses of Danish Low Energy Row Houses for IEA SHC Task 13 'Advanced Solar Low Energy Buildings'*. Innovative Housing Conference '93, Vancouver, Canada.

3.8 The Finnish IEA5 Solar House in Pietarsaari

SUMMARY

This low-energy residence was built to demonstrate new technologies and prototypical solutions. To minimize purchased energy consumption without compromising comfort, a well-insulated, airtight envelope was constructed. High-quality windows are used, and heat is recovered from the exhaust ventilation air. Solar energy is used as effectively as possible by means of solar collectors, photovoltaic panels, and a heat pump.

KEY FEATURES

- super insulation
- airtight construction
- high performance windows
- ventilation heat recovery
- low-temperature floor heating
- active solar heating
- ground-coupled heat pump
- PV system

LOCATION

Pietarsaari is located on the west coast of Finland about 550 km north-west of Helsinki at a latitude of 62°N. The building site is situated by the sea with a sea view from the west side of the house. The main façade of the house faces south. The IEA5 Solar House was built for the housing exhibition held at Pietarsaari, in 1994.

CLIMATE

The climate is cold maritime (Figure 3.8.1).

CONCEPT

Traditionally, Finnish detached houses are constructed of wood and are well insulated. Modern houses are

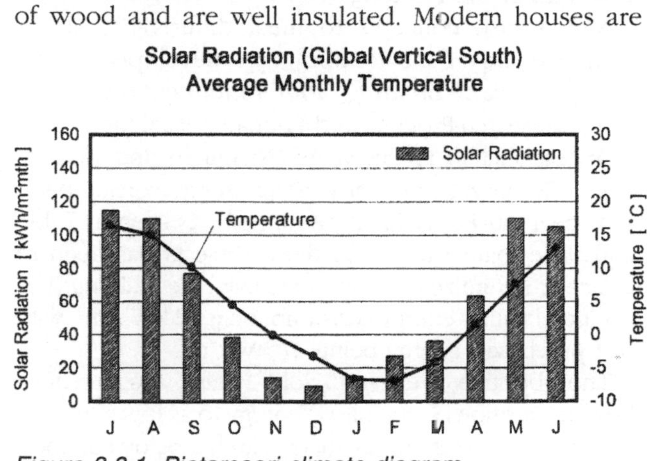

Figure 3.8.1. Pietarsaari climate diagram

Table 3.8.1. Typical Finnish small houses and low energy demonstrations

Building type	U-value/component (W/m²K)	(ac/h)	Ventilation (ac/h)	Infiltraton %	Heat recovery	Heating energy consumption (kWh/m²)
Standard house[a]	wall	0.28	0.5	–	–	160
	floor	0.36				
	roof	0.22				
	window	2.1				
	door	0.7				
Existing houses[b]	wall	0.28–0.17	0.5	0.2–0.3	60	120
	floor	0.24				
	roof	0.22–0.13				
	window	1.8–2.1				
	door	0.7				
Demonstration[c]	wall	0.17–0.12	0.5–0.7	0.05–0.1	60 - 80	40–80
	floor	0.20–0.15				
	roof	1.2–0.85				
	window	0.7–0.5				
	door	0.12–0.09				
IEA5[d]	wall	0.12	0.6	0.05	80	12
	floor	0.11				
	roof	0.09				
	window	0.7				
	door	0.5				

[a] Corresponds to the National Building Code requirements
[b] Existing houses built from l983–88 with average heated floor area of 159m²
[c] Houses built from l991–95 with a heated floor area of 125–300m²
[d] Built from l993–94 with heated floor area of 166m²

generally insulated to a somewhat higher standard than is required by the Building Code. The average annual heating energy consumption of a small house complying with Finland's Building Code is about 160 kWh/m², while the annual consumption in new houses averages 120–140 kWh/m² in the climatic conditions of central Finland.

A light-weight wooden construction with a wooden façade was chosen for the IEA5 Solar House in order to tie solar technologies to traditional ways of building. Spatial flexibility and temperature zoning by means of modular design are design features. The house was first designed for a hypothetical site. Thermal zoning had to be decreased owing to considerations of spatial functionality and the actual building site. In the final design the living area is one temperature zone, while storage rooms and the garage on the north side of the house have a lower-temperature zone and perform as buffers to the living area. For the final design of the house it was imperative to take advantage of the sea view on the west side of the building.

New house concepts to meet different levels of heating energy consumption were developed in the *Energy-Efficient Buildings and Building Components* (ETRR) Research Program and a number of solutions have been test-built in recent years (Nieminen and Kouhia, 1994). The target levels of heating energy consumption have been 40–80 kWh/m² per annum, as seen in Table 3.8.1. The main strategy in these projects has been to reduce or minimize the heat losses from the building. The resulting heating energy consumption is 12 kWh/m², with total purchased energy being 27 kWh/m².

The objective of the IEA5 Solar House was to reduce the consumption of purchased energy to as low a level as possible, using technology that is either already commercially available or at the prototype stage of development.

CONSTRUCTION

The plans and elevation of IEA5 Solar House are shown in Figure 3.8.2.

Building envelope

A section through the envelope is presented in Figure 3.8.3. The timber-framed building envelope is extremely well insulated with between 315 and 500 mm of mineral wool. The envelope was assembled from large prefabricated units. The wooden exterior of the house is covered with an extra layer of non-woven faced mineral woolwind proofing. The 70 mm thick non-woven faced windproof panel attached to the outer surface of the wall unit extends beyond the bottom of the wall as far as the foundations. Also, the concrete foundations are insulated on the inside. The foundation of the house stands on piles because of the poor load-bearing capacity of the soil. The ground floor of the living area has an unventilated 400–600 mm high air space. Extruded polystyrene foam panels were used to insulate the foundations.

When the building was designed, there were no superior insulating windows on the market. New windows were therefore designed for the building in collaboration with industry. The target was a window with a U-value of 0.6 W/m²K, but this standard was not achieved because of the construction schedule. The windows actually used in the building have a total thermal transmittance coefficient of 0.7 W/m²K. The glazed unit of the inner frame is a krypton-gas-filled unit with two selective films. There is also a selective film on the outer pane of glass. The frame and casement of the windows are thermally insulated.

Airtight construction was one of the key features of the house. The target level set for the airtightness was

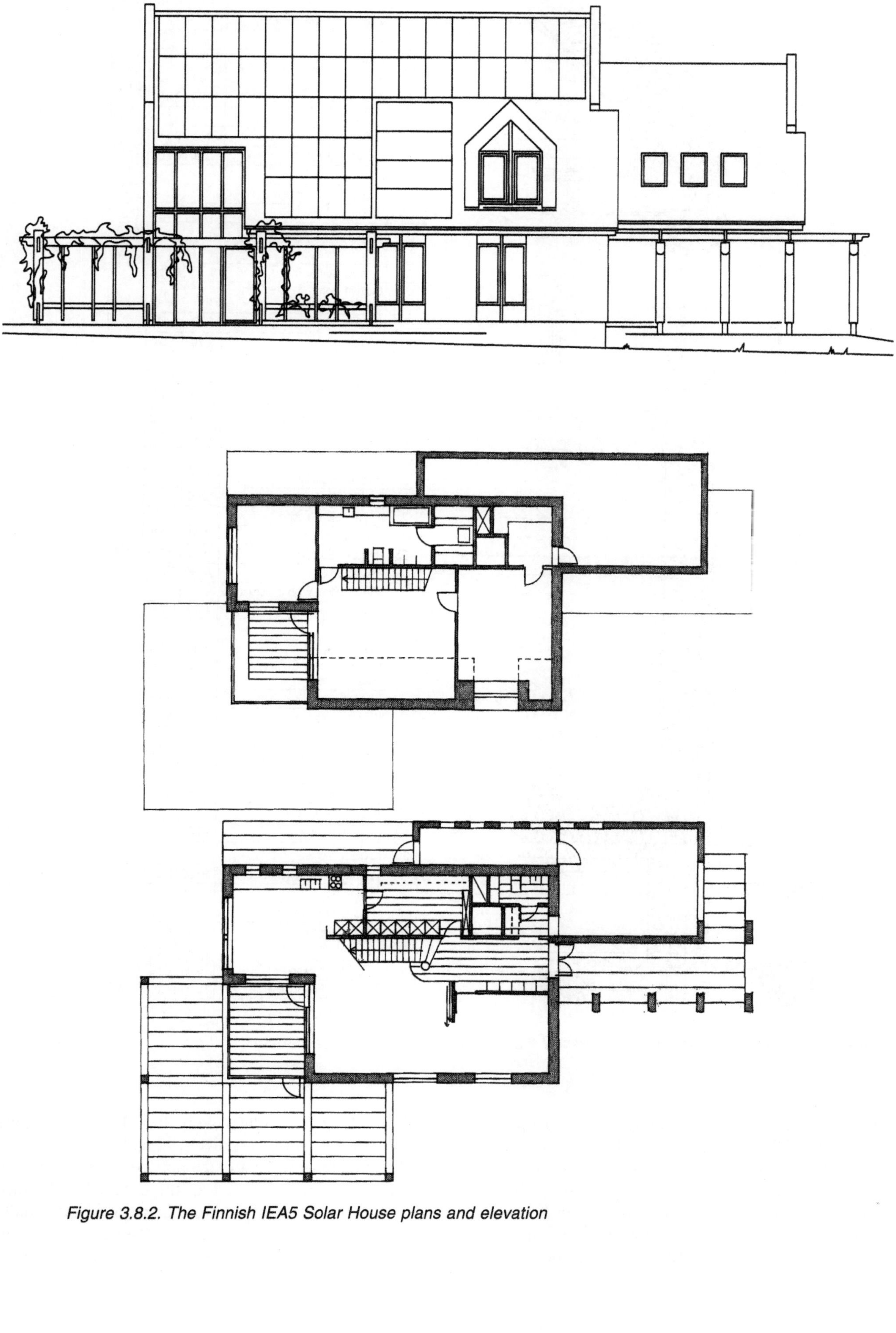

Figure 3.8.2. The Finnish IEA5 Solar House plans and elevation

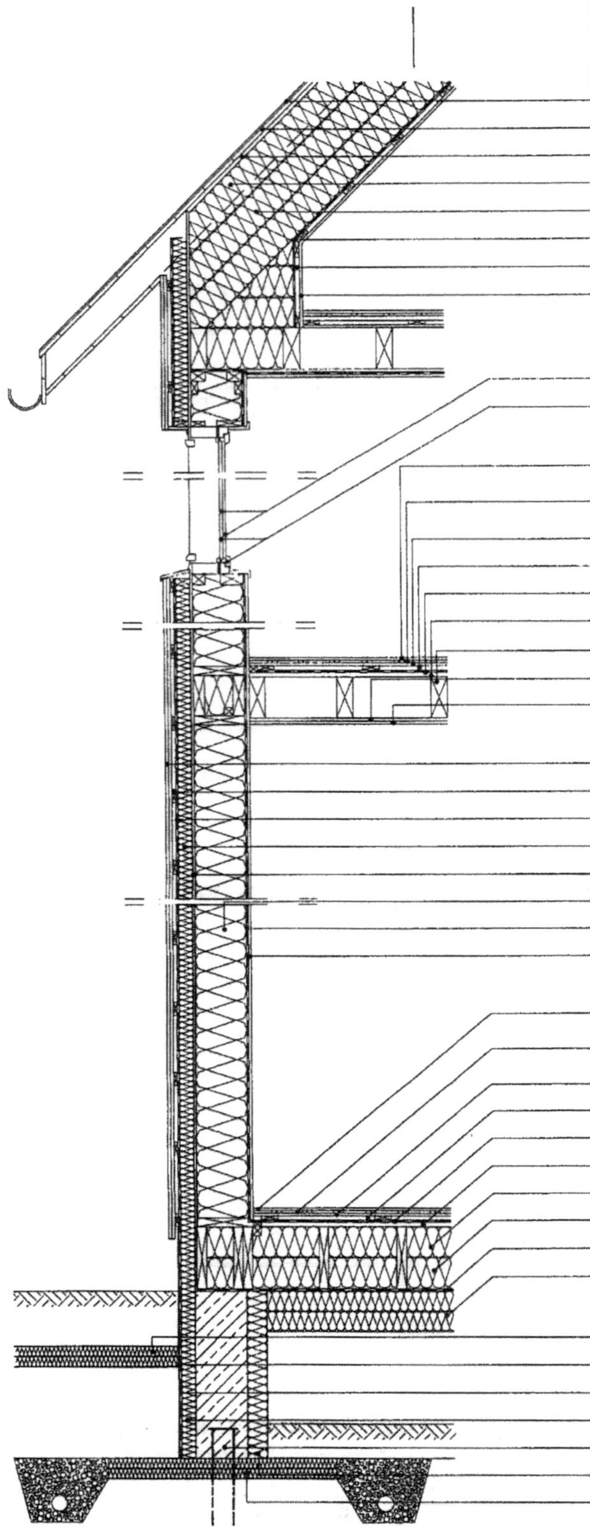

Layer	Description	Width	Thermal Conduct.	U-Value
		cm	W/mK	W/m²K
Roof				
1	Sheet Steel Waterproofing	–	–	
2	Boarding	1.90	–	
3	Ventilation Gap	5.00	–	
4	Wood	10.00	0.12	0.09
5	Mineralwool	45.00	0.035	
6	Vapour Barrier	–	–	
7	Lattice	2.20	0.120	
8	Gypsum Board	1.30	0.230	
Window				
9	Frame Aluminium/Wood			0.70
10	Glazing			0.70
11	Glazing and Frame			0.70
Ceiling				
12	Gypsum Board	1.30	0.230	
13	Gypsum Board Stripes/ Plastic Piping	1.30	0.230	
14	Gypsum Board	1.30	0.230	
15	Boarding/Airspace	2.20	0.120	–
16	Plastic Film Vapour Barrier	–	–	
17	Wood Frame Structure	2.20	0.120	
18	Wood Composite Beams	20.00	0.014	
19	Wood Frame Structure	2.00	0.120	
20	Gypsum Board	1.30	0.230	
Exterior Wall				
21	Wooden Cladding	3.00	–	
22	Lattice/Airspace	2.00	–	
23	Plastic Fibre Fabric	–	–	
24	Rigid Mineralwool	7.00	0.035	
25	Gypsum Board	0.90	0.210	0.12
26	Mineralwool	24.50	0.035	
27	Plastic Film Vapour Barrier	–	–	
28	Gypsum Board	1.30	0.230	
Floor				
29	Gypsum Board	1.30	0.230	
30	Gypsum Board Stripes/ Plastic Piping	1.30	0.230	
31	Gypsum Board	1.30	0.230	
32	Boarding/Airspace	2.20	0.120	
33	Plastic Film Vapour Barrier	–	–	0.11
34	Wood Frame Structure	2.00	0.120	
35	Mineralwool	30.00	0.035	
36	Wood Composite Beams	30.50	0.014	
37	Plastic Fibre Fabric	–	–	
38	Rigid Mineralwool	20.00	0.035	
Foundation				
39	Extruded Polystyrene	10.00	0.030	
40	Rendering	0.50	–	
41	Plastic Fibre Fabric	–	–	
42	Rigid Mineralwool	7.00	0.035	
43	Concrete	25.00	1.500	–
44	Rigid Mineralwool	10.00	0.035	
45	Extruded Polystyrene	10.00	0.030	
46	Steel Columns	11.00	60.000	

Figure 3.8.3. Section through the structures of the building envelope (Finnish house)

air leakage rate of 1.0 ac/h measured at 50 Pa pressure difference. All the technical installations are located as far as possible inside the building's envelope in order to promote airtightness. Rubber seals are used in the joints of all the building components, and vapor barriers are taped together at their meeting points. The number of penetrations through the building envelope was minimized. Flanged conduits were installed in envelope penetrations in order to prevent air leaks.

The first floor units are 220 mm deep beams, which extend to the outer surface of the external wall units. The load-bearing roof structure consists of industrially manufactured nail-plate trusses. The roof was constructed on the building site. The underlay for the roof covering is tongue-and-groove boarding, which was assembled into panels at the factory. The thickness of the roof's thermal insulation is 450–500 mm.

SOLAR SYSTEMS

Heating and ventilation

Energy provided by the solar collectors and the ground heat pump is stored in a hot water storage tank. The solar collector system consists of four 2.5 m² modules, connected in parallel on the south-facing side of the roof. To satisfy the winter heat demand a ground heat pump with a capacity of about 7 kW is included in the system. Approximately 400 m of ground pipes are installed in the garden at a depth of about 1 m. The ground pipe system consists of two 200 m long circuits, either one of which can be switched off.

Photovoltaic system

The photovoltaic (PV) system consists of 45 solar panels (800 mm x 1.55 mm) installed on the south-facing side of the roof. The panels are connected in series in groups of three and the groups are connected in parallel. The voltage produced by the panels varies between 100 and 160 V and the maximum power output is about 2 kW. The panels are connected directly to the building's electrical system. This improves efficiency, while the absence of batteries reduces maintenance and replacement costs. In the summer, when sunlight is in abundance, the system supplies energy to the house mains.

HVAC

A 3 m³ water-filled heat storage tank is the heart of the building's heating system. The temperature of the storage tank can be kept at a low level during the winter, so that the heat losses from the storage tank are small and the temperature coefficient of the ground heat pump is as large as possible. Heat from the storage tank is distributed to the rooms with the aid of low-temperature floor heating. The floors are made from three layers of gypsum board, with plastic floor-heating tubes installed in the middle layer. Separate floor-heating pipe circuits for each of the rooms allows the temperature in them to be regulated independently.

The possibility of using the floor heating system for cooling is being studied in the building. The ground heat pipe system includes a heat exchanger which can be used to cool the water circulating in the floor heating system. A diagram of the building's heating system is given in Figure 3.8.4.

The building is ventilated by two separate fans, each being equipped with heat recovery. The design value for the building's main ventilation system is 0.6 litre/h (approximately 240 m³/h). The main ventilating system is fitted with a rotational heat exchanger with a thermal efficiency of about 80%. The outlet of the kitchen stove hood is handled separately with a heat recovery efficiency of about 60%. The stove hood fan can back up the main ventilation system. The building has operable windows, which can be used to increase ventilation in situations of overheating.

Control system

The ventilation system is manually controlled. Four power settings on each of the fans allow the air volume flow to be regulated. Heat recovery in the rotational ventilating heat exchanger can be switched on and off. The fan in the kitchen stove hood has a plate-type heat exchanger, so activating the heat recovery means exchanging the cell.

The heating system is regulated by a programmable control unit which regulates the system based on outside temperature, the feed and return temperatures of the solar collector, heat pump, floor-heating and domestic hot-water circuits, the temperatures of the storage tank at different levels, and the power output of the photovoltaic system. Energy from the solar collectors is stored in the heat-storage tank whenever possible. The operation of the heat pump is restricted to the daytime, and its start-up occurs, depending on the output of the photovoltaic system, at different temperatures of the heat storage tank. The temperature in the floor heating system is regulated according to the outside temperature, and individual rooms are controlled by thermostatically controlled valves. When the floor heating system is to be used for cooling purposes, the minimum temperature of the water pumped to the floor heating system is two degrees lower than the room temperature. This restriction prevents the formation of condensation on the floor surfaces.

ENERGY USE

The savings in heating energy in the building are based on minimizing heat losses with the use of generous thermal insulation as well as exploiting free energy. In addition super windows were specified with total

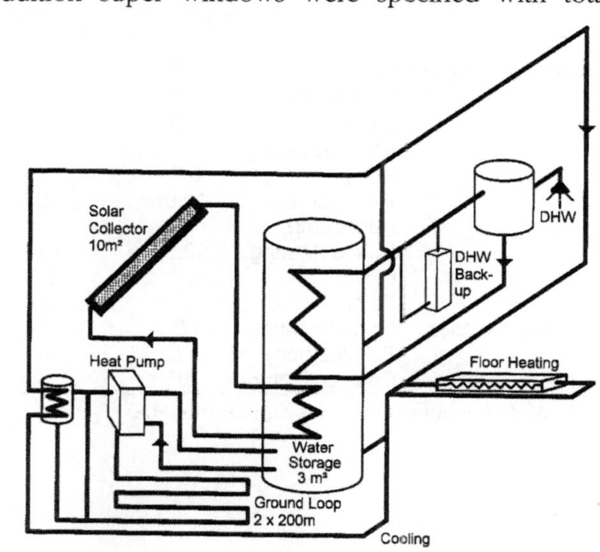

Figure 3.8.4. Heating system of the IEA5 Solar House, Finland

U-value three times better than conventional windows used in Finland.

Solar energy is exploited directly in the house's PV system as well as through solar thermal collectors. Moreover, solar energy is exploited with a ground heat pump. A small sunspace was built on the south-west corner of the house more as a solar amenity than as a means of saving energy.

Low-energy lamps were used for the house's lighting system. Furthermore, energy was a major criterion in the selection of commercially available household appliances.

MONITORING AND MEASUREMENTS

After the building's envelope was completed, its airtightness was measured and all leaks were located and marked for repair. In addition, the heat loss coefficient and effective heat capacity of the building were measured for a full week.

A value of 0.87 litre/h was obtained for the air infiltration n_{50} of the completed building. The air leaks were located by two-stage thermography (at normal room temperature and a negative pressure of 50 Pa).

Short-term measurement of the building's thermal performance was made in March 1994. The measurement results were used to calculate the building's heat loss coefficient, which is approximately 85 W/K. A value of approx. 6.5 kWh/K was obtained for the effective heat capacity. The measured heat loss coefficient is 10–15% lower than that calculated with the Tase program.

Monitoring was conducted to check that the building performed as planned.. Data collection started at the beginning of 1995 and runs to the end of 1996.

The energy consumption of the building has also been calculated. Table 3.8.2 shows the measured and predicted energy consumption and production by the solar systems. The measured consumption has been somewhat higher than predicted. This is mainly because the energy needed to run the fans and pumps is nearly three times higher than predicted. The measured energy consumption for heating is quite low, even though the COP of the heat pump is 2.4 instead of the assumed COP of 3.3. The solar collectors proved to be very efficient, although 6 of the measured 19 kWh/m² were non-usable solar energy.

Table 3.8.2. Measured and simulated energy consumption and production of the Finnish IEA5 Solar House

	Consumption/ production	Results 1995–6 (kWh/m²)	Simulated (kWh/m²)
Energy consumption	Lights	6.5	6
	Household	10.0	7
	Sauna	1.5	3
	DHW Back up	6.5	4.5
	Pumps	5.5	2
	Ventilation	10.0	5
	Heat pump		
	Space Heating	12.5	13.0
	DHW	6.5	7.5
	Total	59	48
Energy production	Grid electricity	48	34
	PV electricity	11	14
	Solar collectors	19[a]	12

[a] 13 kWh/m² usable solar, 6 kWh/m² non-usable

CONCLUSIONS

The house achieves its aim of drastically reducing heating energy consumption. This has been achieved through highly insulating and airtight construction of the building envelope and direct solar gains through high quality specially fabricated windows. Also, the photovoltaic system of solar panels directly connected to the building's electrical system proved effective. Furthermore, the measured heat loss coefficient was 10–15% lower than calculated with the Tase program, which demonstrates the success of the project.

ACKNOWLEDGEMENTS

- *Project manager.* Ilpo Kouhia, VTT – Building Technology, P.O. Box 18011, FIN-02044 VTT, Finland.
- *Architect.* Andreas Walterman, ALV-ARK Oy, Lastenkodinkatu 5, FIN-65100 Vaasa, Finland.
- *Author.* Jyri Nieminen, VTT – Building Technology, P.O. Box 18011, FIN-02044 VTT, Finland.

PUBLICATIONS

ETRR *Energy Efficient Buildings and Building Components (1993). Final report on the energy research programme 1988–1992.* Reviews B:163, Ministry of Trade and Industry, Helsinki.

Kouhia I, Nieminen J (1995). *IEA5 Solar House, VTT– Research Notes 1674*, Espoo. (ISBN 951-38-4829-9)

Nieminen J (1994). Low-energy Residential Housing. *Energy and Buildings*, 21, 187–197.

Nieminen J, Kouhia I (1994). Low-energy Residential Housing: Case Studies on Energy Efficient Buildings. *Proceedings of the International Conference of HVAC in Cold Climate, Cold Climate HVAC' 94.* Finvac. (ISBN 952-90-5367-3)

Nieminen J, Kouhia I (1994). Solar house IEA5: The Finnish IEA Task 13 Demonstration House. *Proceedings of the International Conference of HVAC in Cold Climate, Cold Climate HVAC' 94.* Finvac. (ISBN 952-90-5367-3)

Nieminen J, Kouhia I (1994). Solar house IEA5: The Finnish IEA Task 13 Demonstration House. *Proceedings of the International conference Solar energy at High Latitudes, North Sun' 94.* Scottish Solar Energy Group, James & James (Science Publishers) Ltd, London. (ISBN 1-873936-33-8)

3.9 The German Zero Heating Energy House, Berlin

SUMMARY

This house is the result of a progression from the low-energy house to the zero-heating-energy house, i.e. a house that requires no auxiliary space-heating energy. This is achieved by consistently reducing the transmission losses and by active and passive solar strategies. In order to achieve zero heating energy, the house has to be constructed as an ultra low-energy house. The house is built on the corner of two rows of terraced houses, running perpendicular to one another. The house has a surface to volume coefficient of 0.60/m. The heated floor area is 170 m^2 and the heated volume 670 m^3. Solar collectors with seasonal storage provide the auxiliary space and domestic hot-water heating energy.

KEY FEATURES

- active solar collectors
- seasonal storage
- super glazing
- heat recovery
- low-energy appliances

LOCATION

The house is built in the Berlin satellite town of Spandau as part of a new residential zone. The geographical location is 52°N and 14°E.

CLIMATE

While the average monthly air temperature is warmer than in Rottweil, the location of the other German Task 13 house, the global solar radiation on a south vertical surface is significantly less. The Berlin climate diagram is given in Figure 3.9.1.

CONCEPT

Owing to the positioning of the house, the exposure to the south is large, while the exposure to the north is minimal. Including the cellar, the northern side of the house has four storeys while the southern, eastern, and western sides have three storeys. An important feature of the house is a storage system located in the centre of the house. The collectors are integrated in the heat-

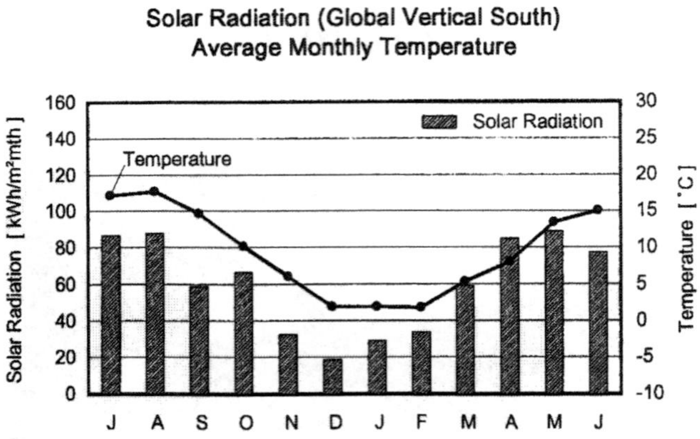

Figure 3.9.1. Berlin climate diagram

exchanging envelope area of the large living space. The floor plans, cross-section, and site plan of the Zero Heating Energy House are given in Figure 3.9.2.

CONSTRUCTION

The windows are triple-glazed with the voids filled with xenon gas, and with two surfaces having low-emissivity coatings. The frames are wooden and account for approximately 25% of the windows' gross area. The glazing U-value ranges from 0.4 (north) to 0.7 (south) W/m²K, while the entire window U-value ranges from 0.68 to 0.88 W/m²K.

The collector field has a U-value of 0.10 W/m²K. Other important surface areas are the external wall with a U-value of 0.19 W/m²K. The main components are a 24 cm thick layer of porous lightweight brick and a 16 cm thick layer of mineral wool. The roof has a U-value of 0.12 W/m²K with a main insulation layer of 30 cm mineral wool, and the cellar ceiling has a U-value of 0.23 W/m²K. In Figure 3.9.3 the construction details with the corresponding component U-values can be seen. The use of high-performance insulation materials has made it feasible for the solar collectors to provide all the space heating.

SOLAR SYSTEMS

Figure 3.9.4 presents a diagram of the heating system. The main features include solar collectors, a seasonal storage tank, heat exchangers, heating of ventilation air, and extreme low-temperature radiators. One solar collector field is divided by the glazing of the living space. The solar collectors are built into the sunspace on the southern side of the building and form a part of the external cladding. They have a combined absorber surface area of 54 m² at a slope of 52°. The flow in the collector field is 10–15 litre/m²h.. A small pump is used to compensate for the pressure drop over the heat exchanger. Heat is transferred to the storage system via a counterflow heat exchanger.

The main component of the seasonal storage system is a 20 m³ water tank. It is cylindrical in form, insulated with 30 cm insulation, and located in the middle of the house to make use of any thermal transmission losses through the wall. There are windows built into the roof above the tank for ventilation in the summer, to prevent space overheating.

Space heating is achieved by direct connection from the main storage tank to wall-mounted radiators. The radiators are flat plates constructed from aluminum. They have a high thermal conductivity. This means that even if the storage tank temperature drops to 30°C, the system will nevertheless be able to maintain a space temperature of between 20°C and 24°C.

The domestic hot water is stored in a 350 litre tank on the top floor. Heat is transferred to this tank by means of a plate heat exchanger. The temperature is maintained at 60°C. Incoming cold water is first circulated through the heat exchanger before entering the tank.

Back-up gas heating can also be used both for space heating or domestic hot water heating in case the storage tank temperature should become too low. It is intended, however, to run the building without a back-up system during the monitoring phase.

The hot-water storage system is also used to heat ventilation air which, despite already passing through a heat-recovery ventilator, may not be at 20°C. A small, thermostatically controlled radiator is built into the ventilation duct. One further application for the heat storage system is that during periods of excessive heat, two of the neighboring houses will receive heating for domestic hot water.

HVAC SYSTEMS

A diagram of the ventilation system can be seen in Figure 3.9.5. The system includes two heat exchangers. The first is a plate heat exchanger where heat is recovered from the exhaust air. The second is described in Section 2.9. The ventilation system will only operate during the heating period. During the summer, the living room, kitchen, and toilets can be ventilated by windows. When in operation, the ventilation system delivers 130 m³/h of fresh air to the living room, while the three bedrooms and the storeroom each receive 30 m³/h. The stale air is exhausted at the rate of 120 m³/h from the two bathrooms and 110 m³/h from the toilets. There are remote controls in the bathrooms which will enable the extraction fan to increase the volume of expelled air during periods of heavy moisture loading. As the cooking facility in the kitchen is a gas range, the kitchen is not connected to the ventilation system. There is, however, an extraction hood over the range so that the air can be exhausted directly outside.

ENERGY USE

Table 3.9.1 shows the predicted annual energy consumption for the zero-energy house. The major appliances used have all been selected for their energy efficiency.

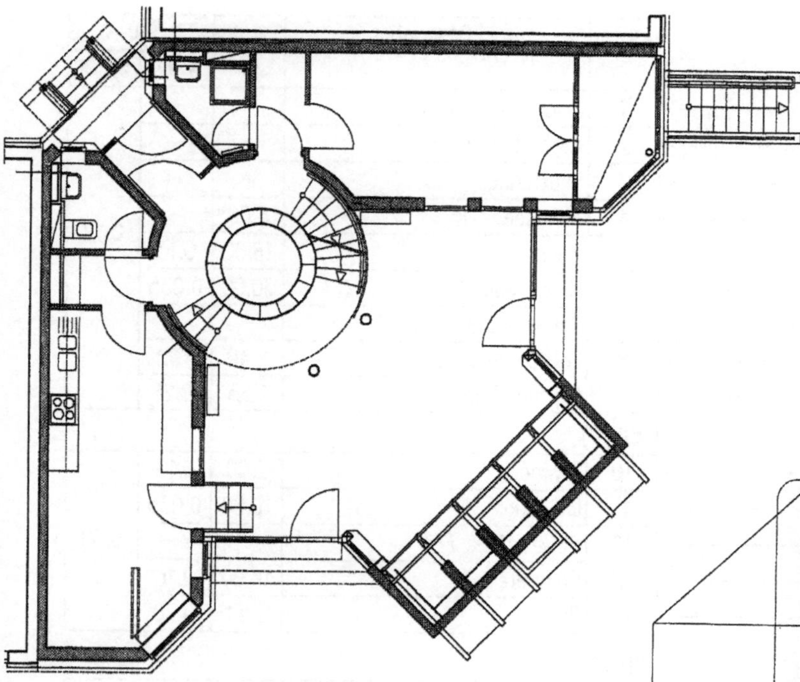

Ground floor

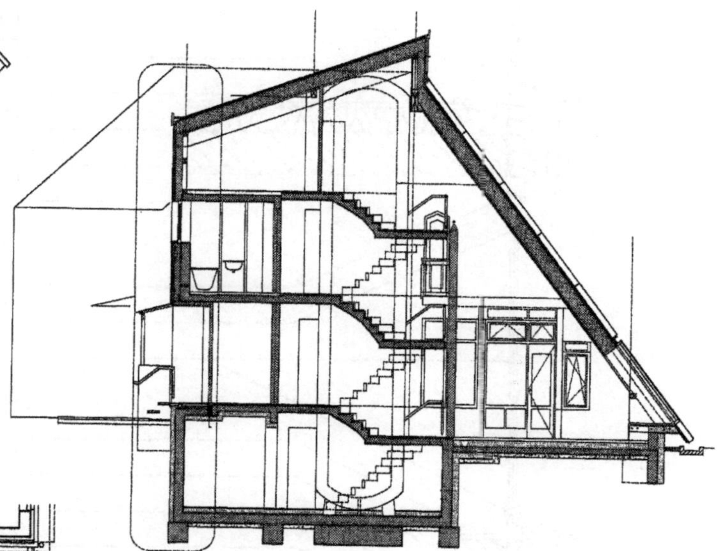

Section

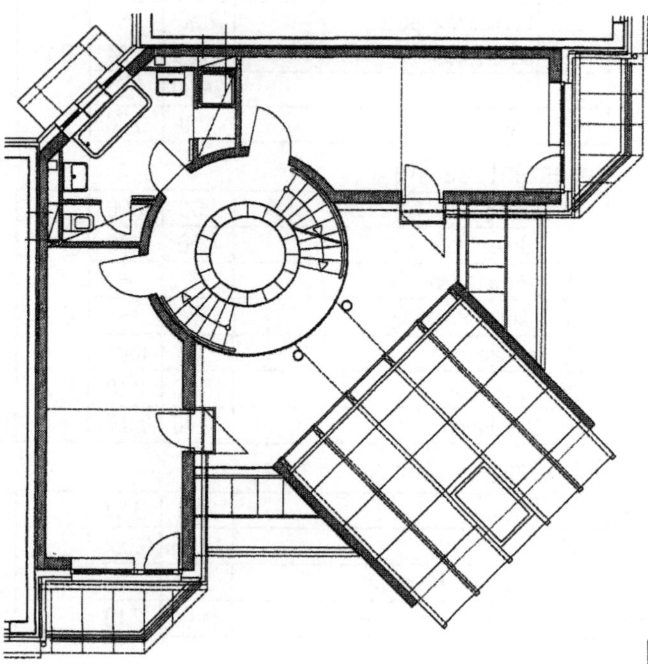

First floor

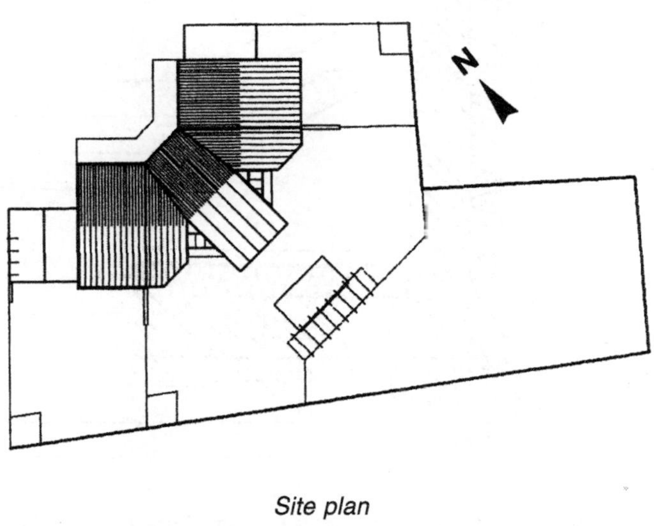

Site plan

Figure 3.9.2. Floor plans, cross-section, and site plan of the Zero Heating Energy House, Berlin

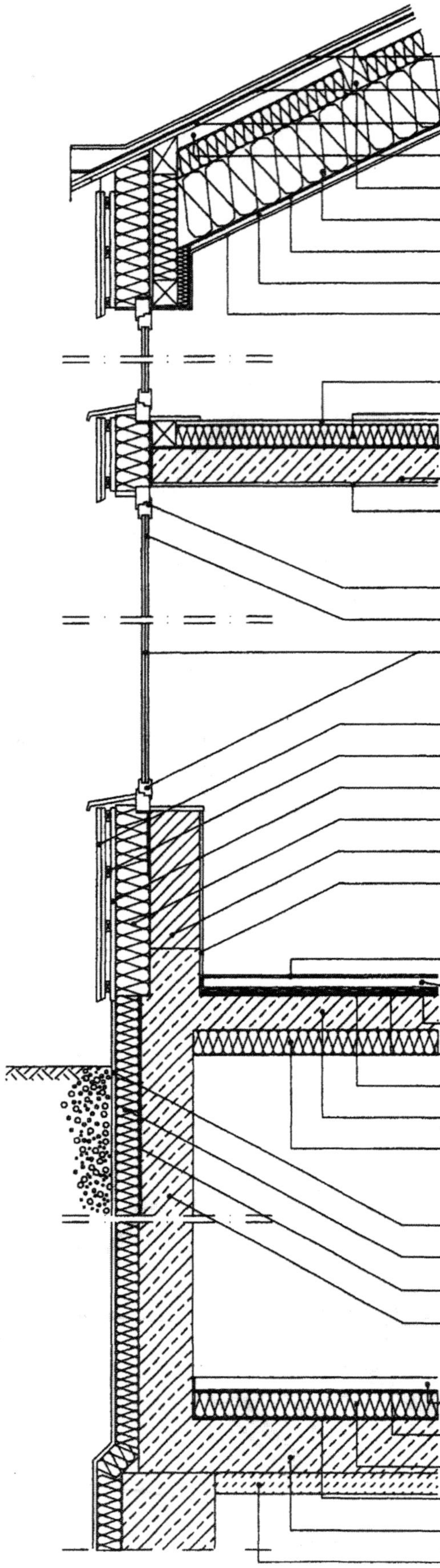

Layer	Description	Width	Thermal Conduct.	U-Value
		cm	W/mK	W/m²K
1	Zinc metal sheet	–	–	0.12
2	Asphalt Bitumen	–	–	
3	Shuttering	–	–	
4	Airspace	6.00	–	
5	Wood	18.00	0.13	
6	Mineralwool	30.00	0.035	
7	Polyethylene Foil	–	–	
8	Lattice/Airspace	2.40	λ=0.17	
9	Gypsum Board	1.25	0.21	
Ceiling				
10	Plywood	2.00	0.13	0.30
11	Insulation	10.00	0.035	
12	Polyethylene Foil	–	–	
13	Concrete	16.00	2.10	
14	Plaster	1.50	0.70	
Window				
15	Wood Frame			1.8
16	Triple Glazing			0.40
17	Glazing and Frame			0.68
Exterior Wall				
18	Timber Shuttering	4.00	–	0.19
19	Lattice/Airspace	2.00	–	
20	Lattice/Airspace	2.50	–	
21	Mineralwool	16.00	0.04	
22	Brick	24.00	0.21	
23	Plaster	1.50	0.70	
Basement Ceiling				
24	Tiling	1.00	1.00	0.23
25	Screedfloor	4.50	1.40	
26	Polyethylene Foil	–	–	
27	Water Proofing	–	–	
28	Insulation	2.00	0.04	
29	Concrete	16.00	2.10	
30	Insulation	12.00	0.035	
Below Grade Wall				
31	Plaster	2.00	0.87	0.27
32	Insulation	12.00	0.035	
33	Asphalt Bitumen	–	–	
34	Watertight Concrete	24.00	2.10	
Basement Slab				
35	Screedfloor	6.00	1.4	0.26
36	Polyethylene Foil	–	–	
37	Polyurethane Insulation	12.00	0.035	
38	Water Proofing	–	–	
39	Watertight Concrete	25.00	2.10	
40	Concrete	10.00	2.10	

Figure 3.9.3. Construction detail for the Zero Heating Energy House, Berlin

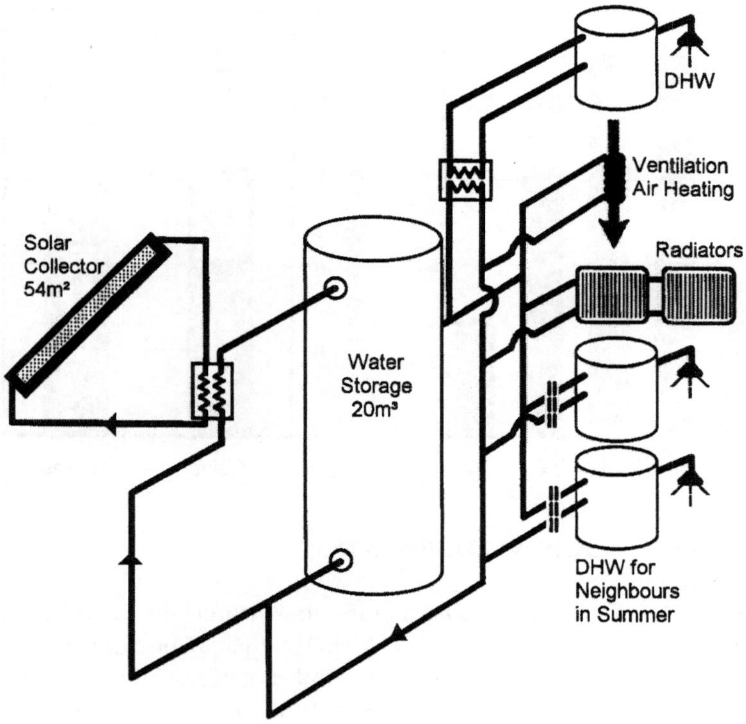

Figure 3.9.4. Diagram of the space and domestic hot water heating systems in the Berlin house

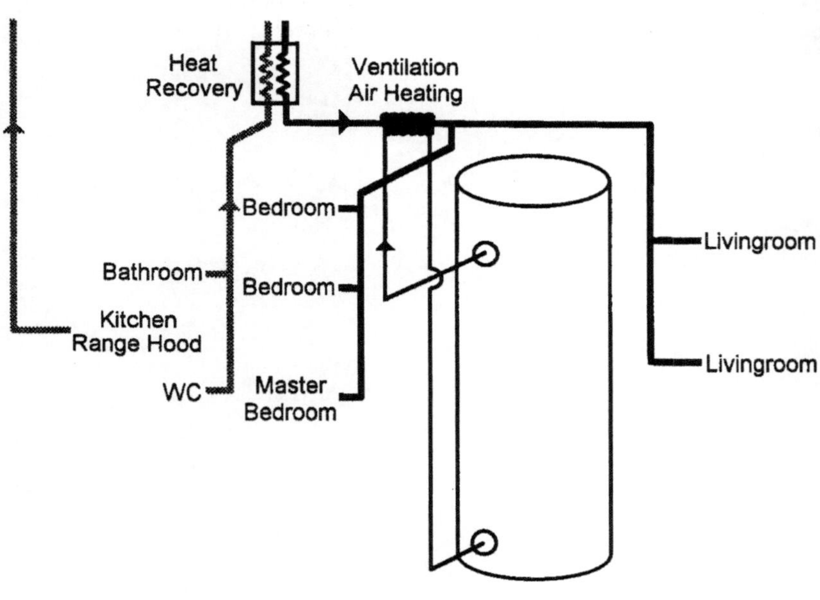

Figure 3.9.5. The ventilation system used in the Zero Heating Energy House, Berlin.

Figure 3.9.6 shows the monthly space-heating energy components. It can be seen that even without the solar collectors, the house can still be classified as a low-energy house because the total auxiliary heating (active solar heating) required is only 19 kW/m²a. Figure 3.9.6 also shows the complete energy balance over the heating season.

Table 3.9.1. Predicted annual energy consumption for the Berlin house

Energy usage	Active solar gains (kWh/m²a)	Purchased energy needs (kWh/m²a)
Space heating	19	–
Domestic hot water	15	–
Lights and appliances	–	12
Fans and pumps	–	3
Total	34	15

CONCLUSIONS

The zero-heating energy house can only be built if the building design shows an extremely low heating-energy demand. The storage has to be integrated in the

Figure 3.9.6. Calculated monthly heating energy and total energy for the Berlin house

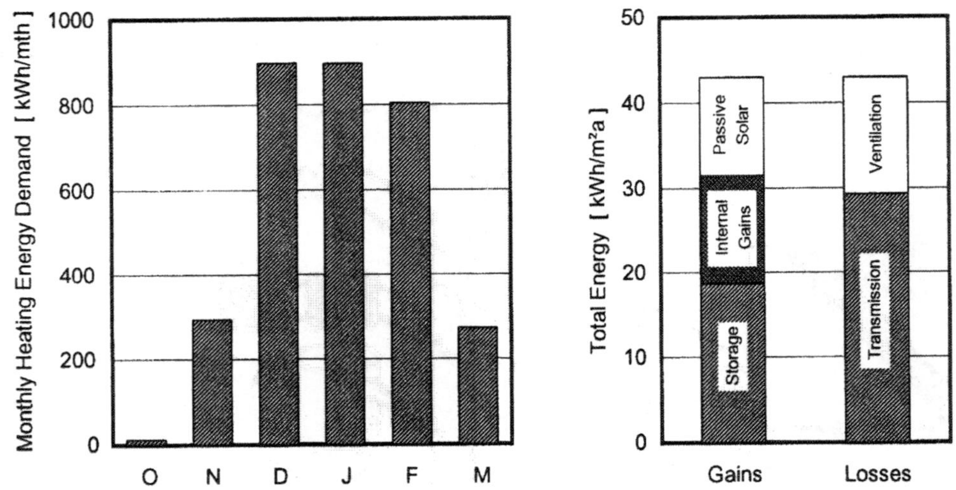

heated area, otherwise the losses will amount to about 1000 kWh/a. The installations (heating, ventilation) should be switched off as soon as the windows are opened to avoid increasing heating energy losses.

An important lesson learned in this project is that extreme energy concepts like this can only be put into practice by a close collaboration among the architect, building physicists, and installation experts working together in a team.

ACKNOWLEDGEMENTS

- *Design Team.* Fraunhofer-Institut für Bauphysik, Nobelstrasse 12, D-70569 Stuttgart, Germany; Institut für Bau-, Umwelt- und Solarforschung GmbH, Caspar-Theyss-Strasse 14A, D-14193 Berlin, Germany; and Institut für Thermodynamik und Wärmetechnik, Universität Stuttgart, Germany.
- *Authors.* Michael Beckert, Hans Erhorn, Gustav Hillman, Heike Kluttig, Johann Reiss, Hans-Martin Schmid, and Hasso Schreck.

3.10 The German Ultra House in Rottweil

SUMMARY

The main strategies employed in this house are to minimize space-heating losses through thermal transmittance and ventilation and to use passive solar energy. The aim was to reduce the annual purchased space heating to less than 20 kWh/m²a. The house is constructed such that all living space is supplied with daylight from the south side. Because of the split-level construction, and because the house is built like a fan with the largest surface area facing south, daylight and solar energy penetrate deep into the building. The house is compact, with a 0.52/m ratio of exterior surface to volume.

KEY FEATURES

- super-insulation
- optimal building surface area south orientated
- superglazing
- heat recovery
- ground air preheat
- sunspace

LOCATION

Rottweil is located 48°N and 10°E. It nestles in the foothills of the Black Forest in the province of Baden-Württemberg. The house is built on higher ground in a new suburb on the southern outskirts of town.

CLIMATE

In mid-winter average monthly temperatures fall below 0°C while summer temperatures are moderate. In winter the solar global radiation is sufficient to allow solar gains to be an important part of the energy balance equation (Figure 3.10.1).

CONCEPT

The Ultra House is half of a two-family house. The other house half was constructed as a low-energy unit. It was also monitored as a control for the Ultra House. A sunspace on the south side acts as a buffer against large temperature swings in the living area as well as a source of daylight. The sunspace, however is not required to

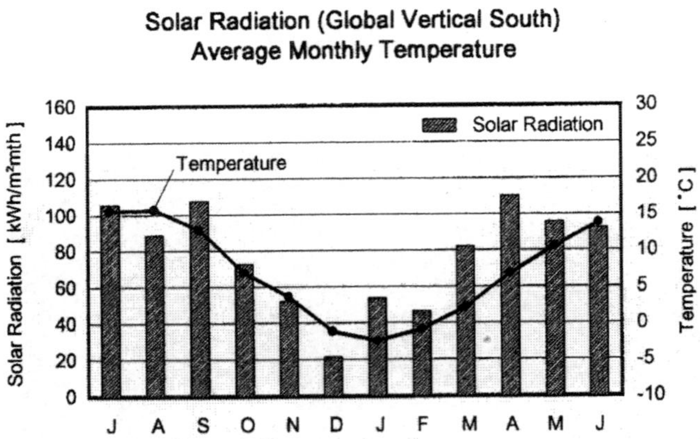

Figure 3.10.1. Rottweil climate diagram

achieve the low energy demand. Other design strategies were to decrease the proportion of surface area to volume and to maximize the surface area facing south, while minimizing the surface area facing north. To achieve these objectives, the building was given the shape of a quarter segment of an octagonal ring, with each of the house units occupying half of the segment. The ground floor plan and a cross section of the building can be seen in Figure 3.10.2. All the living spaces are situated in the southern section of the building and have access to the sunspace, which extends over the height of the building. Bathrooms, toilets, stairwell, and kitchen are all situated in the northern zone of the building. A cellar, located under the northern half of the building, is not heated. The roof has a grass covering, which was only installed as a decorative feature. The heated living area is 175 m².

The heating energy for the building is provided by a local power plant. The hot water arrives at about 60°C and is also used for heating the domestic hot water. Space heating is provided by wall-mounted convection radiators. The hot water for space heating is first cooled slightly in a plate heat exchanger by returned water from the radiators. This allows better control of the room temperatures. The domestic hot water is heated by a coil heat exchanger and is continuously circulated in insulated piping.

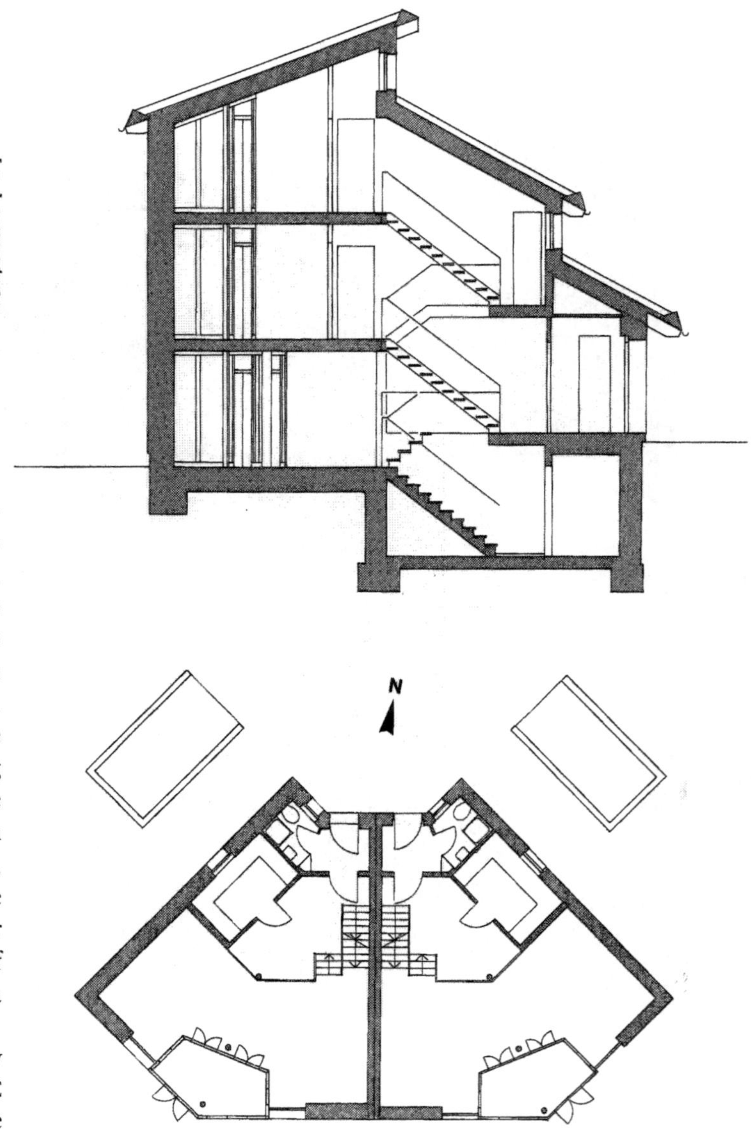

Figure 3.10.2. Cross-section and ground floor plan of the Ultra House, Rottweil

CONSTRUCTION

The details of construction are shown in Figure 3.10.3. The exterior walls of the Ultra House are constructed from 17.5 cm limestone blocks with 40 cm rockwool insulation (thermal conductivity 0.040 W/mK) on the outer side. The inner and outer surfaces are sealed with plaster. The outer plaster coating is strengthened by a layer of plastic webbing (Figure 3.10.4). The insulation is glued to the wall construction in order to avoid thermal bridges in the anchoring.

All the cellar walls were constructed from cellular concrete blocks of 36.5 cm thickness with an outer layer of 10 cm insulation (thermal conductivity 0.035 W/mK).

Between the roof rafters 25 cm of mineral wool is packed, while over the rafters there is a layer of wooden panelling which supports an unbroken layer of 10 cm thick insulation. Under the rafters, there is a vapor barrier of polyethylene foil. The inner surface is gypsum board supported on a lattice. The exterior of the roof is sealed with bitumen, earth-covered, and planted with grass.

SOLAR SYSTEMS

Passive solar gains are an important strategy in the Ultra House. To achieve maximum solar coverage a superior glazing was specified. All external glazing, including the sunspace glazing, is triple pane with two low-emissivity coatings, and both cavities are filled with xenon gas. The glazing has a U-value lower than 0.5 W/m²K. The pine wood frames account for approximately 28% of the window. The resulting window's total U-value is 0.75 W/m²K and 90% of the house's total glazing surface area is south-facing. The glazing has a total solar transmittance of 48% and a light transmittance of 61%. The glazing between the living space and the sunspace is double glazing with a U-value of

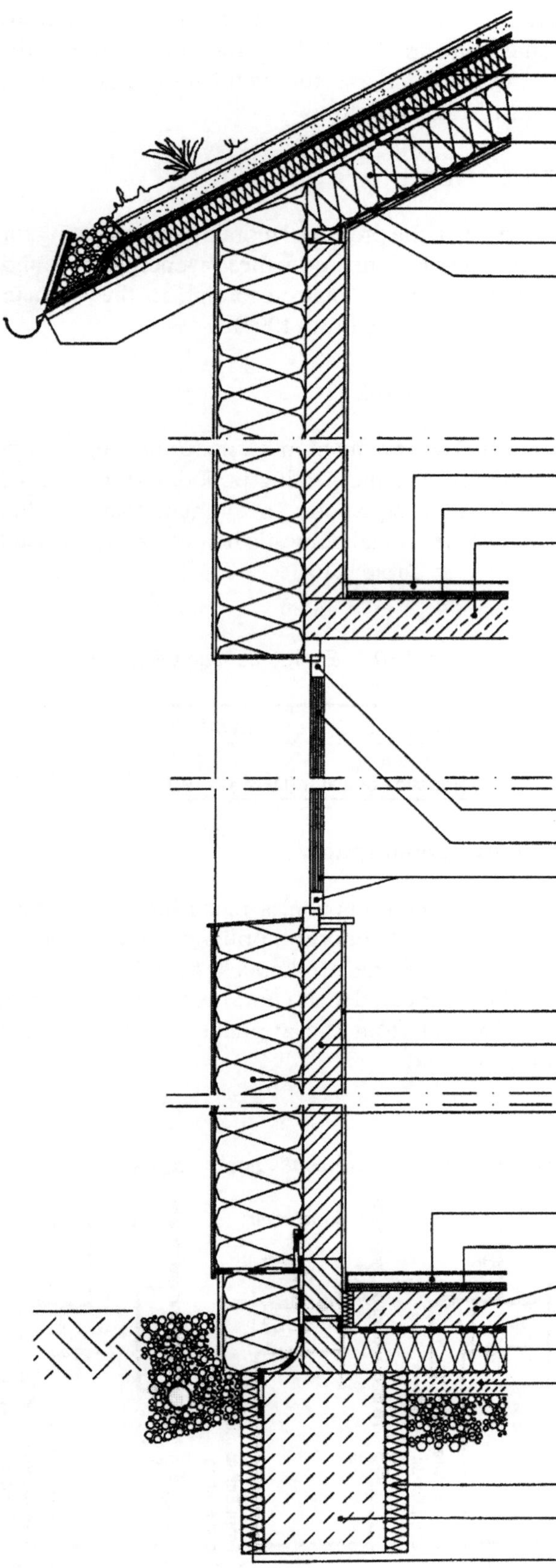

Layer	Description	Width cm	Thermal Conduct. W/mK	U-Value W/m²K
Roof				
1	Earth + Vegetation	11.50	–	0.12
2	Water Proofing	0.02	–	
3	Insulation	10.00	0.035	
4	Wood	2.00	0.130	
5	Insulation + Wood	24.00	0.049	
6	Vapour Barrier	0.02	–	
7	Lattice	2.40	–	
8	Gypsum Board	1.50	0.700	
Ceiling				
9	Screedfloor	4.00	1.400	0.83
10	Insulation	3.00	0.040	
11	Concrete	18.00	2.100	
Window				
12	Sealed Frame			1.40
13	Glazing			0.50
14	Glazing and Frame			0.75
Exterior Wall				
15	Plaster	1.50	0.700	0.11
16	Limestone	17.50	0.990	
17	Insulation	40.00	0.045	
18	Plaster	2.00	0.870	
Basment Slab				
19	Screedfloor	4.00	1.400	0.13
20	Insulation	3.00	0.040	
21	Concrete	18.00	2.100	
22	Water Proofing	0.02	–	
23	Insulation	20.00	0.030	
24	Concrete	10.00	2.100	
Foundation				
25	Insulation	10.00	0.035	–
26	Concrete	60.00	2.100	
27	Insulation	10.00	0.035	

Figure 3.10.3. Construction details of the Ultra House, Rottweil

Figure 3.10.4. Insulation layer on the exterior wall on the south side of the Rottweil house

2.6 W/m²K, which allows the sunspace to receive some heating during the winter.

HVAC SYSTEMS

To ensure a continuous supply of fresh air, a ventilation system has been built into the house, the main components of which are a ground storage system, a plate heat recovery ventilator (HRV), and a fan for bypassing the HRV. As can be seen in Figure 3.10.5, there are four operational modes which are thermostatically switched. For temperatures below 8°C (winter), the ground storage preheats the air and then the HRV completes the process of heating the air. For temperatures between 8 °C and 15 °C (spring and autumn), the fresh air bypasses the ground storage system and is only heated by the HRV. For temperatures from 15°C to 25°C (summer), the fresh air is distributed directly to the living space, while for temperatures greater than 25°C (summer), the fresh air is first cooled by the ground-storage system before being distributed to the living space.

The air flow into the building is approximately 90 m³/h, which is just under 0.2 air changes per hour. Fresh air is delivered to the bedrooms and living room, while stale air is extracted from the toilet, bathroom, and kitchen. There are two remote switches, one in the kitchen and one in the bathroom, which increase the ventilation rate during periods of high humidity.

ENERGY USE

In Figure 3.10.6 the predicted monthly space heating for the Ultra House is compared to the low-energy house and to a reference house built according to the German Standard (Reiss and Erhorn, 1994).

MONITORED RESULTS

Figure 3.10.7 shows the monthly space-heating energy requirements over the heating period and the annual energy balance equation. The annual total heating energy and the annual domestic hot water energy used can be seen in Table 3.10.1.

Table 3.10.1. Energy consumption in the Rottweil house

Component	kWh/m²a
Space heating	18.7
Domestic hot water (DHW)	9.1

Electricity consumption

The electricity consumption is much higher than predicted because low-energy appliances were not selected. Many of the appliances with high energy demand were located in the non-heated cellar, hence, the waste heat from the appliances did not offset heating demand.

Figure 3.10.5. Ground-coupled heat recovery ventilation operation modes in the Rottweil house

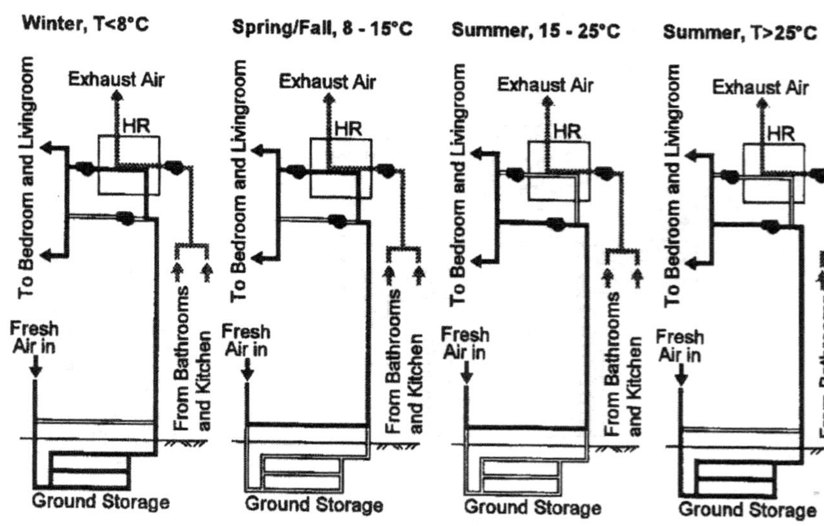

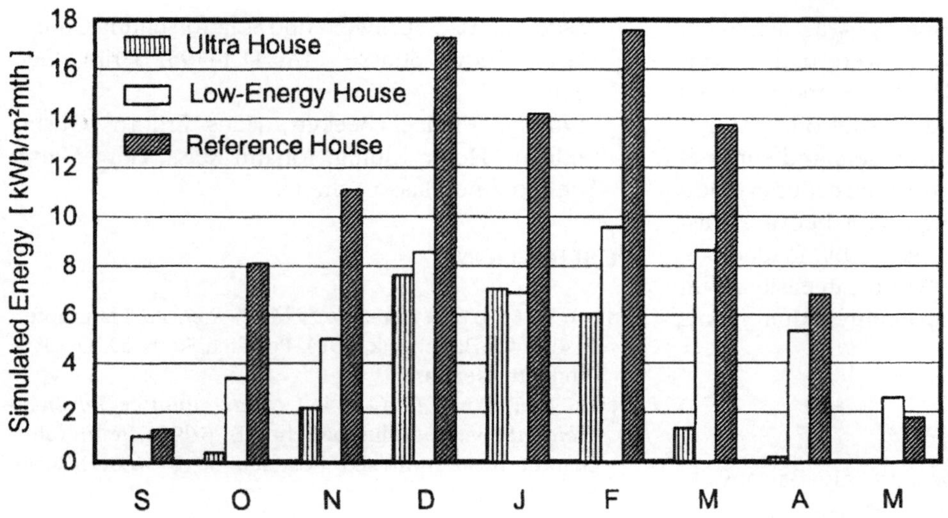

Figure 3.10.6. Predicted space heating in the Rottweil house compared to a low-energy house and a reference house

Domestic hot water

3250 kWh/a was predicted for water heating; however, only 1600 kWh/a was required. This is much less than estimated. The reason for this large difference is that the occupants had the water temperature set at 41°C and not the normal 60°C. The water storage tank was poorly insulated and showed circulation losses; hence the large difference between purchased and required energy.

Air infiltration rate

Because the heat-recovery–ventilation system (HRV) ran continuously in the Ultra House, infiltration tests were carried out in the naturally ventilated low-energy house, which has an identical building form. The annual average natural air infiltration rate in that house is 0.18 ac/h. Blower door tests carried out in the Ultra and the low-energy houses showed 1.6 ac/h and 1.4 ac/h at 50 Pa pressure difference. The difference is due to the ventilation system in the Ultra House.

Thermography

Thermographs taken of the north and south sides of the building showed that the Ultra house's wall and window outer surface temperatures were significantly cooler than those of the low-energy house.

Thermal Comfort

An important consideration for the occupants is whether they feel comfortable in their house, whether it is too cold or too warm. Thermal comfort tests have shown that for both winter and summer the predicted dissatisfaction levels (PPD) in the living area are all less than the expected 10%. However, in the sunspace, where air temperatures sometimes rose above 35°C, blinds have been added.

CONCLUSIONS

The main aim of this project, which was to build a passive solar house in a cold climate with an annual space heating energy requirement of less than

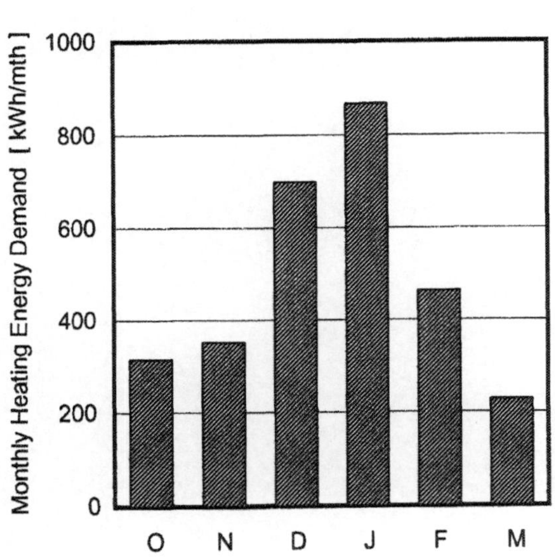

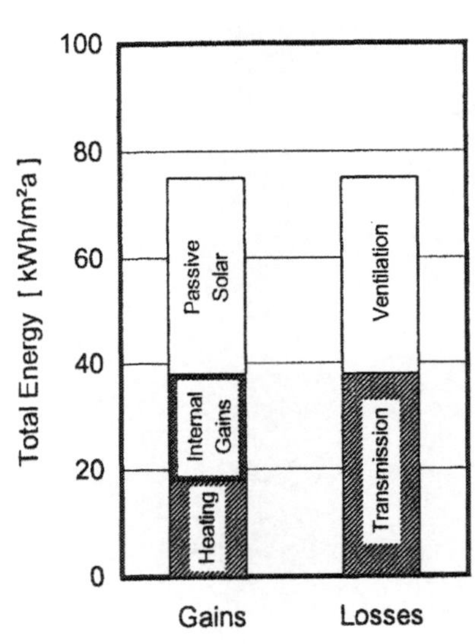

Figure 3.10.7. Space heating components for the heating period and the total energy balance for the Rottweil house

20 kWh/m^2 with current technology, was achieved. Important lessons from this project were that, if windows are open for long periods of time, the radiators and heat recovery system should be shut off, and that for new technologies informed installers and supervision during special installations are a must. Supervision during the building phase is important because mistakes caused by wrong materials or by inadequate execution of the building construction can easily result in a doubling of the space-heating energy requirements in very low-energy houses.

ACKNOWLEDGEMENTS

- *Design Team.* Fraunhofer-Institut für Bauphysik, Nobelstrasse 12, D-70569 Stuttgart, Germany, and Institut für Bau-, Umwelt- und Solarforschung GmbH, Caspar-Theyss-Strasse 14A, D-14193 Berlin, Germany.
- *Authors.* Michael Beckert, Hans Erhorn, Gustav Hillman, Heike Kluttig, Johann Reiss, Hans-Martin Schmid, and Hasso Schreck.

PUBLICATIONS

Erhorn H, *Ultra and Low-Energy House Rottweil.* Fraunhofer-Institut für Bauphysik (IBP), Postfach 80 04 69, D-70504 Stuttgart, Germany.

Reiss J, Erhorn H (1994). *Niedrigenenergiehäuser Heidenheim,* Abschlussbericht WB 75/1994, Fraunhofer-Institut für Bauphysik, Postfach 80 04 69, D-70504 Stuttgart, Germany.

3.11 The Italian prototype apartment building

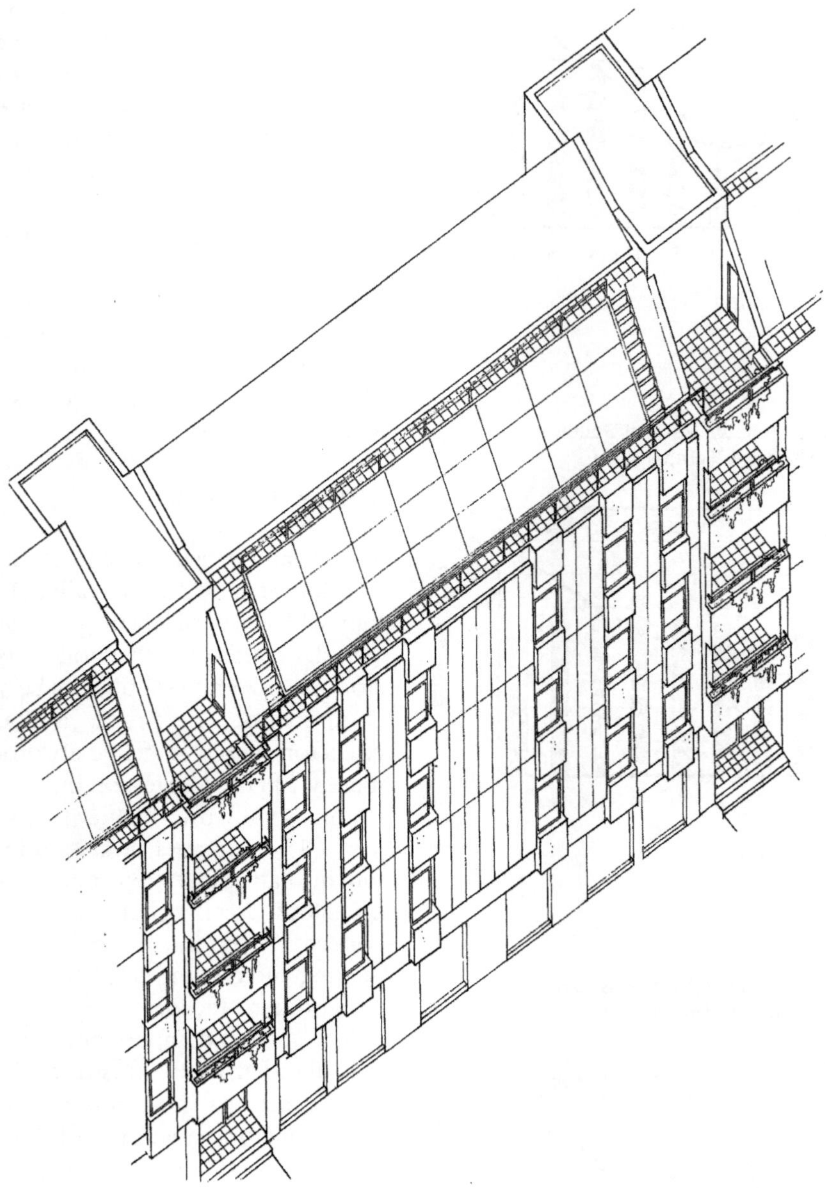

SUMMARY

An apartment building design was dynamically simulated to quantify how much energy consumption for heating and cooling could be reduced by conservation and solar strategies. To reduce the heating demand, the building envelope is assumed to be insulated far beyond normal practice in Italy and a solar air system is incorporated. Solar energy is also used to heat domestic hot water. To provide cooling with less energy input, an innovative solar absorption system was investigated. The calculated energy savings for this design over a conventional building are: Milan 44%, Rome 62%, and Trapani 62%. The building has not yet been built.

KEY FEATURES

- solar heating system
- solar cooling system
- active solar system for domestic hot water

LOCATION

A major aspect of the project presented here has been to adjust energy-saving strategies and technologies to the specific conditions of three representative Italian climates, namely Milan: 45°N, Rome: 42° N, and Trapani: 40° N.

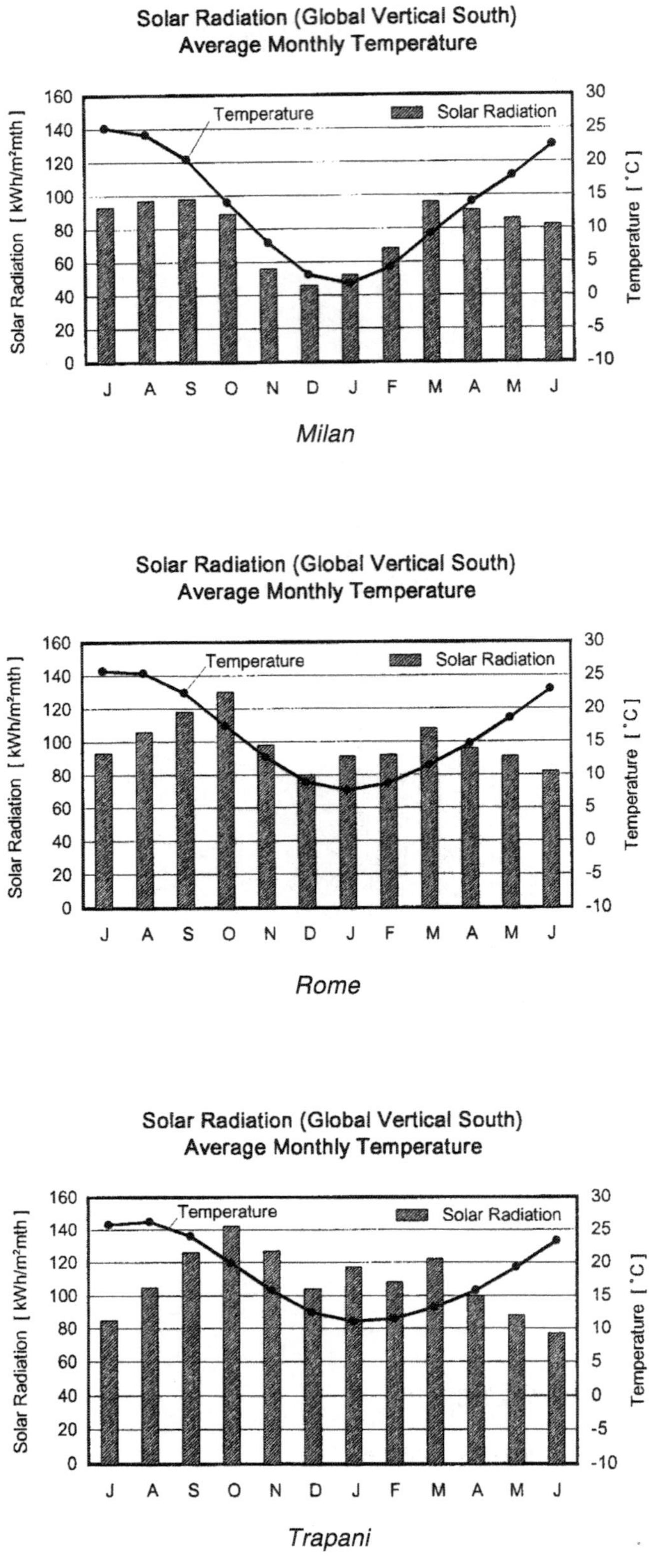

Figure 3.11.1. Diagrams of the three representative Italian climates

CLIMATE

The issue of climate is critical for a country like Italy, where temperatures span from 37°N to 47°N. Diagrams of the three representative climates are given in Figure 3.11.1.

CONCEPT

An apartment building was selected for this study because such building types are widely used in the country. The building block has a total of twelve apartments on three floors and is served by two staircases. The ground level is occupied by a garage. Most of the building services are located in the attic. Each apartment, comprising a net floor area of 90 m², houses a family of four persons. The main axis of the building runs east-west as shown in Figure 3.11.2. The pitch of the south-facing roof has two different slopes: the first, 40° from horizontal, contains solar water collectors for domestic hot water, and the second, at 26°, is a solar absorbing surface to regenerate the water–lithium bromide solution of the cooling system. Most of the southern façade is occupied by vertical solar air collectors.

SOLAR SYSTEMS

The solar heating system of the house is a further development of the so-called Barra–Constantini system, which is basically the thermal coupling of a vertical air solar collector with a hollow-core floor used as heat storage and air distribution. The system has three main components. The first is a double-flow vertical air collector, with a selective absorbing plate, of approximately 108 m², located on the southern façade. A detail of this is shown in Figure 3.11.3. Behind the collector are: a 15 mm panel of glass-reinforced concrete, a 50 mm insulating layer, a 50 mm cavity, and a 100 mm clay-brick wall. Air is circulated through the cavity and the collector to the mechanical system. A fan coil for both heating and cooling is located beneath each window in the southern façade.

The second component of the system is the hollow-core thermal floor, illustrated in Figure 3.11.4. This floor performs the double function of both storing and circulating heat or cold. There are 14 ducts, each 500 mm x 150 mm (*w* x *h*) in the floor of each apartment.

The last components of the system are the venting units located below each window sill in the northern façade. From here room air is blown into the return ducts located in the floor according to the following four operation modes and illustrated in Figure 3.11.5:

- *Mode 1.* Return air is blown through the solar collector if the outlet temperature is greater than 25° C and the room air temperature is below the thermostat set- point. The valves A and B are open, while valve C is closed
- *Mode 2.* If, as in the first mode, the room-temperature set- point is not achieved, part of the return air is blown through the solar collector and part through the internal cavity. Valves A, C, and E are open, valve B is closed

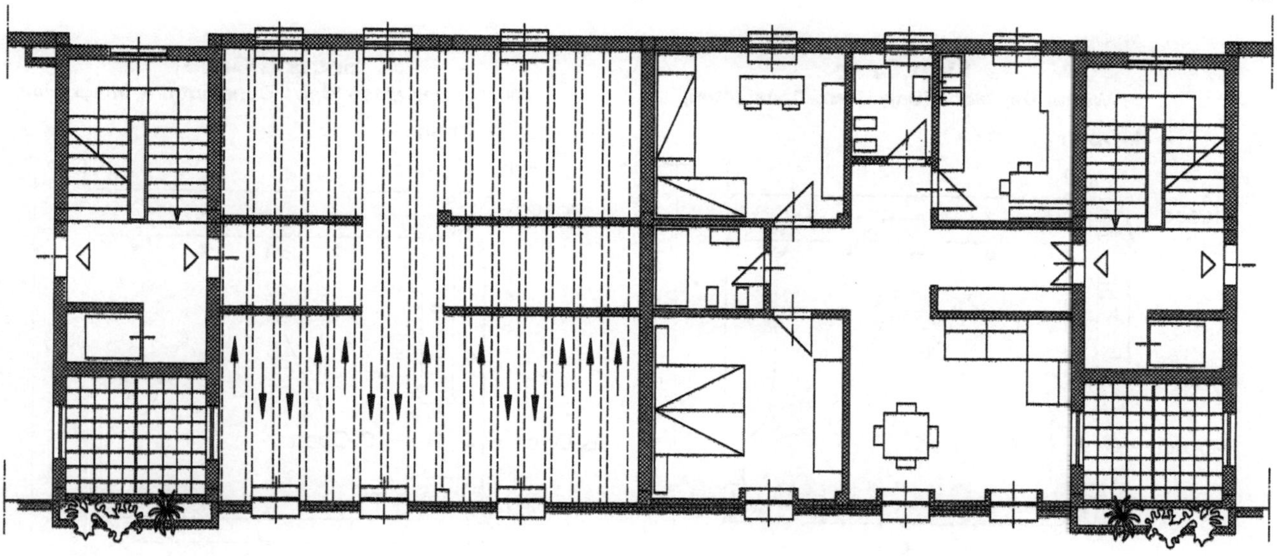

Figure 3.11.2. Floor plan of the Italian apartment

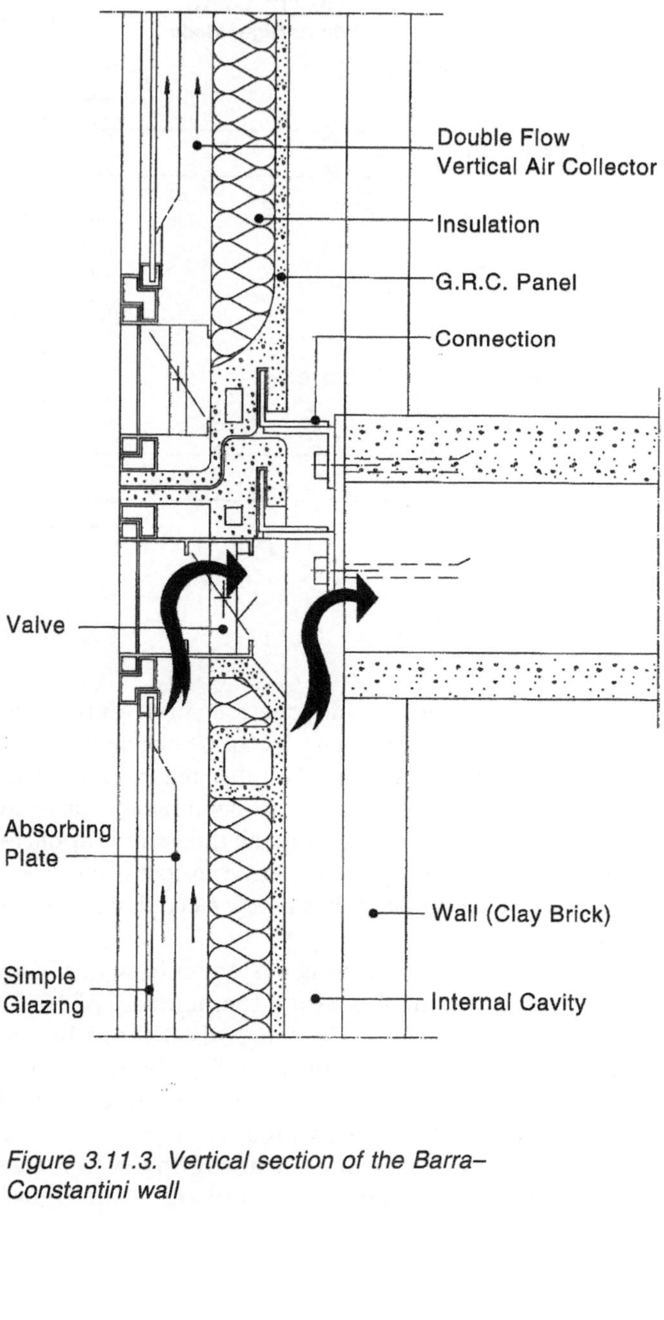

Figure 3.11.3. Vertical section of the Barra–Constantini wall

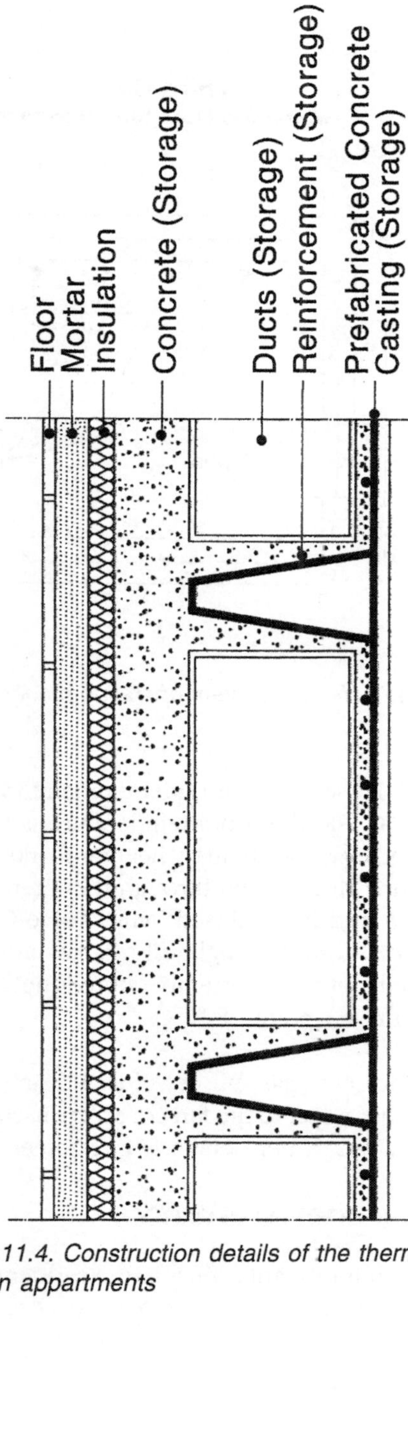

Figure 3.11.4. Construction details of the thermal floor of the Italian appartments

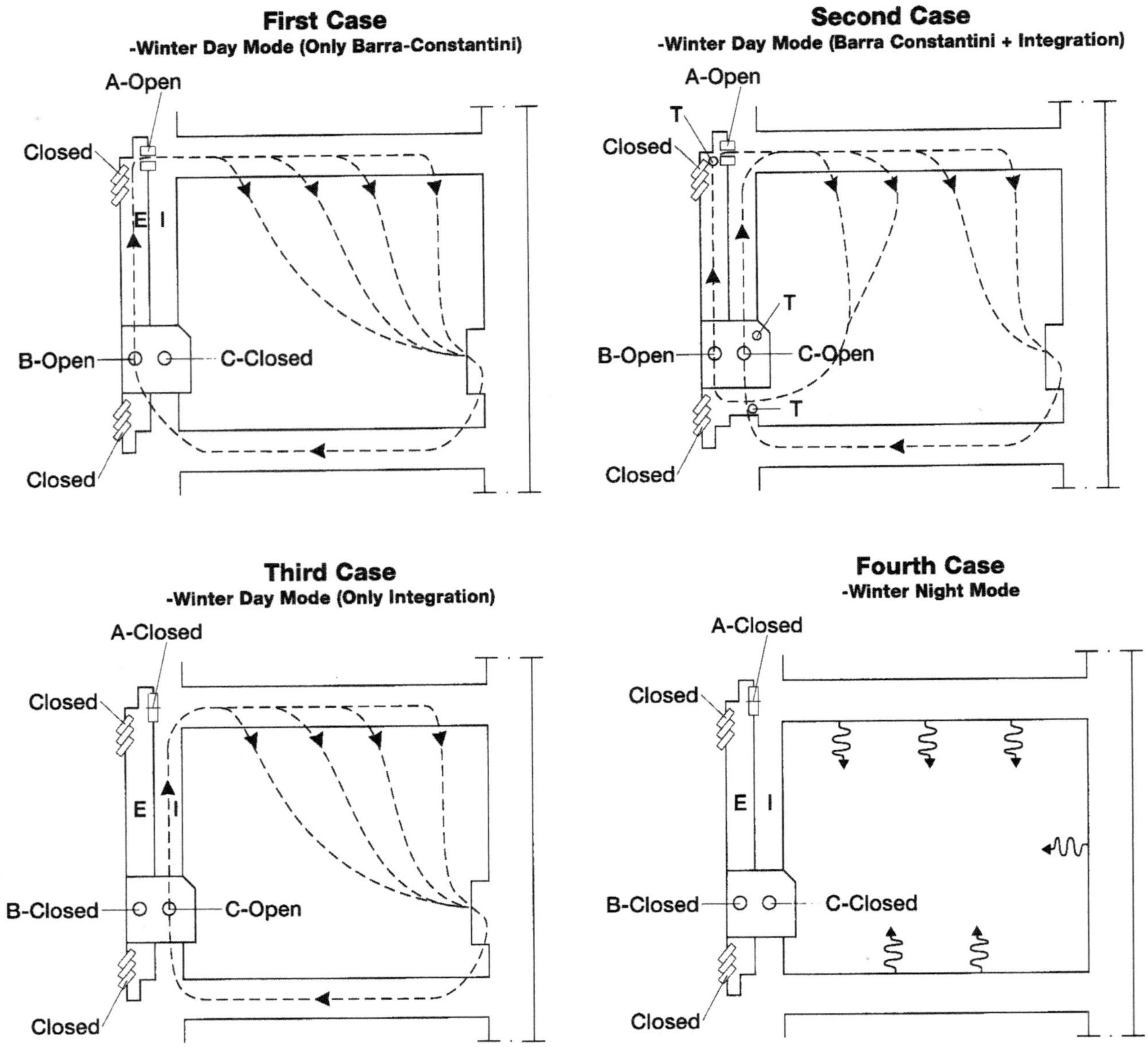

Figure 3.11.5. Heating system operating modes in the Italian apartment building

- *Mode 3.* If the collector outlet temperature is lower than 25°C and the room temperature is below the set-point, only the auxiliary heating works and all of the return air is blown through the internal cavity. Valves A and B are closed, while valve C is open
- *Mode 4.* During the night all systems are shut off, provided that room air temperature is not below the night-time set-point of 16°C.

A high-efficiency gas burner, located in the attic, distributes hot water to the fan coils. An air-change rate of at least 25 m^3/h per person is guaranteed.

THE SOLAR COOLING SYSTEM

Cooling is an important aspect in the residential sector all over the country. In Italy passive cooling alone cannot readily provide adequate comfort during the hot season, which in certain locations can be as much as one third of the year. For this reason an active cooling system is necessary, but a semi-passive one can still meet the demands. Among all the various applications of solar energy for cooling purposes, the open cycle absorption system has been selected, because of its special advantages.

In the proposed cooling system cold water from the absorption machine (at a temperature of 10°C) is circulated through the fan coils located below the window sill in the southern façade. Cool air is distributed to the rooms through the same vertical and horizontal ducts, which form part of the auxiliary system in winter. The cooling system has three modes of operation, as illustrated in Figure 3.11.6:

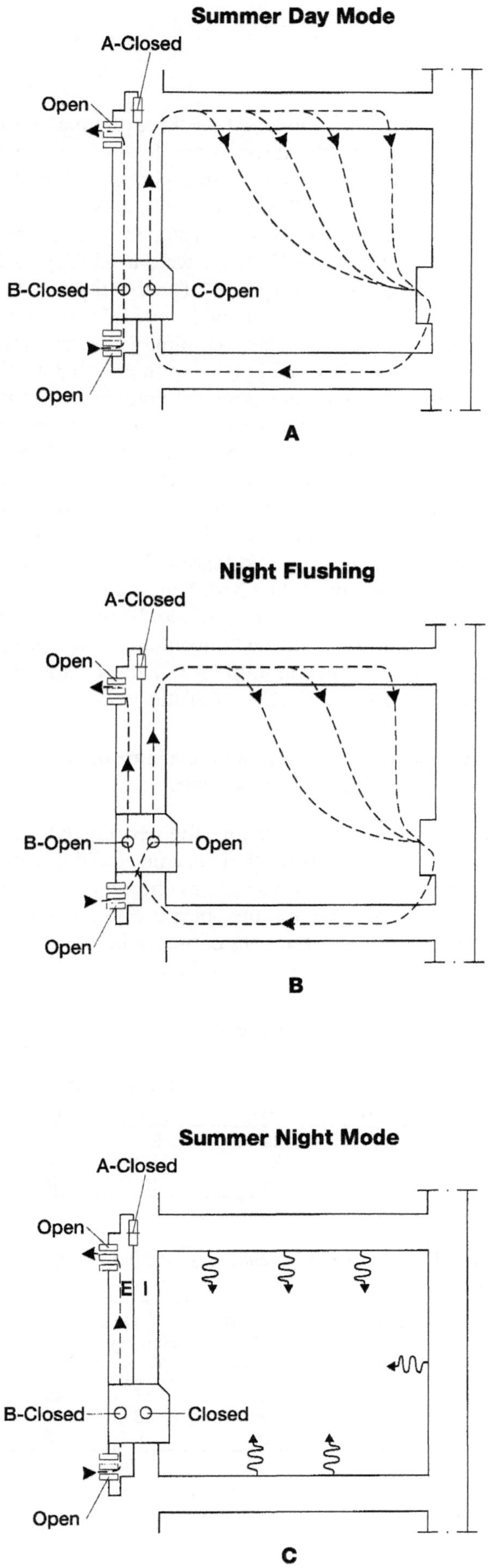

Figure 3.11.6. Cooling system operating modes in the Italian apartment building

- *Mode 1.* Return air is cooled through the cold coil and then blown through the internal cavity of the southern façade if the room air temperature is above the set point. Both the lower and the upper valve of the external cavity of the collector are kept open to induce a chimney effect
- *Mode 2.* If the absorption machine is not working, either because it is night or not a sunny day, the system can still be used to provide mechanical ventilation, both during day and night (night flushing). This is accomplished by blowing the return air back through to the external cavity. In this case the valve on top of the internal cavity is kept closed while both the lower and the upper valves of the external cavity of the collector are open
- *Mode 3.* During the night all fans are off and the thermal floor extracts heat from the apartment underneath through the irradiation effect.

THE SOLAR OPEN-CYCLE ABSORPTION MACHINE

The cold water necessary to operate fan coils in summer is produced in a solar open-absorption machine located in the attic. With absorption-cycle cooling, as with the saturated-vapour compression cycle, cooling is achieved through evaporation of a fluid at a low temperature. Instead of being mechanically compressed, a vapour is absorbed, in this case by a solution which has a strong affinity with the refrigerant. The most commonly used absorbent–refrigerant in building applications is lithium bromide water. The advantage of an open-cycle absorption system is that the regeneration of the weak solution is obtained by means of evaporation of the refrigerant, i.e. water in the atmosphere, rather than in a condenser as in the case of the closed

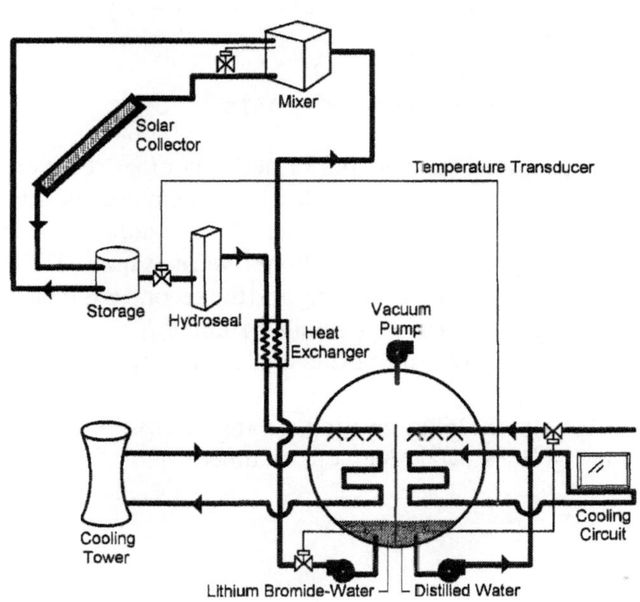

Figure 3.11.7. Diagram of the open-cycle absorption machine

system. To achieve this phenomenon, the weak solution passes as a fluid film over a black-painted flat collector, adequately tilted to be strongly irradiated by the sun. In this application the collector is the upper tilt surface of the roof at an angle of approximately 26° to the horizontal. The main components of an open-cycle absorption machine are indicated in Figure 3.11.7.

Compared to other types of absorption machines the open cycle offers important advantages:

- the collector is simply a black surface, at a suitable inclination above the horizon
- it needs neither a glass cover nor selective coatings
- low regenerating temperatures are possible, thanks to the low partial vapour pressure in the surrounding air
- no condenser is required; the collector itself serves as both the condenser and regenerator, resulting in lower investment costs.

There are also some disadvantages, such as those of a vacuum pump continuously operating to extract air from the tank, or the problem of dust and other chemicals dragged into the concentrated solution.

AN ACTIVE SOLAR SYSTEM FOR DHW

Each of the two apartment buildings has an active solar system with a single-glazed collecting surface of 32 m^2 for heating domestic hot water. The absorber of the collectors has a selective coating. A 1200 litre storage tank, located in the attic, has to satisfy the domestic hot water demand of six families of four persons each. Back-up of the system is provided by the same high-efficiency gas burner used by the heating system.

ENERGY USE

The performance of the systems described above has been computer-simulated in detail with the program TRNSYS (Klein et al, 1990).

HEATING AND COOLING LOADS

Initial simulations were performed to compare the benefits of the proposed solar heating system and its side-effect on summer performance. Two buildings, each containing six flats on three floors, were examined. The first is a conventional building and the second is the same building with the solar heating system (the effect of the solar cooling system is evaluated separately in the next paragraph).

For the building envelope components the following U-values (W/m^2K) were assumed:

vertical walls:	0.49
direct gain windows:	2.80 and 1.50
top and lower floors:	0.66

The building was modelled as three thermal zones, corresponding to the three floors. Set-point temperatures were 20°C in winter with night setback to 16°C between 22:00 and 07:00 h, and 26°C in summer. For the simulation of the Barra–Constantini System, algorithms from the TRNSYS library were used (Type 1 for the air collector and Type 37 for the thermal floor system). The ventilation rate was set to 0.5 ac/h in both seasons. For internal gains a continuous occupancy of four people was assumed, corresponding to 360 W of sensible and latent load. Artificial lighting output was assumed for the hours from 17:00 to 24:00 h. Of the 500 W released, 25% occurs by radiation and 75% by convection. During simulations of the summer the solar air collectors work like ventilated walls, because of the chimney effect. An air flow of 50 kg/hm^2 was estimated to occur. The sites selected for simulation were Milan, Rome, and Trapani. Tables 3.11.1 and 3.11.2 summarize winter and summer peak load and specific energy consumption for both the conventional and the solar buildings (all values in Table 3.11.2 are means of the results for the three zones simulated).

Performance of the solar open-cycle absorption machine of the cooling system

In a second set of simulations the performance of the solar open-cycle absorption machine of the cooling system was evaluated using a mathematical model of the system written for this purpose (Lauritano and Marano, 1995). The fractions of the cooling load met by the system have been estimated for the three locations

Table 3.11.1. Heating and cooling peak loads for six flats

Location	Conventional building		Solar building	
	Heating (kW)	Cooling (kW)	Heating (kW)	Cooling (kW)
Milan	24	22	24	17
Rome	18	23	16	18
Trapani	12	24	10	25

Table 3.11.2. Specific heating and cooling energy consumption

Location	Conventional building		Solar building	
	Heating (kWh/m^2a)	Cooling (kWh/m^2a)	Heating (kWh/m^2a)	Cooling (kWh/m^2a)
Milan	51	42	38	42
Rome	23	48	13	47
Trapani	9	61	3	60

Table 3.11.3. Seasonal cooling load fractions for the Italian apartment building

Location	Seasonal cooling fraction of total load (7–19h) (%)	Seasonal cooling fraction of total load (1–24h) (%)
Milan	61	40
Rome	65	44
Trapani	65	44

for the whole cooling season. The results of this simulation are listed in Table 3.11.3.

PERFORMANCE OF THE SOLAR DHW SYSTEM

The solar fraction provided by the domestic hot-water system has been calculated with TRNSYS. The assumed daily hot water demand at 40°C was 400 litres per family. A yearly average inlet temperature of 15°C was also assumed. Calculation results, rounded to whole numbers, are summarized in Table 3.11.4.

Table 3.11.4. DHW solar system performance for the Italian apartment building

Location	Annual solar fraction (%)	Residual load (kWh)
Milan	66	8666
Rome	85	3823
Trapani	99	255

These values are only the first results of the calculation analysis on a preliminary project which was not site oriented; further design optimizations and new simulations have been used for tuning the envelope and the solar systems to the local specific climate conditions. These results show, however, that the base case used in the sensitivity study is correctly dimensioned for the latitude of Milan.

Finally, the overall performances of the building with regard to each end use of energy and each location are summarized in Table 3.11.5.

CONCLUSIONS

The predicted overall energy savings (for heating, DHW, and cooling) in this project are substantial compared to the traditional building standard in Italy. For heating, peak load analysis shows that the Barra–Constantini system can make an important energy contribution. Moreover the façade does not worsen the energy balance either as an envelope during winter or as a ventilated wall during summer.

A reasonable fraction of the daily cooling load can be met by the solar absorption machine, throughout the country. This fraction would be higher (60% or more) if only the daytime cooling were to be covered or if the comfort conditions were fixed slightly above the standard values (28°C and 60% relative humidity instead of 26°C and 50%).

The feasibility of integrating three solar systems (heating, cooling, and DHW) into the structure of the building has been demonstrated (i.e. heating and cooling share the same storage and distribution). The thermal floor, used as diurnal heat storage, is correctly sized because its temperature during the cycle is always a few degrees above room temperature.

Thermal comfort is nearly always guaranteed during winter, and overheating conditions are narrowed to a very short period of time if, according to ISO 7730, higher thermal comfort indices are assumed (25°C air temperature and 35% relative humidity for sedentary activity).

Given that the degree of thermal insulation (of both the glazed and the opaque envelope) of the reference building is far better than required by local building standards or used in common practice, the performance of the innovative building appears to be particularly promising. The energy-saving potential of the innovative building, according to Table 3.11.5, is 44% for Milan, 61% for Rome, and 68% for Trapani.

ACKNOWLEDGEMENTS

- *Design Team.* Stella Harangozo, M.T. Portfiri, D. Marano, ENEA (National Agency for New Technology, Energy and Environment), Energy Department, 0060 S. Maria di Galeria, Rome, Italy, and Aldo Lauritano, Universita di Palermo, Dipartimento di Energetica, I- 90128 -Palermo, Italy.

Table 3.11.5. Comparison of specific heating,cooling, and DHW energy consumption for the conventional (C) and innovative (I) buildings

Location	Heating (kWh/m^2a)		Cooling (kWh/m^2a)		DHW (kWh/m^2a)		Total (kWh/m^2a)	
	C	I	C	I	C	I	C.	I
Milan	51	38	42	25	47	16	141	79
Rome	23	13	48	27	47	7	118	46
Trapani	9	3	61	34	47	1	117	37

PUBLICATIONS

Harangozo S, Lauritano A et. al. (1994). Edilizia a basso consumo energetico Low Energy Building. *Proceedings of the I.S.E.S. Conference: Public Building and Dwelling, Renewable Sources, New Technologies and Innovative Materials*, Lecce, Italy, July 1994.

Klein S A et al. (1990). *TRNSYS. A Transient System Simulation Program, version 13.1.*Solar Energy Laboratory, University of Wisconsin-Madison, Madison, WI 53706, USA.

Lauritano A, Marano D (1995). *Frigoriferi solari ad assorbimento a ciclo aperto. Possibilità di impiego nel settore residenziale in Italia (Solar open cycle absorption refrigerative system. possible utilization in the buildings sector in Italy).* Condizionamento dell'aria, May 1995. Editoriale PEG SPA, Via F.lli Bressan 2, I-20126 Milan, Italy.

3.12 The Japanese WISH House

SUMMARY

The Japanese WISH House uses a hybrid collector and phase change material (PCM) storage to reduce the heating and cooling loads. A grid-connected PV array of 4.2 kW peak power also serves as a solar air heater. In winter the warm air, heated by cooling the PV cells, is used as the heat source for a heat pump unit designed for space heating. Heating and cooling are provided via PCM thermal-storage panels embedded in the ceilings and floors. When the heating or cooling load is greater than the capacity of the PCM thermal storage panels, a heat pump which includes a heat recovery circuit is operated as a backup. Because the PCM thermal-storage panel cooling cannot provide dehumidification, a dehumidifying unit impregnated with a lithium chloride (LiCl) solution is also used. The system is computer controlled to optimize performance while ensuring comfort.

KEY FEATURES

- insulation
- airtightness
- photovoltaic–thermal hybrid collector
- PCM thermal storage panels
- dehumidifying unit with lithium chloride

LOCATION

The house is built in the city of Iwaki, located 180 km north of Tokyo at latitude 37°N.

CLIMATE

The climate in Iwaki is warm in winter and hot and humid in summer (Figure 3.12.1).

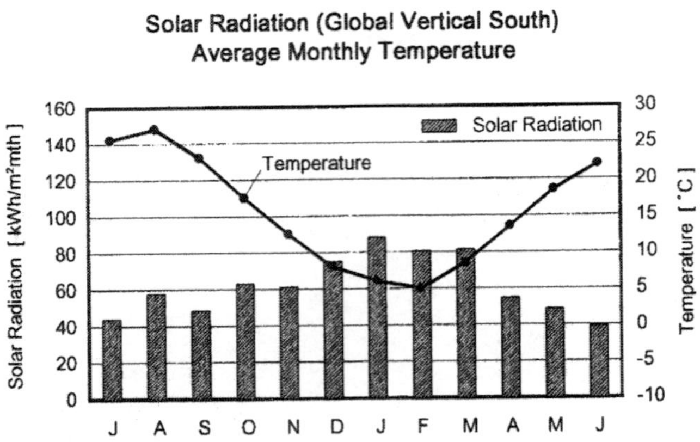

Figure 3.12.1. Iwaki climate conditions

CONCEPT

The house was built with industrialized construction methods using factory-made steel frames. As it is well insulated and airtight, the heating load is estimated to be one third of that of a house with standard insulation for the locality. It is part of the research and development project of MITI called the WISH house project.

The house consists of linked southern and northern units (Figure 3.12.2). The southern unit is designed for the aged parents, the northern unit for a family with two children. The central unit serves as an entrance hall for both families. Several innovative systems are included in the northern unit, while the southern unit only has a simple passive system. The northern unit is discussed here.

CONSTRUCTION

The northern unit has two storeys and a total floor area of 150 m^2. The house is built using an industrialized construction method employing factory-made steel frames. The exterior walls are made of lightweight concrete panels for fire protection. The walls are insulated with 50 mm urethane foam, the floor with 100 mm polystyrene foam, and the roof with 220 mm blown mineral wool. All windows are made of low-E coated and argon-gas-filled double glazing with plastic frames. The airtightness of the house is 2.0 cm^2/m^2 at 1mm Aq. The overall heat loss coefficient of the house is 1.55 W/m^{2}K. The overhangs on the south façade cut out 70% of the solar radiation into the rooms in summer.

The space heating and cooling load of the WISH house has been calculated by simulation. The heating load is 4620 kWh and the cooling load is 2360 kWh. As shown in Figure 3.12.3, the estimated heating load of the WISH house is one third of a house model with the insulation standard for the locality.

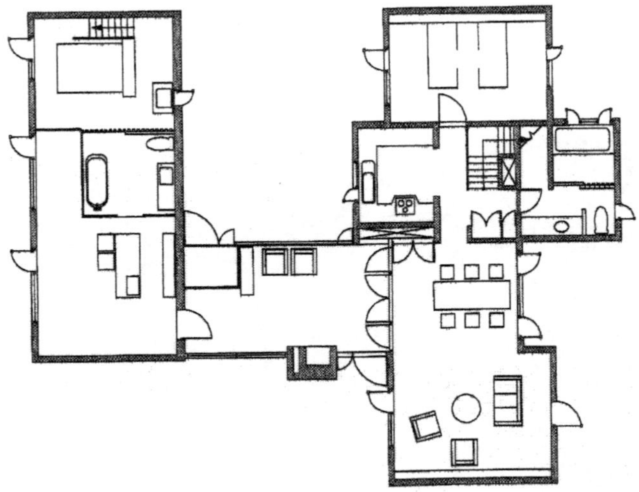

Ground floor

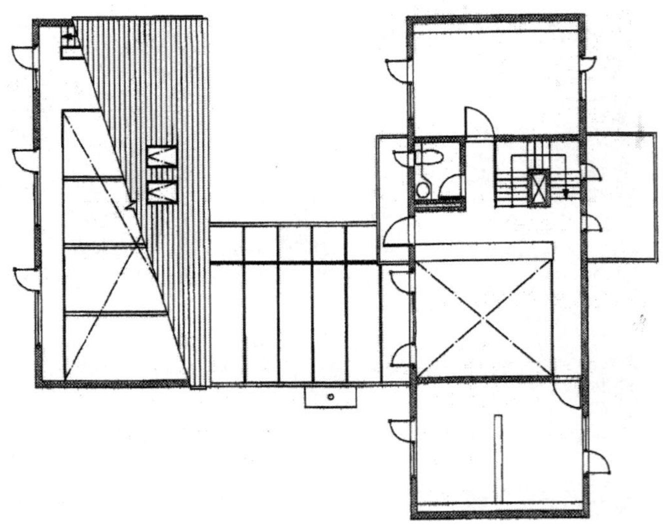

Upper floor

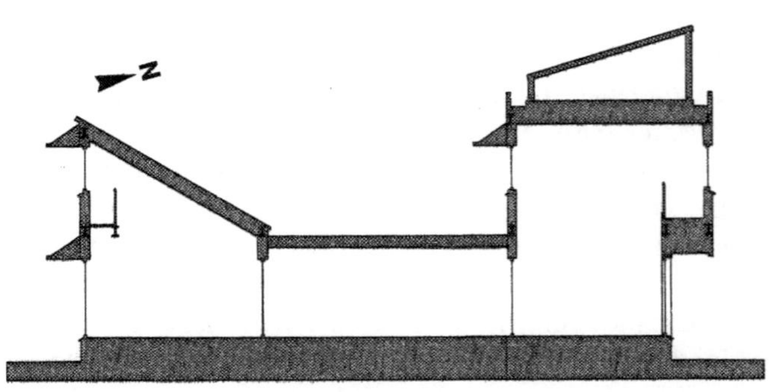

Section

Figure 3.12.2. Floor plans and section of the Japanese WISH house

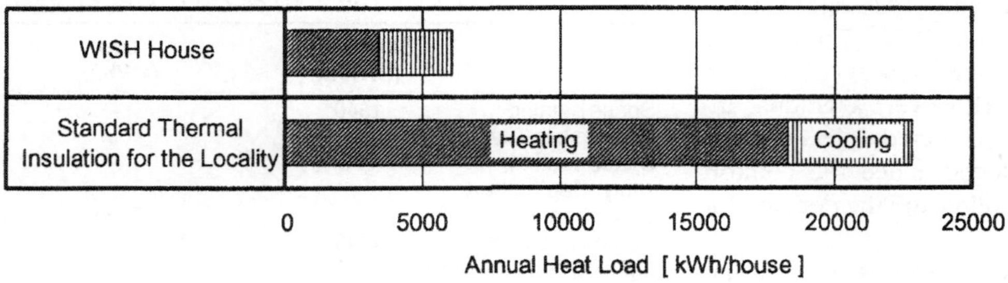

Figure 3.12.3. Annual heat load of the Japanese WISH house

Figure 3.12.4. Photovoltaic–thermal hybrid collector in the Japanese WISH house

SOLAR SYSTEM

A photovoltaic–thermal hybrid collector system of polycrystalline PV modules backed up by aluminium honeycomb air collectors is used (Figure 3.12.4). The array area is 44.8 m² for 4.2 kW power generation at the peak of solar radiation (14% generation efficiency). The power generation efficiency of the PV array is expected to be increased by more than 10% by the cooling effect of the air flow at the rear surface of the PV modules. When the generated electric power of the PV system is greater than the power demand of the house, the excess electricity is sent back to the utility grid. The heated warm air for cooling the PV cells is exhausted during summer. In winter the heated warm air becomes the heat source for a heat pump unit used for space heating.

HVAC

The system is shown in Figure 3.12.5.

PCM thermal-storage panel and dehumidifier

Basically, the heating and cooling loads are covered by the PCM thermal-storage panels embedded in the ceiling and floor. Each thermal storage panel is 260 × 300 mm with a thickness of 36 mm. The capacity of the thermal storage is about 550 kJ. The phase change temperature is 22°C for both cooling and heating. The total panel area is 22 m² on the first floor, 12.5 m² on the second floor, and 26 m² on the ceiling of the second floor. The PCM thermal storage panels are heated by a heat pump unit during the night using off-peak electricity.

When the heating or cooling load exceeds the capacity of the PCM thermal storage panel, a heat pump serves as a backup. The heat pump is capable of supplying both chilled or warm water and cool or hot air. The efficiency of the heat pump in winter is improved by using the air from the thermal-storage space under the floor as a heat source. Since the PCM thermal-storage panel cooling cannot work for dehumidification purposes, a dehumidifying unit impregnated with lithium chloride (LiCl) solution is used. The unit is regenerated using an electric heater operated by off-peak electricity. The dehumidifying module uses an osmotic sheet to prevent LiCl solution from being lost. As the osmotic sheet allows only vapour to be transmitted, the circulated room air is not in direct contact with the LiCl solution.

ENERGY USE

According to a statistical investigation of the energy consumption of residential houses in Japan, the average annual purchased energy for a housing unit, including both single and multi-family houses, is estimated to be 11.6 MWh (Nakasami et al, 1990). Another investigation

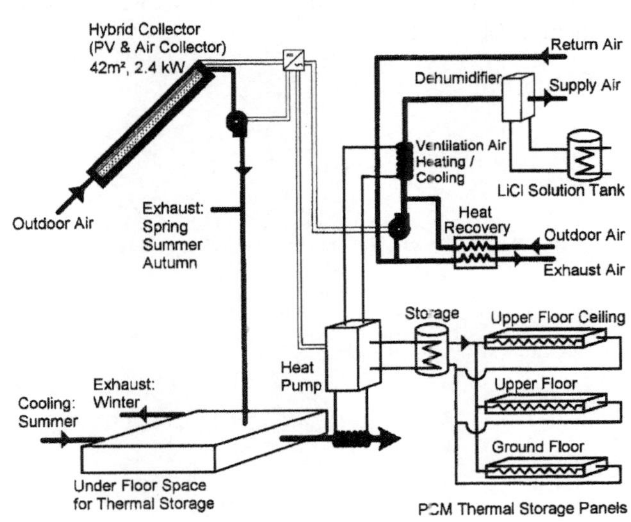

Figure 3.12.5. Heating, cooling, and ventilation system in the Japanese WISH house

of single family houses built after 1985 showed the value to be 14.5 MWh (Ishida et al, 1992). These values are rather small compared to those in most countries in Europe and North America. However, a gradual increase in purchased energy is predicted unless energy-efficient methods are introduced, since the comfort level for space heating and cooling and the demand for DHW is increasing in Japan.

The usage of the purchased energy is divided into three categories: household appliances, space heating, and DHW heating. While cooling is required in most areas except the northern district, the annual total amount of energy consumption is still small. Averaging the daily profile of electric power consumption is also important for the effective use of electric power generation plants. Therefore, the strategies for energy-efficient houses in Japan are:

- thermal insulation and airtightness of the building envelope to minimize space heating
- shading and natural ventilation to minimize cooling energy
- solar DHW heating system to reduce the amount of DHW energy
- PV system to reduce electricity purchased
- thermal storage to avoid peak electricity loads.

The domestic hot-water heating load for an average household in Japan is 3750 kWh per year. If a conventional flat-type collector of 6 m^2 had been installed on the roof, solar energy would have provided about 50% of the annual load, thus requiring 1700 kWh of auxiliary energy. In fact, however, a solar domestic hot water system was not installed, as it was not regarded as an innovative system to be tested in the WISH house project.

Various types of sensors are installed to monitor temperature, humidity, occupancy and comfort in and around the house. The system operation is computer-controlled so that comfort can be achieved with energy conservation.

The predicted performance of the WISH house is given in Table 3.12.1.

Table 3.12.1. Predicted performance of the WISH house

	Electricity (kWh/a)	**Gas (kWh/a)**
Space heating (November–March)	1540	–
Space cooling (July–September)	786	–
Water heating		2131
Pumps and fans	1749	–
Lighting and appliances	6682	–
PV	–5510	–
Total	4247	2131

CONCLUSIONS

In the case of such a narrow site with limited solar radiation and breeze, an active solar system needs to be introduced for living comfort. The high levels of insulation of the WISH house reduce the heating and cooling energy consumption to one third of the values of conventionally built houses in the neighbourhood. The PCM panel system was coupled to both a heating and a cooling system, driven by off-peak electric power during the night. Electric power for lighting and appliances can be provided almost entirely by solar energy using a photovoltaic-thermal hybrid collector, which also helps meet part of the heating load by means of the heat pump.

ACKNOWLEDGMENTS

- *Building Design*. Tokyo planning office, Sekisui House Ltd., Yoyogi 2-1-1, Shibuya-ku, Tokyo, 151-70 Japan.
- *Authors*. Ken-ichi Kimura, Waseda University; Mitsuhiro Udagawa, Kogakuin University; and Ken-ichi Ishida, Sekisui House Ltd.

PUBLICATIONS

Nakagami H et al. (1990). *Annual Statistics of Household Energy*, Jyukankyo Research Institute Inc. (in Japanese).

Ishida K, Ohta A, Kosaka S, Kimura H (1992). Energy Consumption for Residential Houses in Japan, *Transactions of AIJ (Architectural Institute of Japan)*, 1349–1352 (in Japanese).

3.13 The Urban Villa 'Licht en Groen' in the Netherlands

SUMMARY

The Urban Villa 'Licht en Groen' is a 42-unit apartment building. Sixteen apartments with a 100 m² floor area have been developed as advanced solar low-energy dwellings. The goal of the project was to demonstrate that advanced technologies can achieve extremely low energy consumption. The purpose was to design a comfortable dwelling, pleasant to live in, and with a high architectural quality. The dwellings were designed to use less than 40 kWh per m² floor area for space heating, hot water, and fans and pumps. This is a reduction in energy use of about 60 to 70% compared to conventional design. Preliminary monitoring results during summer show that the performance regarding comfort is very satisfactory.

KEY FEATURES

- super-insulation
- high-performance windows
- advanced ventilation and heat recovery
- advanced control system
- solar hot water
- passive cooling
- sunspace

LOCATION

Amstelveen is located 20 km south of Amsterdam in a maritime climate at 52°N latitude (Figure 3.13.1).

CLIMATE

Two 42-unit apartment buildings with a curved shape are located at the entrance of a new urban settlement.

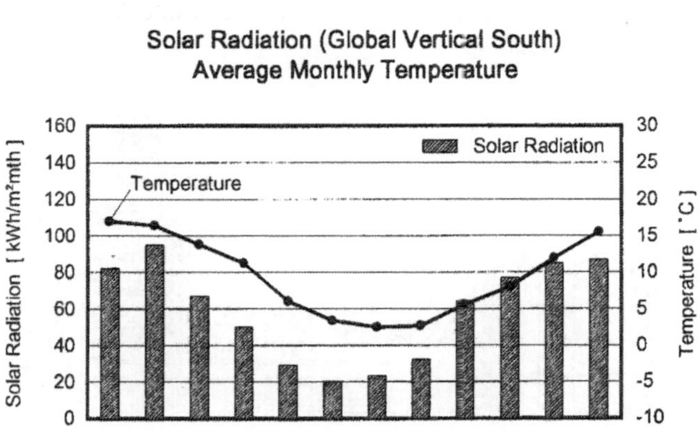

Figure 3.13.1. Amstelveen climate diagram

In one of the buildings 16 apartments have been designed with an advanced solar low-energy concept.

The building consists of three blocks joined by a large sunspace that serves as an entrance and is intended for communal use. The 16 Task 13 houses dwellings are located in the curved blocks facing south. The floor plan is shown in Figure 3.13.2 and the building cross-section in Figure 3.13.3.

CONCEPT

Dutch government policy addresses the need both for higher housing densities in urban environments and for energy saving. To harmonize energy savings, comfort, and architectural quality, a detailed study was carried out by the research team. A major topic was the transparency of the envelope, because the architect desired a high degree of transparency. Key issues were then energy conservation, passive solar use, and overheating avoidance. From this study the following conclusions were drawn:

- Extremely low energy consumption for space heating can be achieved through super-insulation, airtightness, and high ventilation heat-recovery efficiencies, combined with advanced passive solar technologies
- The control of the mechanical system should ensure optimal functioning of all systems and also help to prevent overheating. Heating of domestic hot water dominates and the electricity demand for fans and pumps is relatively important. This is all the more true in the case of low-efficiency electricity generation from fossil fuels. The more sophisticated control system needed in such installations costs money but is essential
- A collective system is more energy- and cost-effective than individual systems. The public, however, prefer individual systems, so a compromise was found by designing the solar DHW system as partly collective, and by collectively preheating ventilation air in a large sunspace.

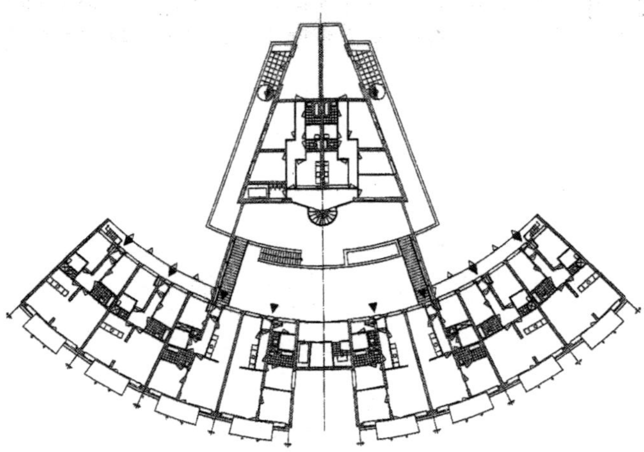

Figure 3.13.2. Building floor plan of the Netherlands apartment building

Figure 3.13.3. Building cross-section of the Netherlands apartment building

The energy concept consists of:

- heat-loss reduction
  - compact building
  - super-insulated airtight envelope
  - heat recovery from ventilation air
- use of solar energy
  - direct gain through transparent envelope
  - modest preheating of ventilation air, additional to heat recovery
  - solar hot-water system
- heating
  - high-efficiency boiler for space heating and DHW supply
  - rapidly reacting convectors controlled per room
- ventilation
  - balanced ventilation and heat recovery during heating season, mechanical, exhaust and natural supply during the rest of the year
- passive cooling
  - controlled shading devices
  - high-efficiency ventilation through vents.

It became clear from the various simulations that the principal problem was to bring the various measures into balance in such a way that comfort can also be achieved during summertime.

CONSTRUCTION

The building consists of a concrete load-bearing structure closed at the south and north side with wood-frame façades. The concrete structure has at least 20 cm exterior insulation. The transparency of the south façade has been brought to the maximum of 40%.

The apartment floor plan is shown in Figure 3.13.4. In order to achieve adequate summer cooling by natural ventilation, all the rooms have vents and windows to the outside air. A structure with a dual function has been designed for the south side of the block functioning as a separate load-bearing structure for the

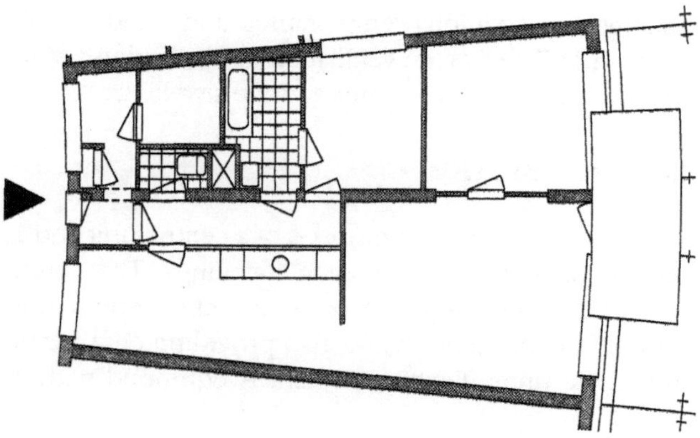

Figure 3.13.4. Apartment floor plan of the Netherlands apartment building

balconies and as support for the shading devices for the summer.

The thermal properties of the building shell (W/m²K):

floor:	0.15
wall:	0.17
glazing:	0.7
south façade:	0.9
roof:	0.15
sunspace (single glazing):	5.7

Airtightness

The dwelling envelope with vents sealed has an air leakage of 1.27 air changes per hour at a pressure difference of 50 Pa.

Shading devices

Movable shading is operated by a control system or manually outside the windows of the dwellings and under the roof of the atrium.

South façade

The south façade (Figure 3.13.5) consists of integrated wall elements comprising windows, doors to the balconies, opaque wall parts, vents in the lower and upper parts, and convector elements for space heating. Much attention was paid to developing this wall element. The requirements of insulation level, airtightness, solar gain, venting, burglary prevention and noise transmission had to be fulfilled. The thermal aspects of the façade element were tested in the PASSYS Test Facility at TNO-Bouw in Delft. The thermal performance met the expectations.

Special care was taken to minimize the thermal bridges. The elevator and staircase section between the curved blocks is not connected to the dwellings. The load-bearing structure of the balconies partly has its own foundation. The structural connection of the balconies to the concrete floors has minimal conduction losses. Also the penetrations through the roof insulation (collector supports, roof trusses of the sunspace) are thermally insulated. Even so, the thermal bridges increase conduction losses by 16%.

SOLAR SYSTEMS

Passive solar gain through windows covers about 41% of the space-heating demand. Internal gains cover 42%, while 17% is provided by the heating system.

A solar hot-water system provides about 60% of the hot-water demand. A communal solar collector arrangement was selected. The collector area is 18 m² per unit of four dwellings. Back-up heating consists of high-efficiency boilers. Each dwelling has its own storage with a heat exchanger in the top for back-up heating to a minimum of 65°C. If a storage tank can be charged with solar energy, the 18 m² collector area will serve this storage. The principle of the control is that the collector serves any storage that can be charged. Washing machines and dishwashers are supplied directly from the solar hot-water tanks.

HVAC SYSTEMS

Space heating

Super-insulated dwellings have a minimum heat requirement. Ultimately the dimensions of the installation are determined by the heating power needed to warm up a cooled-down dwelling. Attention was given to the possibility of a collective system and/or connection with the solar water-heating installation (Figure 3.13.6). The final choice was for an individual heating installation with a high-efficiency boiler of about 10–22 kW, with an efficiency of 90%. Integrated air-feed and flue-gas ducts are used for the boiler. Heating is delivered to the rooms by rapidly reacting convectors, integrated in the parapet walls of the exterior wall elements. This provides a quick response when heat is needed, as a result of the low water volume of this system and the efficient heat transfer of the convector elements. A buffer tank prevents the boiler from being switched on and off too frequently. In this way the high efficiency of the boiler can be fully achieved.

Ventilation of the dwelling

Winter ventilation

Air exchange in the dwelling is by balanced ventilation. The incoming air flows from the sunspace through two cross-flow heat exchangers which are connected in series with a combined efficiency of about 80%. The heat exchanger was tested at the Delft University of Technology. Total air quantities were established according to the Netherlands standard (250 m³/h for the dwelling in total). In winter the residents are advised not to use natural ventilation, but all bedrooms are provided with windows that can be tilted open. However, any opening of windows automatically shuts off the heating element.

Summer ventilation

Outside the heating season a red light on each room control indicates that natural ventilation can be used. Occupant opening of the vents is detected by the sensors of the control system and the inlet ventilator is switched off. This reduces the fan energy consumption. The exhaust ventilator always remains on.

Summer cooling

The upper and lower vents were designed so that a proper flow through the rooms (about 6 air changes per hour) is guaranteed at various temperatures and wind speeds. Figure 3.13.7 shows the ventilation modes and use of the shading device of the atrium and the dwellings.

CONTROL SYSTEM

The control system is comprised of a central unit and 16 individual units (one for each dwelling). The central unit controls the large sunspace and other central functions of the building, while also providing data for the individual units. Each apartment is equipped with an

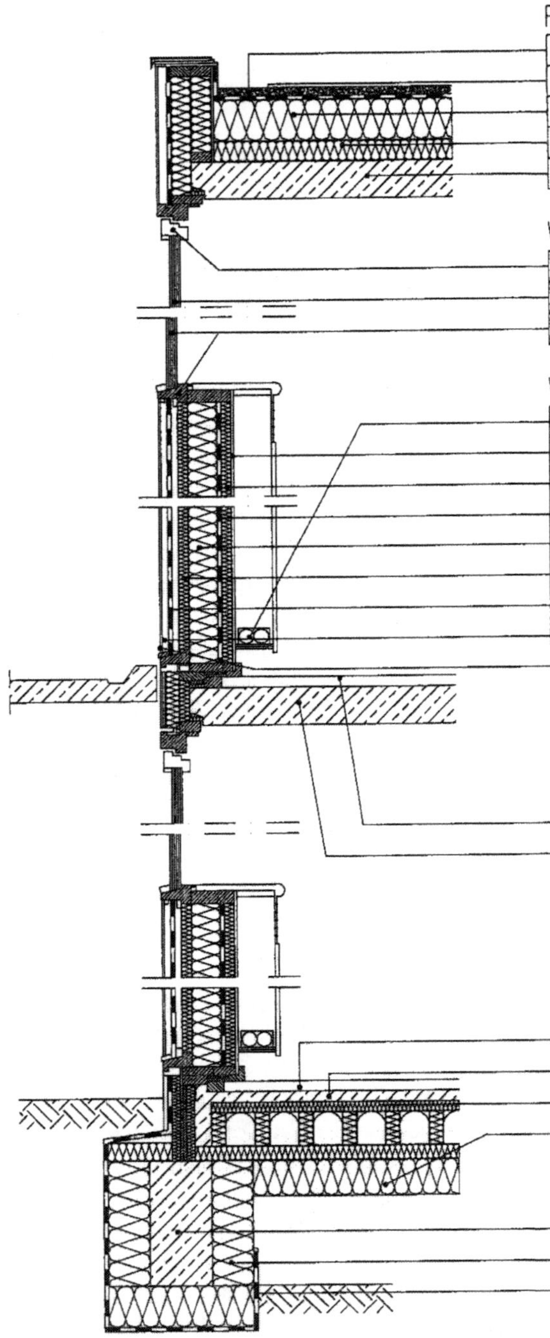

Layer	Description	Width (cm)	Thermal Conduct. (W/mK)	U-Value (W/m²K)
Roof				
1	Ballast	4.00	-	0.14
2	Bituminous Cover	-	-	
3	Expanded Polystyreen	20.00	0.035	
4	Expanded Polystyreen	4.00 to 10.00	0.035	
5	Concrete	18.00	2.000	
Window				
6	Wood Frame			2.27
7	Triple Glazing			0.69
8	Glazing and Frame			1.35
Wall				
9	Heating Elements (Convectors)	-	-	0.17
10	Gypsum Board	1.25	0.250	
11	Glas Wool	5.00	0.037	
12	Airlight Vapour Barrier	-	-	
13	Mineral Wool	14.00	0.037	
14	Mineral Wool	5.00	0.037	
15	Wind Barrier	-	-	
16	Air (vented)	-	-	
17	Plywood Sheeting	-	-	
Ceiling				
18	Finishing Floor (Cement)	5.00	1.000	-
19	Concrete	18.00	2.000	
Floor				
20	Finishing Floor (Cement)	5.00	1.000	0.17
21	Ribbed Concrete Slab	5.00	2.000	
22	Hollow Core Exp. Polystreen	28.00	0.035	
23	Expanded Polystyreen	17.00	0.035	
Foundation				
24	Concrete	30.00	2.000	-
25	Expanded Polystreen	20.00	0.035	
26	Polyethylene Foil	-	-	

Figure 3.13.5. Cross-section of the south façade of the Netherlands apartment building

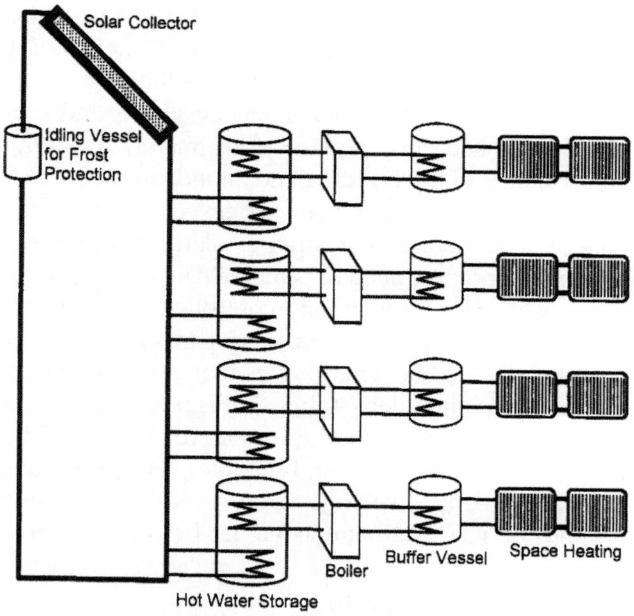

Figure 3.13.6. Heating system in the Netherlands apartment building

individual independent system, based on locally measured data and data sent from the central unit, which controls the apartment.

The central unit prevents the sunspace from overheating by means of shading devices and vent openings. To conserve the energy in the heating season, the roof shading devices are used only during the night. The ventilation openings are opened if smoke is detected. The central unit also controls the distribution of the solar heat from the collectors to the dwellings and the lighting in the central areas of the building, while providing climatic data to the individual units.

The ventilation, the shading and the space and hot water heating are controlled in each dwelling. Space heating is controlled by valves in each room. They are operated by time-switch, programmed by day of the week with manual override. If a window is opened, the space heating is switched off. The control unit switches

Table 3.13.1. Energy performance for one dwelling

	Energy consumption	**Primary energy consumed (taking into account a power plant efficiency of 40%)** per dwelling	per m² floor area
	(kWh)	(kWh)	(kWh/m²)
Space heating	920	920	9.2
Domestic hot water	1630	1630	16.3
Electric fans/pumps	450	1125	11.3
Total consumption	3000	3675	36.8

the ventilation system to the day or night mode based on a time-switch program with a facility to increase the ventilation in the kitchen and bathroom as required. The control of the sunshades is linked to the dwelling temperature and outside climate conditions (wind speed, rainfall, temperature); again there is the facility for manual adjustment. The functioning of the control system was tested in the laboratory of the Delft University of Technology.

ENERGY USE

The energy performance for these apartments (Table 3.13.1), taking into account the thermal bridge effect, was calculated using the SUNCODE-PC model. Electricity consumption has been determined based on the equipment characteristics and the anticipated operating times.

MONITORING RESULTS

The air leakage was measured in four of the 16 dwellings. The average value was 1.3 air changes per hour (ac/h) at 50 Pa pressure difference, close to the target of 1.27 ac/h.

As far as overheating in summer is concerned, the results are very satisfactory. During an exceptionally warm and sunny summer period with temperatures far

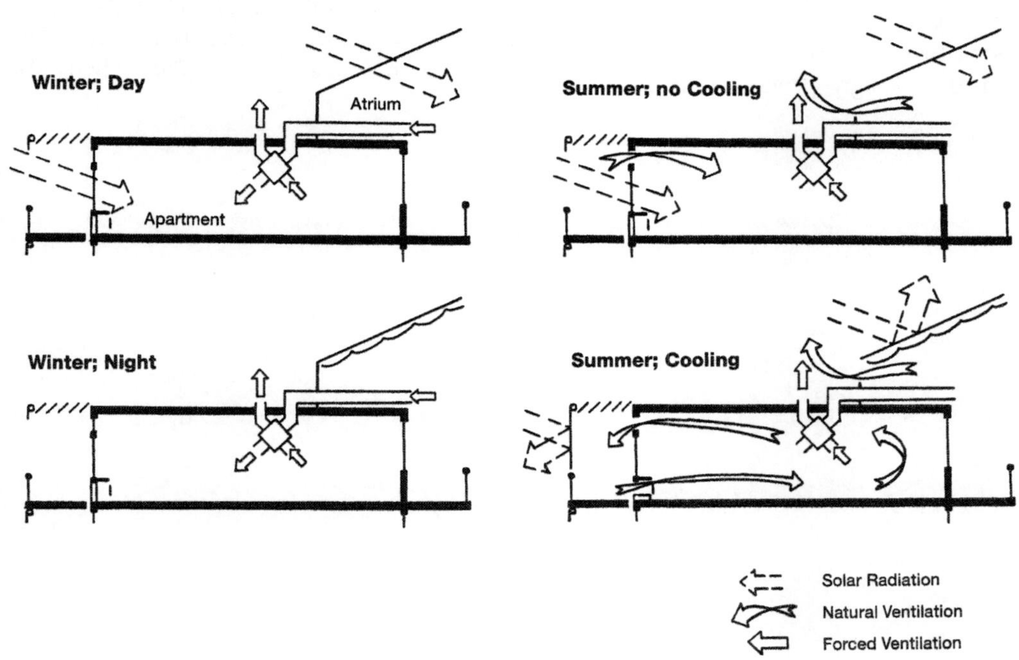

Figure 3.13.7. Ventilation modes and shading devices of the atrium and dwellings in the Netherlands apartment building

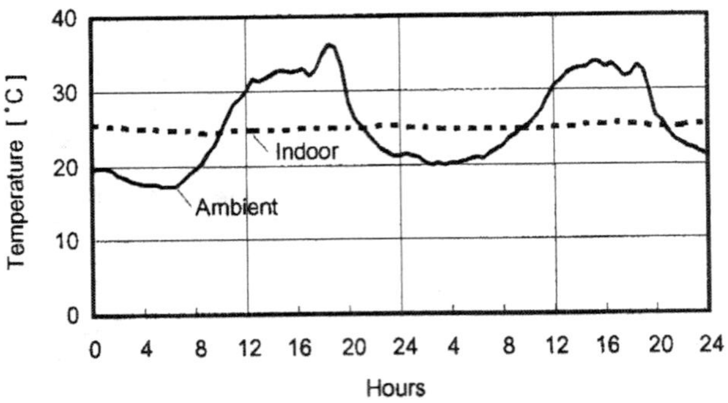

Figure 3.13.8. Indoor and outdoor temperatures in the Netherlands apartment building over two warm summer days

above 30°C, the indoor temperatures were about 25°C in the living room. Figure 3.13.8 shows indoor and outdoor temperatures measured by the control system over a two-day period. The indoor temperature is measured at the thermostat. The wall behind the thermostat stabilizes the temperature swings in the air to a certain extent. These results show that the passive cooling provisions are being used properly and function adequately.

Interviews with the occupants indicate that they were very content with the thermal comfort during summer. Some even stated that the apartment was cooler than their previous house. The control system is thought to be complex, especially for the elderly. More information about the use of the system was desired.

Apart from some teething problems, the concept seems to work quite well under summer conditions, although the handling of the control systems is complicated.

CONCLUSIONS

This project concentrated on finding a proper balance between conservation and passive solar measures, taking into account passive cooling options. Considerable attention was given to thermal comfort, especially during summertime. Monitoring results proved that the passive cooling approach is successful. The interaction of the various disciplines in the design process was sometimes difficult to organize but essential to create a well-balanced design. The choice of a multi-storey building has made it subject to various limiting factors (town planning requirements, selling points and others) that caused design concessions, mainly regarding compactness of the building and an individual installation per dwelling. Thermal bridges can reduce the energy performance of a super-insulated building significantly, and they need to be taken care of in a proper way. Advanced installation, controlled by a home automation system, is a necessary part of the energy concept. The Urban Villa proves that extreme energy saving within high density housing is possible without being an obstacle to good architecture. The possibilities for very transparent and flexible, yet at the same time low-energy dwellings, such as will be required around the year 2010, are as yet far from exhausted.

ACKNOWLEDGEMENTS

- *Architecture.* Zdenek Zavrel, Atelier Z Zavrel Architekten b.v., St. Jobsweg 30, 3024 EJ Rotterdam, The Netherlands.
- *Project management and building physics.* Bart Poel, Damen Consultants, P.O. Box 694, 6800 AR Arnhem, The Netherlands.
- *System design.* Rob Kouffeld and Dolf van Paassen, Delft University of Technology, Faculty of Mechanical Engineering, Mekelweg 2, 2628 CD Delft, The Netherlands.
- *Real Estate Development.* Pieter Hameetman, BAM Woningbouw, P.O. Box 43, 7400 AA Deventer, The Netherlands.
- *Author.* Bart Poel

PUBLICATIONS

Poel A,. Eijdems H H E W (1996). *Urban Villa Amstelveen, energie- en comfortberekeningen.* Damen Consultants; Arnhem.

Liem S H, van Paassen A H C (1995). *Digital Control Systems for Extreme Low Energy Apartments.* Delft University of Technology, Paper 19th Congress IIR/IIF, The Netherlands.

3.14 The Norwegian row house at Hamar

SUMMARY

The Norwegian Task 13 houses is a centre unit of a two-storey row house. The heating demand has been minimized by the use of light-weight, super-insulated constructions in the envelope, windows with super glazings and insulated frames, and a south-facing sunspace with a capillary-type transparent insulation in double-glazed window units. Balanced ventilation is provided by a mechanical ventilation system with a rotating-wheel heat-recovery unit with an efficiency of 90%.

All space and domestic hot-water heating is supplied by a solar-assisted heat pump that uses the ground and the small sunspace as heat sources, with a ground loop and an uncovered absorber in the sunspace. Part of the electricity demand is met by a photovoltaic system. The system is coupled to the grid and is expected to deliver surplus energy to the utility during sunny periods in the summer months.

The total auxiliary energy need is 33 kWh/m^2. This is about 15% of the amount needed in similar dwellings built according to the existing building code.

KEY FEATURES

- super-insulation
- solar-assisted ground-coupled heat pump
- heat recovery
- PV system
- sunspace

LOCATION

Hamar is located 120 km north of Oslo, near the town of Lillehammer, at latitude 61°N.

CLIMATE

The climate is cold, with long cold winters and cool summers (Figure 3.14.1).

CONCEPT

Contemporary dwelling developments in Norway are typically low-rise high-density semi-detached or row house type apartments in wood-frame construction. The high-

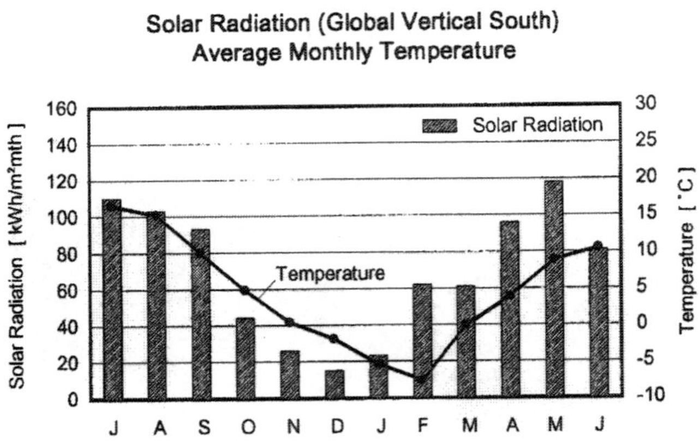

Figure 3.14.1. Solar radiation and average monthly temperatures at Lillehammer

latitude location, with a typically cloudy climate, gives negligible insolation during the most severe part of the winter. In November, December, and January, the daily solar contribution, even in the south of the country, is only 0.9 kWh/m^2 on the most favourable orientation. A high solar fraction for space heating requires some kind of seasonal storage. The high latitude also gives maximum insolation on steeply sloping or vertical surfaces during the heating season, making direct gain windows a favourable solar system. Off-hand, one could conclude that solar energy utilization is rather unlikely to be a success in Norway. However, the heating season is long and stretches into fall and spring, when appreciable amounts of solar energy are available.

The IEA dwelling is the middle unit of a row of houses with sloping roofs and a 125 m^2 heated floor area. The layout is rather conventional. Norwegians prefer cool bedrooms, so they are all situated on the lower floor.

The only unusual feature of the plan is an almost fully glazed sunspace, integrated in the plan on the south side. The staircase is located here, and the sunspace also functions as an air lock and buffer for the fully heated part of the dwelling. The sunspace itself is only heated to 15°C. The intermediate wall between the sunspace and the fully heated part of the dwelling is also glazed, thereby providing additional daylight for the deeper parts of the dwelling. The staircase landing upstairs can be used as an addition to the balcony in summer. Plans and section of the house are given in Figure 3.14.2.

CONSTRUCTION

Thermal insulation

A high level of thermal insulation is required in such a climate. The whole row house is constructed with wood/chipboard I-profile wall studs and floor and roof joists. This type of construction accommodates large thicknesses of insulation with less added cost than conventional dressed timber frame members.

The IEA dwelling is super-insulated, while the two neighbouring dwelling units are built to the thermal standard which will be required in the 1996 building code. The U-values (W/m^{2}K) are given in Table 3.14.1. For the Task 13 house this results in mineral-wool thickness of 300 mm for exterior walls and 400 mm for the roof. The ground floor slab is insulated with 180 mm of recycled expanded polystyrene foam; the perimeter of the foundation is also insulated on the outside.

Windows and glazing

In a super-insulated house, the heating season will be concentrated on the dark, no-sun period. This makes it very important to pay attention to the energy balance of all solar apertures. Even south-facing windows will be net losers over the heating season. Superior thermal insulation is therefore more important than a high solar-gain factor.

An interesting solution to this problem was to use a dynamic system with thermal shutters that add insulation to the windows during the loss period at night. A system that would react automatically to the energy balance of the windows was therefore explored. A smart control system would compare measured radiation to ambient temperature, and decide when to close the shutters.

However, when the performance of this system was compared to that of the new generation of superwindows, it was found that the savings in heating energy were only marginal. With the added costs and mechanical problems taken into account, it was therefore concluded that static superwindows were the best solution. The windows consequently have sealed triple glazing with two low-emission coatings and argon gas, and an extra pane of glass with a hard low-emission coating in an extra frame fitted to the exterior side of the window. The sealed unit has insulating spacers.

The exterior glazing of the sunspace is also triple sealed units with low-emissivity coatings and argon. About half of the wall area, in the staircase end of the sunspace, is transparent insulation panels with 150 mm capillary insulation material fitted between glass panes. This gives an overall U-value of about 0.7 W/m^2 K, while the superwindows, including sash and frame, have a U-value of 0.8.

Table 3.14.1. U-values for the Norwegian row house

	Building (code 1987)	End units (code 1996)	IEA row-house unit
Exterior walls	0.30	0.20	0.14
Roofs	0.20	0.15	0.11
Floor (slab)	0.30	0.13	0.13

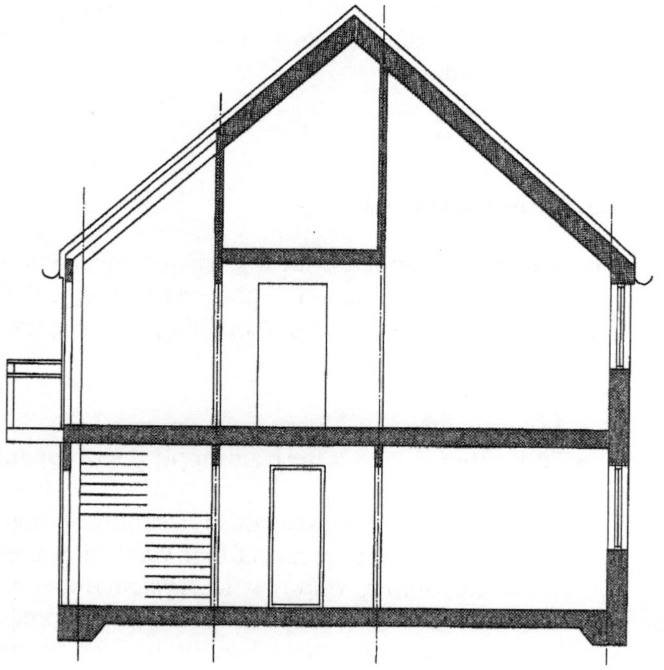

Section

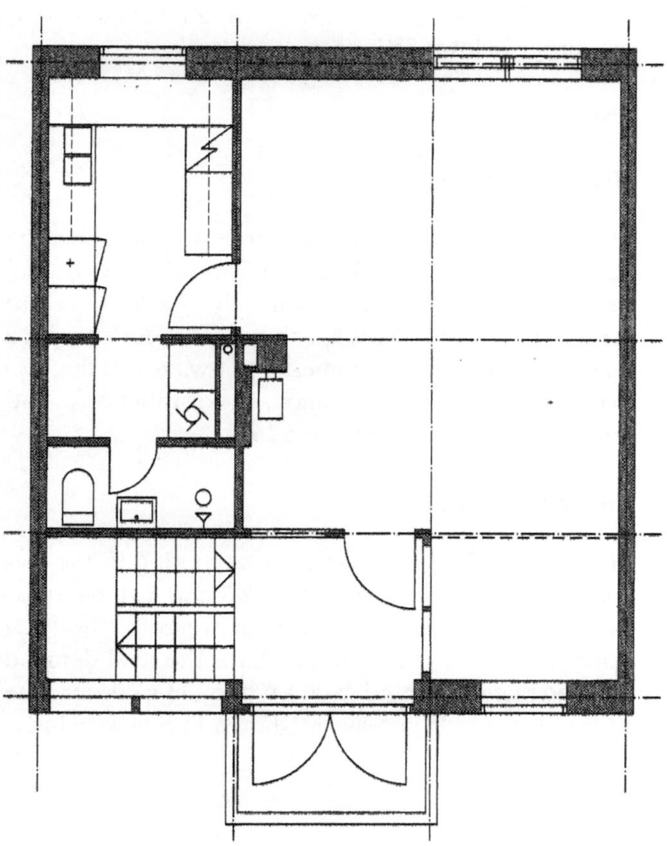

Second floor

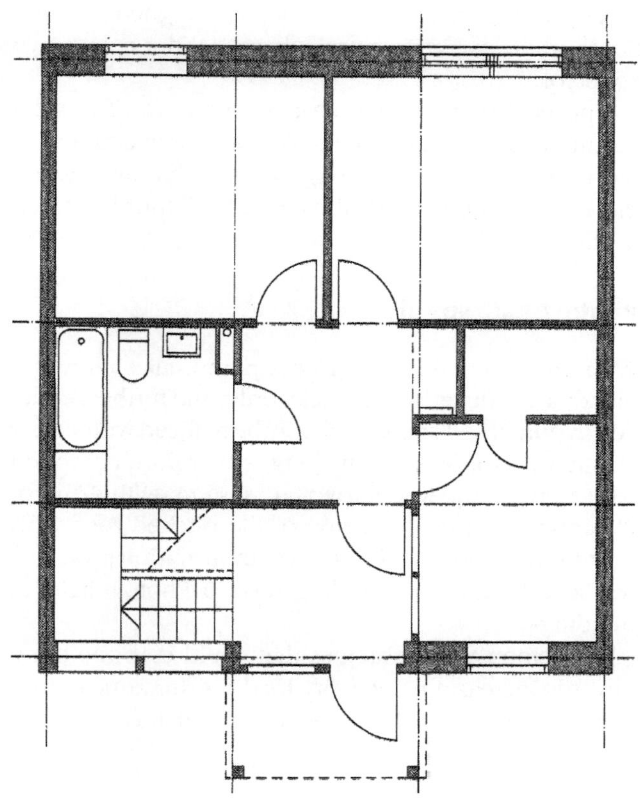

First floor

Figure 3.14.2. Plans and section of the Norwegian row house

SOLAR SYSTEM

Heating system

At the outset, a prime target for the Task 13 house was that it should be self-sufficient in thermal energy (space heating and water heating). As full thermal seasonal storage is not feasible on the scale of one dwelling unit, some purchased back-up heating was later allowed. The system is shown in Figure 3.14.3.

With direct electrical back up, it is logical to employ a heat pump, provided that a good stable heat source is available even in winter. Several options were studied, the final solution being a heat pump that uses both solar surplus heat in the sunspace and ground heat. A coolant circulation system includes both a collector in the upper zone of the sunspace and a 20 m long coil buried at a depth of 2 m in the ground outside the dwelling. The sunspace collector is designed to function both as an active solar collector and a convective collector.

The heat-pump evaporator is always fed from the ground coil. The sunspace collector is only coupled to the loop when excess energy is available and, when there is no heat demand, the energy will thus be stored in the ground.

Photovoltaic supply

With an electrically-driven heat pump, all energy demands are reduced to electricity only, and further reduction in purchased energy can only be realized with a solar electricity supply. The dwelling is therefore equipped with a grid-coupled photovoltaic system, with 22 m^2 of PV panels on the south side of the pitched roof. This system is also used as Norway's demonstration project in IEA Solar Heating and Cooling Task 16 'Photovoltaics in Buildings'.

To demonstrate strategies that could curb the trend towards increasing electricity load for domestic equipment, a very low annual electricity demand target, 25 kWh/m^2, was set. Therefore, very efficient equipment was chosen, including automatically switched compact fluorescent lighting.

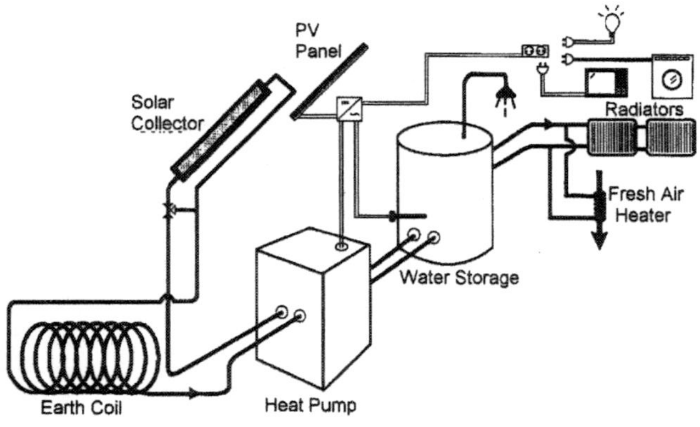

Figure 3.14.3. The heating system in the Norwegian row house

HVAC

Ventilation standard

The house is equipped with a balanced mechanical ventilation system, designed to give an air change that complies with the expected 1996 building code requirements:

0.35 litre/s m^2	base fresh air supply
+7.0 litre/s	fresh air supply per occupant

For the Task 13 house this gives a fresh air supply of about 1.0 ac/h. The code also gives special high ventilation rates for bathrooms without windows. The system therefore takes all return air through the bathrooms, with a forced ventilation rate that operates when the bathrooms are being used, controlled by an occupancy sensor.

Heat recovery

The ventilation system is fitted with a high-performance recuperating heat-recovery unit. With the fan energy added to the heat balance of the ventilation system, there is only a marginal heating load associated with the ventilation of the Task 13 house, when the ambient temperature drops below –5°C.

ENERGY USE

Heating demand

The space-heating load of the house was calculated by simulation with SUNCODE and TRNSYS. The annual load is calculated to be approximately 2700 kWh (Figure 3.14.4), of which 800 kWh is the demand for sunspace heating. In addition, a domestic hot-water heating demand of 4000 kWh/a is assumed. The distribution of end energy uses is given in Figure 3.14.5.

Electricity demand

The output of the PV system is, according to the PV-TRNSYS calculations, about 3000 kWh/a, with a surplus in the summer period which is delivered to the local utility company. On an annual basis the total demand for electricity is around 4000 kWh, or 33 kWh/m^2. The roof-mounted PV panels are shown in Figure 3.14.6.

CONCLUSIONS

The Norwegian Task 13 house is constructed with a super-insulated and airtight envelope, which reduced the space-heating demand to a short period in the middle of winter. The quadruple-pane windows virtually solved the problem of discomfort from downdraft, and therefore

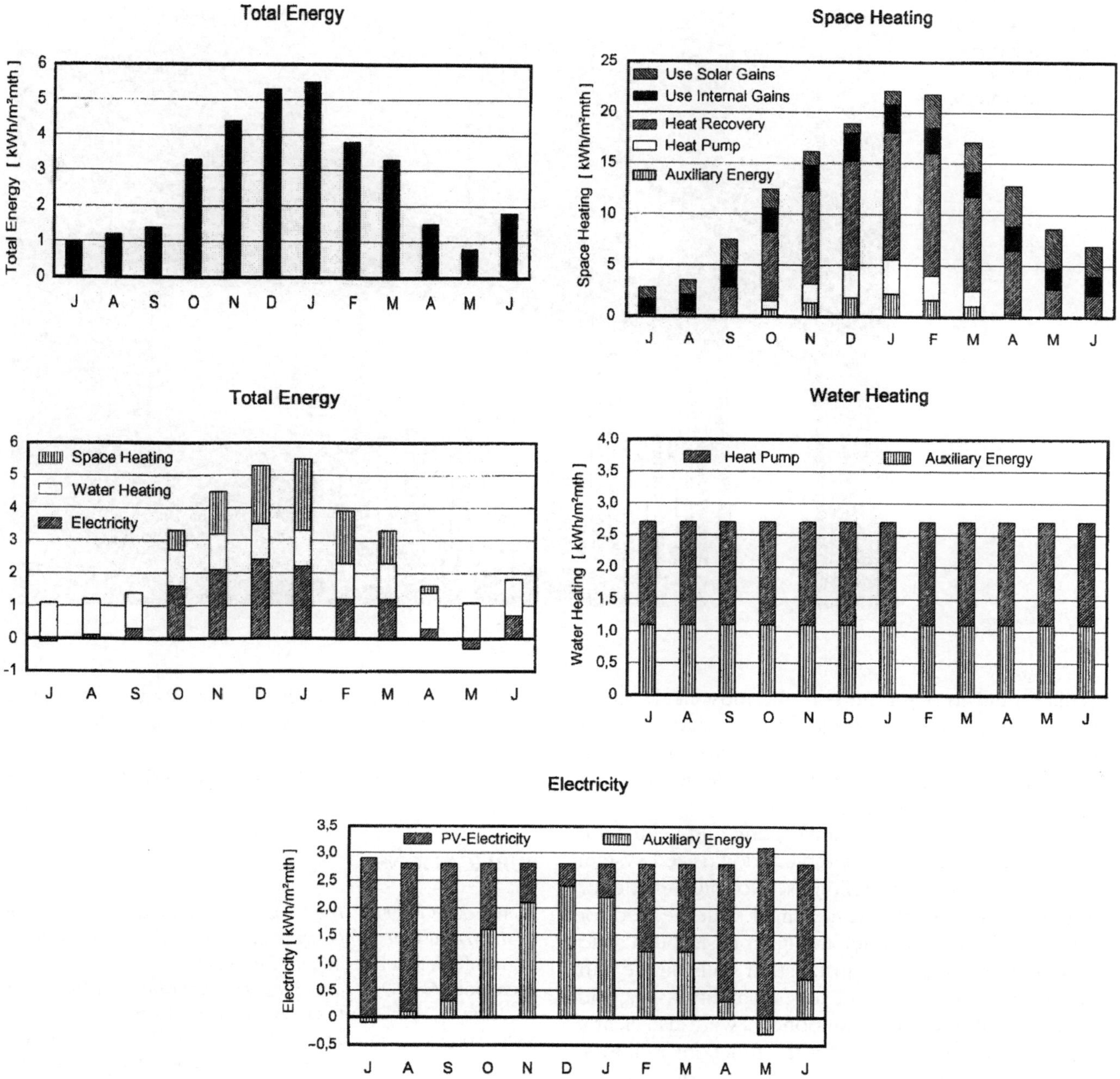

Figure 3.14.4. Predicted annual energy use for the Norwegian row house

hydronic radiators could be placed in the back of the rooms, close to the vertical distribution shaft.

The staircase sunspace on the south side has performed to the satisfaction of the owners. Some rooms gain daylight from this area, which is particularly well lit when direct sunlight is diffused through the transparent insulation panels. At times, this could cause some glare problems, but the staircase is not a space with prolonged occupancy. During short summer periods some overheating in the sunspace occurs.

The house has a rather complex mechanical system, mostly owing to the incorporation of a heat pump for supplying domestic hot water and hydronic space heating. The primary heat source for the heat pump is a 20 m coil buried in a trench with the centre axis of the coil in a horizontal position. It proved too difficult to backfill properly through the top windings of the lying coil. When it was tested later, it was impossible to

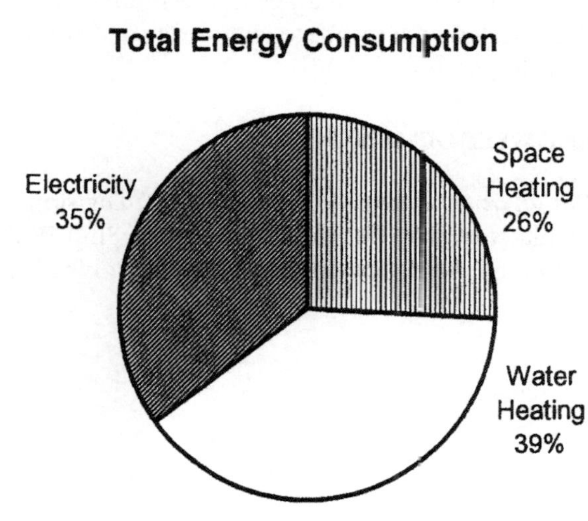

Figure 3.14.5. Distribution of end energy uses for the Norwegian row house

Figure 3.14.6. The roof-mounted PV panels on the Norwegian row house

circulate fluid through the coil, the most likely cause being a bend in the plastic piping used. Later a horizontal zig-zag pipe was installed on top of the coil. It performs as intended, but has a lower thermal capacity than the first storage system.

The mechanical systems for heating and ventilation were custom-built, and resemble a downsized system for large commercial buildings. No high-efficiency equipment is available on the small scale needed for such a small house. This led to higher costs, but also to many problems related to installation skills on the part of local suppliers and contractors, reliability issues, and service availability. Some components were also clearly oversized and too complex for the functions required. A typical example is the controller for the rotating wheel heat recovery unit in the ventilation system. Clearly, there will be a market in the future for such products scaled with regard to complexity and costs for the housing sector.

ACKNOWLEDGEMENTS

- *Design and Analysis Team. SINTEF* Division of Architecture and Building Technology, SINTEF Division of Refrigeration and Airconditioning and Department of Building Technology, Norwegian University of Science and Technology – NTNU, N-7034 Trondheim, Norway
- *Authors.* Øyvind Aschehoug and Anne Grete Hestnes, NTNU.

PUBLICATIONS

Aschehoug Ø (1993). *The Norwegian IEA Task 13 House. The International Symposium on Energy Efficient Buildings,* Leinfelden-Echterdingen, Germany.

Aschehoug Ø (1994). *Low Energy Timber Frame Row Houses in Hamar.* Vision Eureka, Lillehammer, Norway.

Aschehoug Ø, Andresen I (1992). System Design for an Advanced Solar Low Energy Dwelling. *North Sun '92, Trondheim, Proceedings,* pp.85–89.

Aschehoug Ø, Hestnes A G (1994). Snekkerstua Low Energy Dwellings. *North Sun '94,* Glasgow.

Aschehoug Ø, Hestnes A G, Jacobsen T, Raaen H, Thyholt M (1990). *Parametric Studies for a Norwegian IEA Task 13 House,* SINTEF Report STF62 A90018, Trondheim.

Hestnes A G Aschehoug Ø (1992). Transparent Insulation Application: Advanced Solar Low Energy Dwelling in Norway. *Transparent Insulation Technology Meeting TI5, Freiburg Proceedings,* pp. 195–198.

3.15 The Swedish low-cost house at Röskär (and Kullön)

SUMMARY

The Swedish Task 13 house at Röskär saves both energy and natural resources. During the planning stage, high priority was given to features such as effective use of space, recycling, efficient use of materials, and low energy and water consumption. The 54 m² floor area is divided into two living rooms, a kitchen, and a small bedroom. The house represents a suitable alternative for single persons or small families. A large proportion of Swedish households today consists of single persons. This house is an economical alternative for such people. Building materials have been selected that are recyclable and possess non-allergenic qualities.

KEY FEATURES

- super-insulation
- minimal thermal bridges
- direct gain
- solar control
- daylighting
- window night insulation
- heat-recovery systems
- home automation systems
- airtightness and air quality

LOCATION

Röskär is a demonstration site for experimental houses located 30 km north-east of Stockholm city at latitude 60°N. The house is built with south–north orientation and is in a large glen surrounded by large trees.

CLIMATE

The climate is cold temperate. The outdoor air temperature and the solar gain through a single-glazed window pane in the north and south direction are given for each month in Figure 3.15.1.

CONCEPT

The Swedish Task 13 house is built as a prototype for an experimental project at Kullön, which will take place in 1996–2000 and where later such units will be incorporated. The house serves as a demonstration and test facility for building and system concepts and for building details and components. As such, it was freed from normal building code requirements. It is small to keep costs low and save resources, but it can be expanded both vertically and horizontally when a family grows or when a new owner buys the house.

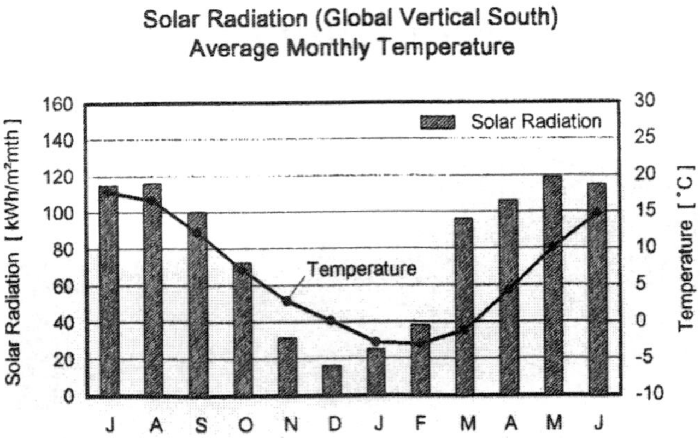

Figure 3.15.1. Outdoor temperature and solar gain through a single-glazed window in the Stockholm climate

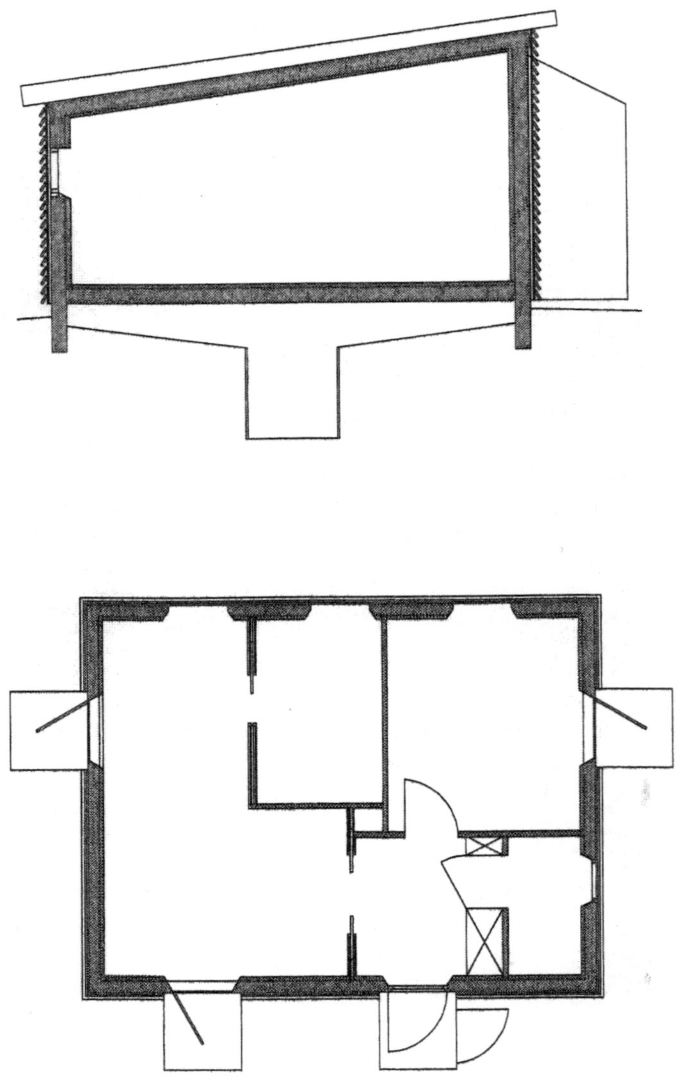

Figure 3.15.2. Floor plan and section for the Swedish Task 13 house in Röskär

The walls are constructed from vertical steel rails integrated in thick polystyrene blocks. This construction allows a free choice of interior and exterior cladding. In the first test house a horizontal panel with overlap was chosen. The windows facing east, south, and west consist of a glazed box on the outside of the wall. Within the box a reflector is positioned with regard to the solar angle, to maximize solar gain and daylight in wintertime and to minimize direct gain in summertime to avoid overheating.

The northern façade of the house has relatively small windows without any reflectors. Fixed exterior blinds made of wooden horizontal strips reduce the long-wave radiation loss to the sky. Some of the design features may be characterized as utopian, both concerning function and performance. In this case the aim is not to create a house for today's market, but to test new ideas and solutions. A floor plan and section are given in Figure 3.15.2.

The electrical installations are based on a BUS system with a high degree of flexibility regarding placement of switches and sockets, both for a 230 V circuit and a low-voltage circuit. The switches can be controlled via infrared sensors in each room. The sanitary systems are also flexible since no common walls are used for the plumbing. The building thus functions as a shell where the owner has total freedom to partition it into spaces. The exception is the wet cubicle, which is finished at the same time as the building structure.

CONSTRUCTION

The building system is made of EPS (expanded polystyrene) blocks and thin-walled steel profiles. Use of rigid foam as a structural element is, of course, not new, but the challenge here was to create a site-built structure using standard structural elements with a minimum degree of prefabrication.

The rigid foam with steel framing was chosen for several reasons:

- EPS is relatively inexpensive and environmentally feasible. It can be cut with relatively high precision. As a consequence, the structural frame has to be an exact and stable form
- wood lacks the stability to be fitted on site to a rigid insulation
- plastic creeps under continuous loads; thin-walled steel profiles were therefore chosen.

The steel construction consists of metal C-profile studs with the web penetrating the wall in order to provide maximum load-bearing capacity. To achieve good insulation values and air and moisture tightness the EPS panels are interlocked with sheet metal 'C' channels. At the top and bottom the walls are locked together with top and bottom sills of perforated thin-walled steel channels (Figure 3.15.3). The thermal bridge effect of these is similar to that of a 38 mm wooden board.

Roof construction

The roof is a cantilever, trapezoid, metal-sheeting construction. The roof slope is 8° with its highest point over the south façade. Under the cantilever roof construction a 200 mm polystyrene section panel is mounted. The panel is of the same type as the walls and the floor. Sheathing paper between the roofing and the EPS panel drains away condensation.

Foundation and floor construction

The house is built over a crawl space. The floor construction is made of profiled steel which serves as the formwork for an *in situ* cast concrete slab. The crawl space, which is highly insulated from the ground, is heated to warm the floor, providing distribution of air to the rooms through a perimeter slot in the floor; it also allows complete freedom for installation work under the floor.

In addition, the crawl space acts as a settlement chamber for dust particles. It is covered with a waterproof membrane with a drain at the centre and can be washed down to clean the inlet air path at regular intervals.

Window construction

The 'magic window' (Figure 3.15.4) is a glass box which allows control of the amount of incoming light and energy, but which does not change the colour of light more than standard clear glass. The objectives behind this window concept were:

- to make use of standard glazing, i.e. without low-E coating
- to increase daylight to the interior by using the outer reflector in winter
- to decrease solar energy to the interior by using the outer reflector in summer
- to provide privacy at night by using the reflectors in a parallel position
- to reduce heat transmission through the window at night as a result of the insulating effect of the reflector's position in front of the window in the façade.

A reflector controls the amount of incoming light. According to the occupant's individual lighting wishes, the reflector can be adjusted into a position that:

- reflects more light into the room
- shades the window from direct sunlight while admitting diffused light
- shades the window completely.

To protect the reflector against damage from rain and wind a glazed enclosure is used. It has a square base

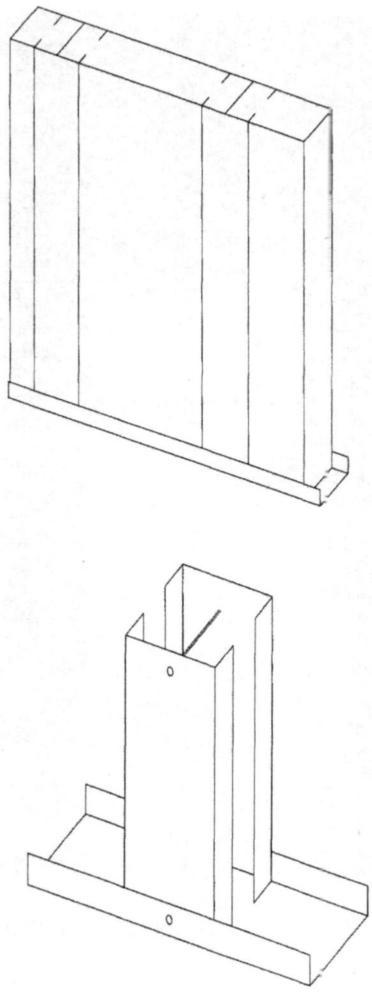

Figure 3.15.3. The EPS blocks locked together at the edges (Swedish Task 13 house)

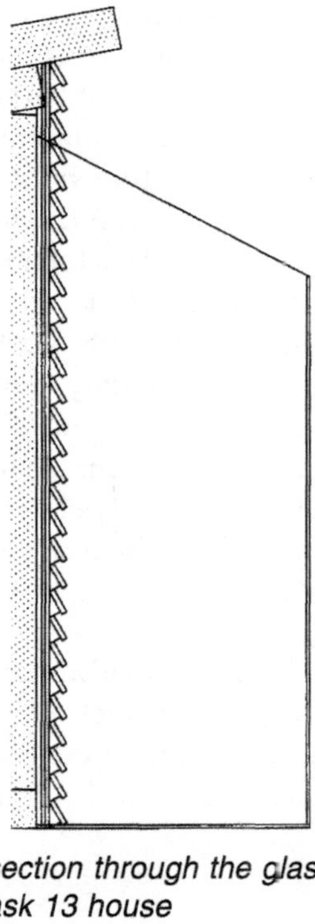

Figure 3.15.4. A section through the glass box attached to the Swedish Task 13 house

Figure 3.15.5. Prototype of the glass box with the reflector used in the Swedish house

with glazing on three sides and the façade and a double-glazed window on the fourth side. The reflector is housed in a glass box which is added to the house on the outside of a standard window. This construction allows for free movement of the reflector and maximizes light reflection and heat reduction. Moreover, this glass box performs as a temperature buffer zone which lowers the temperature difference and, in this way, the energy flow between the inside and outside of the standard window. The temperature inside the glass box can be controlled via the reflector position and a ventilation system. This ventilation system is necessary to avoid overheating during summertime. The reflector itself is a simple slab of polystyrene. The surface can be coated with a material of free choice so that the individually requested performance is achieved. Driven by small wheels on the top, the reflector follows a track on the inside of the glass box. The prototype is shown in Figure 3.15.5.

To keep the amount of shadow on the reflector to a minimum, the supporting track is reduced to corner elements and a connecting steel rod. This steel rod presses the wheels onto the Velcro track so that contact between the track and the wheels is maintained. The Velcro is attached directly to the glass, except where it meets the corner elements of the track. The advantage of the system can be illustrated with a south-facing conventional window in early morning. It receives almost none of the glancing sun's rays from the east. The 'magic window', which projects out from the wall and has a reflector, captures the early morning sun. Measurements on a prototype inside the laboratory at the Division of Building Technology showed local luminance values inside the room which were up to seven times higher than without the reflector.

HVAC

The main elements of the HVAC system are:

- a heated crawl space, which provides floor heating and inlet air distribution
- new type of integrated fresh air inlets
- effective solar screening to avoid overheating
- operational strategies
- solar system for domestic hot water.

The house is heated by a low temperature hydronic floor heating system. One goal of the project was to develop methods of using low-temperature (< 30°C) heating to reduce storage losses. There will also be a 2 m³ water tank for thermal storage and this will be shared between the house system ands a solar-powered car.

Minimum ventilation is provided with an air-to-air heat exchanger unit. This is a compact air-handling unit mounted under the floor construction in the crawl space in the kitchen. It achieves up to 75% heat recovery. A separate natural ventilation system can be used in summertime to save fan energy and temporarily increase ventilation when demanded by the user. The air inlets are specially designed dampers with high-porosity polycarbonate insulation as filters.

An intermittently operated open-flow solar collector is mounted on the roof to preheat tap water in a 300 litre storage tank.

THE ELECTRICAL SYSTEM

An interactively operated BUS system controls the electrical system. Signals are carried by a two-wire low-voltage BUS that connects to all components in the system. A sensor detects a process and forwards the signal on the BUS. Buttons, presence detectors, IR detectors, a wind meter, temperature sensors, and an operational clock all transmit signals to the BUS. There are receivers on the BUS that take appropriate action to put lights off or on, start a fan motor, call the fire department etc. The receivers are programmed to react on signals from different sensors in different ways as, for instance, to put the light off when daylight sensors detect adequate daylight. The sensors are connected to the low-voltage system only whereas the receivers are also connected to the high voltage net (230 V).

LABORATORY MEASUREMENTS

The capacity of the wall system for vertical loads was tested at the laboratory of building structural mechanics as part of a master's thesis. A wall element of 2400 mm height was tested standing on a flat surface. The vertical load was pressed on top of the joint with a surface of 110 mm x 200 mm, simulating a normal roof truss. Test results show that this extremely inexpensive and lightweight construction has the potential to serve as a load-bearing wall in at least a one-storey house. Also, since the material thickness of the vertical profiles can be increased without significant thermal bridging, the load-bearing capacity can be adjusted within a wide range.

ENERGY USE

The predicted total energy used by this house is 74 kWh/m²a. It is broken down as follows:

space heating	17
domestic water heating	21
lights and appliances	25
fans and pumps	11

The space heating load of 64.1 kWh/m²a is met as follows:

internal gains	14
passive solar useable gains	14
heat recovery	19
auxiliary heat	17

The domestic water heating load of 41 kWh/m²a is met by:

back-up water heating	21
solar water heating	20

CONCLUSIONS

This compact experimental house achieves very low energy-consumption levels per square metre, despite its small size and the cold climate. The innovative insulation-block construction allows for rapid construction and essentially eliminates thermal bridges. The highly inventive window construction allows the user to 'configure' the window to optimize its ability to deliver daylight, provide passive solar gains, reduce heat losses, or reduce overheating. The interactive control system, low temperature floor heating, and ventilation-air heat recovery further minimize the auxiliary heating requirement. As a result, as often occurs in very low-energy houses, the energy use for domestic water heating and electricity are equal in importance to the space heating.

ACKNOWLEDGEMENTS

- *Project Managers and authors.* Gudni Johannesson & Carl Michael Johannesson, Division of Building Technology, KTH, S-100 44 Stockholm, Sweden.
- *Consultants.* Göran Werner, ABB, Sfserv, Stockholm, Sweden, and Sten Engwall, TEEG AB, Maersta, Sweden.
- *Research team.* Gudni Johannesson, Carl Michael Johannesson, Per Levin and Folke Björk, Division of Building Technology, KTH, S-100 44, Stockholm, Sweden.

PUBLICATIONS

Johannesson G, Levin P, Johannesson C M (1995) *A New Structural System Made of Sheet Metal Profiles Supported by Blocks of Expanded Polystyrene, Nordic Steel Construction Conference*, Department of Building Science, Division of Building Technology, KTH Stockholm.

Löfgren P (1995). *Underfloor Heating*, Thesis work No 250, Department of Building Science, Division of Building Technology, KTH Stockholm.

Trinius W (1995). *The Magic Window*'. Master Thesis Work, Department of Building Technology, KTH Stockholm.

3.16 The Swiss duplex in Gelterkinden

SUMMARY

This project is noteworthy because of its very low energy consumption by means of mostly conventional features and construction, including superior window frames and glazing to minimize heat losses, light frame exterior walls and roof construction to accommodate thick insulation easily, masonry interior walls and concrete floors to provide thermal mass, and heat recovery from exhaust air. Once the building was well insulated and optimized to profit from passive solar gains (60% solar coverage of heat losses), additional solar measures (active or hybrid) were increasingly difficult to justify economically because the energy demand was so drastically reduced.

KEY FEATURES

- direct gain
- superior glazing and insulation
- heat recovery
- PV system

LOCATION

The duplex is located at 403 m above sea level, latitude 47°N, as part of the Chienbergreben Housing Development in Gelterkinden near Basel. It is built on the site of an old vineyard with a slight slope to the south.

CLIMATE

Temperate middle European climate with long overcast winter periods (Figure 3.16.1).

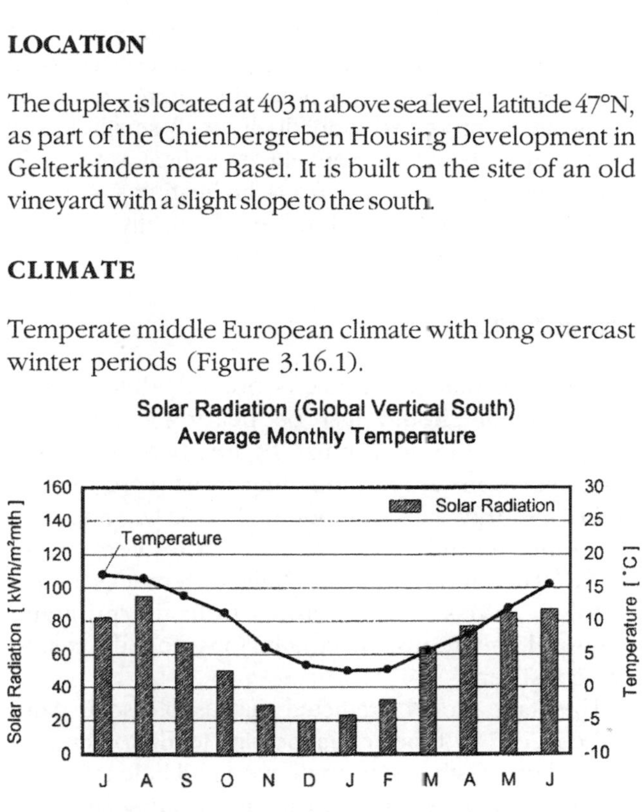

Figure 3.16.1. Climate data for Gelterkinden

CONCEPT

This residence is half of one of eight duplex houses constructed in 1995. It has 180 m² of heated floor area divided over three storeys. To profit from the optimal south exposure, the south façade has windows extending from the ground to the roof (36% of the façade). Open floor plans of the ground and upper levels avoid the overheating which often occurs in small closed rooms with large south-facing windows. Because the house shell is of light wooden construction, interior floors and walls are massive to provide the needed thermal mass. Section and floor plans are given in Figure 3.16.2.

CONSTRUCTION

The properties of the building envelope and interior are given in Table 3.16.1. Selected construction details are illustrated in Figure 3.16.3.

SOLAR SYSTEMS

Passive solar heating is provided by large south-facing windows and internal massive floors and walls. Deep penetration and distribution of passive solar gains are facilitated by the open floor plan.

A standardized grid connected 3 kW peak photovoltaic system is mounted on the roof.

Domestic water heating by an active solar system was considered but omitted for economical reasons when a central heating plant with a distribution net was chosen.

HVAC

A central wood-chip heating unit provides back-up heat distributed to each house by a conventional system with two hot-water pipes. In summer domestic hot water is produced individually with small electric heat pumps using ambient air as a heat source. In winter the water is heated from the heating system via a heat exchanger. Ventilation is mechanically provided with heat recovery from exhaust air.

ENERGY USE

At the time that this book is printed, this housing development should just be nearing completion. Accordingly, the energy use figures here are based on calculation. Three modelling tools were used: a static monthly based model: LESOSAIX, and two dynamic models: SUNREL (a new version of SUNCODE) and TRNSYS. While differences in the absolute values of the outputs from the programs occurred, the trends and conclusions from the results are consistent.

The static model predicted that usable solar gains will cover 57% of the annual heating load. Rotating the house from south east to south west had little influence on the heating demand. Finally, the heating season is reduced to the five months, November through March,

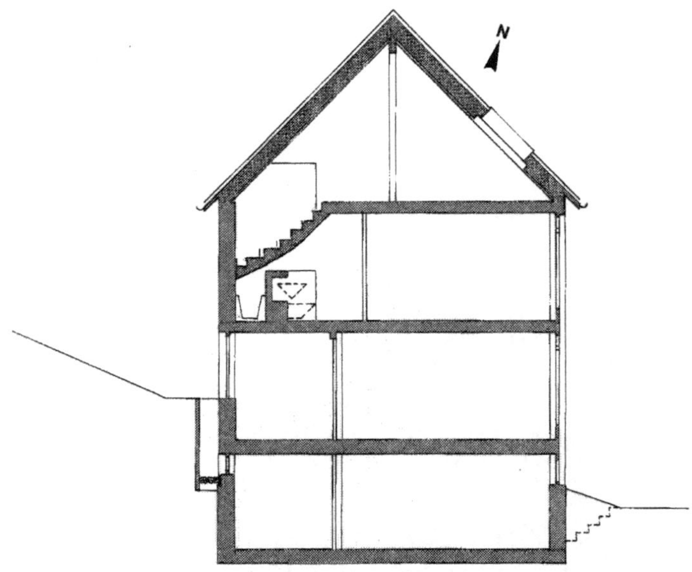

Section

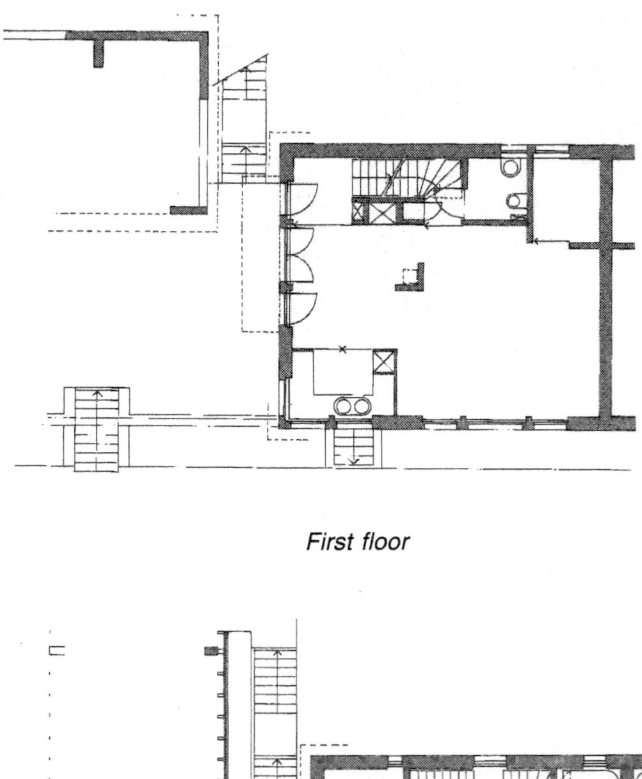

First floor

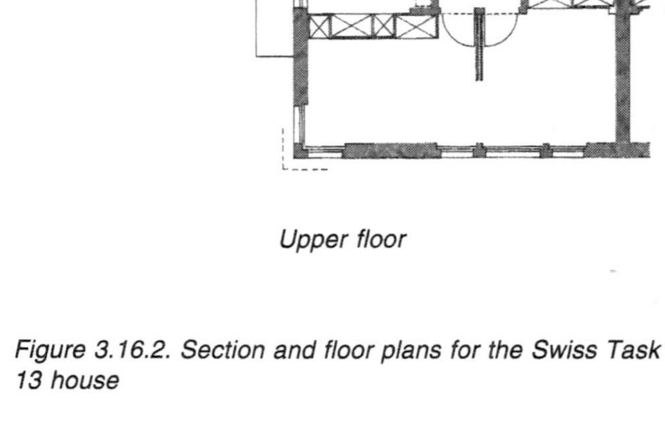

Upper floor

Figure 3.16.2. Section and floor plans for the Swiss Task 13 house

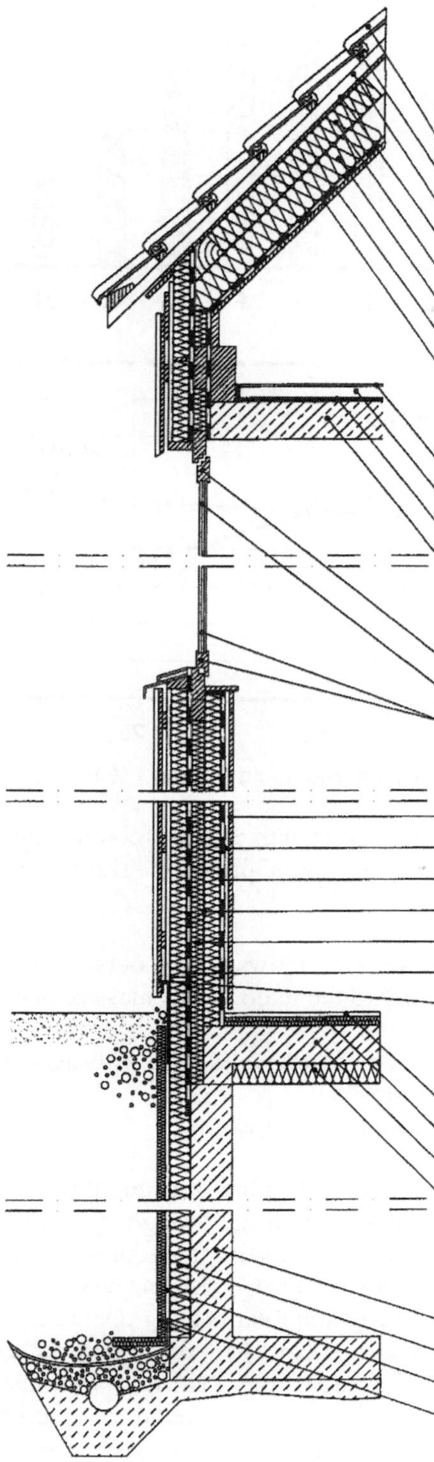

Layer	Description	Width (cm)	Thermal Conduct (W/mK)	U-Value (W/m²K)
Roof				
1	Roofing Tile	-	-	0.16
2	Lattice	2.40	-	
3	Lattice	5.00	-	
4	Hardboard	1.50	-	
5	Insulation (90%)	26.00	0.04	
6	Wood (10%)	26.00	0.14	
7	Vapour Barrier	-	-	
8	Lattice	2.00	0.14	
Ceiling				
9	Parquqet	1.00	0.21	1.54
10	Lower Floor	5.00	1.80	
9	Insulation	2.00	0.04	
10	Concrete	18.00	2.40	
Window				
13	Wood Frame			1.4
14	Glazing			1.1
15	Glazing and Frame			1.2
Exterior Wall				
16	Lattice	0.02	0.14	0.15
17	Airspace	0.02	0.12	
18	Vapour Retarder	-	-	
19	Insulation (90%)	26.00	0.04	
20	Wood (10%)	26.00	1.40	
21	Wind proofing	-	-	
22	Lattice	-	-	
Basement Ceiling				
23	Strand Board Subfloor	3.00	0.21	0.27
24	Insulation	3.00	0.04	
25	Concrete	18.00	2.40	
26	Insulation	10.00	0.04	
Below Grade Floor				
27	Concrete	20.00	2.40	0.33
28	Foamglas	10.00	0.036	
29	Airspace	-	-	
30	Polystrene Insulation	-	-	

Figure 3.16.3. Selected construction details for the Swiss Task 13 house

Table 3.16.1. Properties of the construction of the Swiss Task 13 house

	Description	(W/m²K)
Exterior		
Walls & roof	Vented wooden siding and 26 cm insulation	0.18
Basement wall	20 cm concrete, 10 cm polystyrene and 6 cm sheathing	0.3
Ground floor	10 cm mineral wool, 18 cm concrete (2400 kg/m^3) and 6 cm screed cement	0.35
Windows	Metal clad wooden frames	1.3 (frame)
	Two glass panes, two low-E films, argon gas ($g = 0.57$)	1.1 (centre of glass)
		1.2 (window average)
Interior		
Intermediate floor	18 cm concrete, 2 or 4 cm insulation and finish floor	
Party wall	15 + 12 cm limestone bricks (1800 kg/m^3)	

Figure 3.16.4. Monthly heating requirements for the Swiss Task 13 house and how they are met

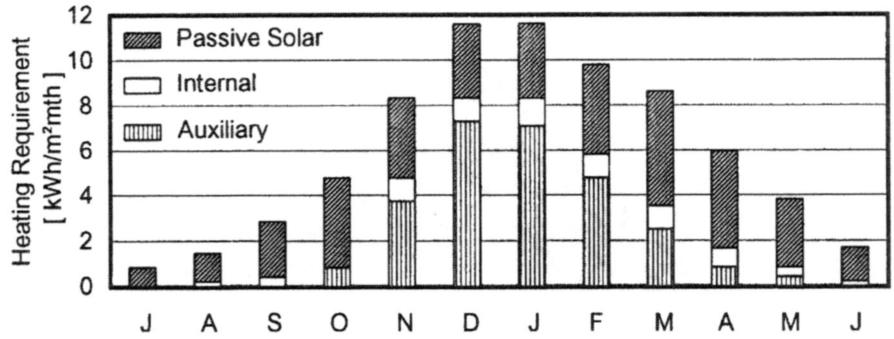

Figure 3.16.5. Effect of window glazing and size on auxiliary heating in the Swiss Task 13 house

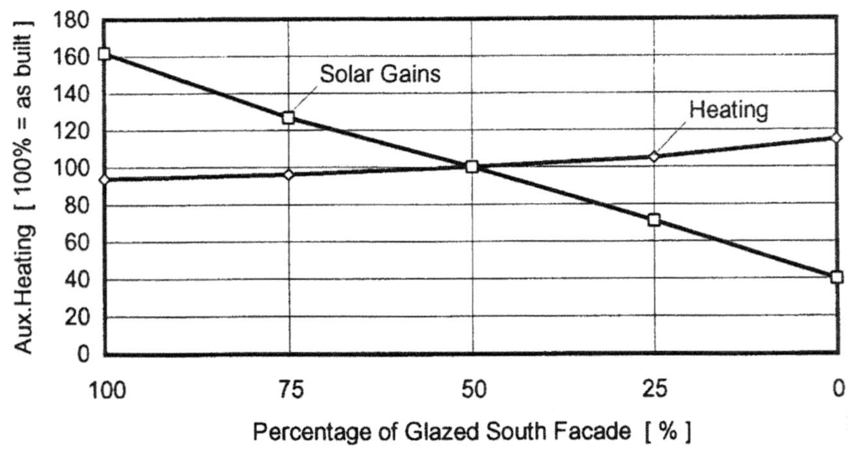

with the largest useable solar contribution occurring in the spring, as can be seen in Figure 3.16.4.

The Program SUNREL was used to investigate in detail the consequences of increasing the amount of south-facing window area and the effect of using a less expensive window (U = 2.0 instead of 1.2 W/m²K). Increasing the window area caused negligible energy savings, given that the reference case already had a large window area, as can be seen in Figure 3.16.5. Larger heating savings could be expected if the internal gains were smaller as a result of more efficient appliances and lighting. On the other hand, using a less well-insulating glazing increased the heating demand by over 40%, even considering its slightly better solar transmittance.

A second variable of interest was the effect of thermal mass. Design guidelines emphasize the importance of insulation being on the outside of exterior walls to minimize thermal bridges. This also provides more thermal mass in the building interior, which should increase the usability of direct solar gain through windows. Simulations with SUNREL confirm this benefit which thermal mass provides. Three variations were studied:

- very massive (exterior walls with outside insulation and masonry to the interior plus massive floors and partitions)
- medium mass – same as (a) but with light weight exterior walls
- light (all wooden construction).

A 12% difference in heating consumption occurred between the massive and light-weight constructions.

The greatest savings occur from the addition of mechanical ventilation with heat recovery. As can be seen in Figure 3.16.6, this reduces the total heating demand by 40%, even if the increased electricity needed to run the fan is included. This reconfirms that, in a well insulated house, reducing the heating demand for heating ventilation air is an important target to save energy.

Parameter	Base case	Variation
a) Heat recovery	without	a) with
b) Glass specs. g=0.57	U=1.1, g=0.42	b) U=0.7,
c) Window/facade[1]	36%	c) 43%
d) Basement floor	U=0.34	d) U=0.25
e) Walls	U=0.18	e) U=0.10
f) Orientation	due south	f) ñ 45°

[1]Ratio for the south facade

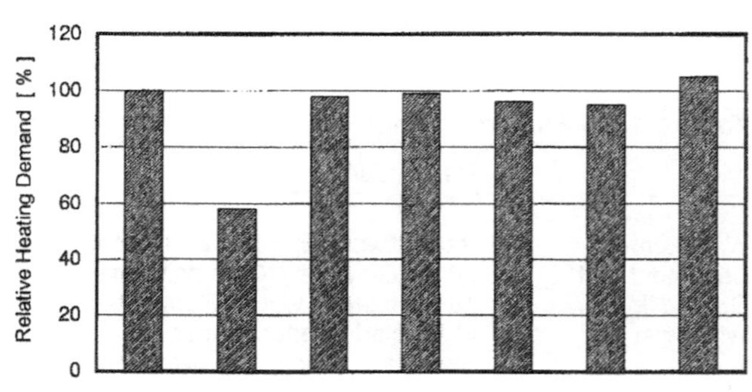

Figure 3.16.6. Energy savings from heat recovery in the Swiss Task 13 house

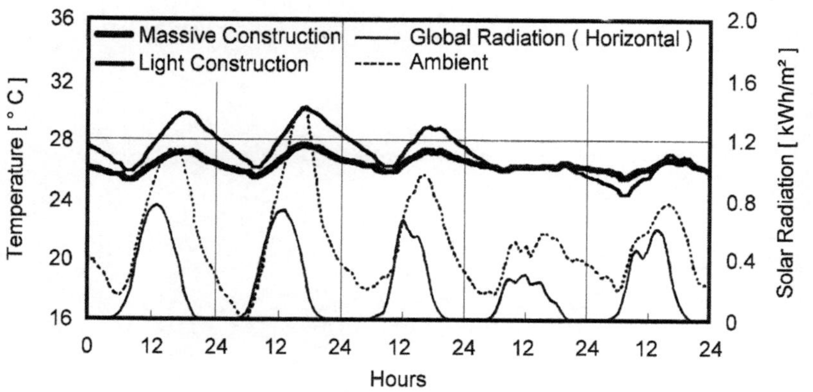

Figure 3.16.7. Computed effect of mass on the room temperature in summer for the Swiss Task 13 house

TRNSYS was also used to calculate the effect of thermal mass and to predict temperature swings. It could be seen that:

- 80% of the effective thermal mass is provided by the inner floors and walls in a massive building such as this
- 10% energy savings (compared with 12% from SUNREL) can be achieved by changing from a light to a massive construction, but it is unimportant whether the outer walls are massive
- temperature swings are drastically reduced if mass is introduced within the building, but again it is unimportant whether the outer walls are massive.

To assess the risk of overheating in summer, five days in August were taken and a worst case computed using TRNSYS. With the windows shut the room temperature quickly rises out of the comfort range (to 31°C) in the case of light construction, but in the case of the massive construction the room temperature stays below 28°C, as can be seen in the curves of Figure 3.16.7.

CONCLUSIONS

The analyses of this project indicate that a simple design consisting of good insulation and deliberate use of passive solar gains can achieve very low energy levels without discomfort problems in middle European climates.

Thermal mass is essential in order for the solar gains to be effective in reducing the auxiliary heating demand but above all to reduce overheating temperatures, which would lead to venting solar gains through window opening.

After passive solar design features are included, heat recovery from exhaust air is the next single most effective investment. Eliminating the small remaining energy demand is extremely expensive, not only with regard to an active solar system, but also with regard to a conventional auxiliary heating system.

ACKNOWLEDGEMENTS

- *Architect.* U. Schäfer, Zollikonstrasse 20, CH-8122 Binz, Switzerland.
- *Energy analysts.* W. Frei and H.P. Rietchle, Solararchitektur, ETH Hönggerberg, CH-8093 Zürich, Switzerland.
- *Author.* R. Hastings, Solararchitektur.

PUBLICATIONS

LESOSAIX is available from LESO-PB, EPFL, CH-1015 Lausanne.

Rietschle P (1994). Erfahrungen mit Planungshilfsmitteln und Simulationsprogrammen bei der Begleitung der Energieplanung einer Niedrigenergie-Siedlung, 8. Status-Seminar: Energieforschung im Hochbau, EMPA-KWH, CH-8600 Dübendorf.

Schäfer U Siedlung Gelterkinden, Schweizer Ingenieur und Architekt, SIA, Postbox 630, CH-8021 Zürich.

SUNREL is available from R. Judkoff, Buildings Systems Research Branch, NREL, 1617 Cole Blvd, Golden, CO 80401, USA.

TRNSYS is available in Europe from several sources, one being TRANSSOLAR, Nobelstrasse 15, D-70569 Stuttgart, Germany.

4

Lessons learned

Anne Grete Hestnes

Lessons learned

- **The Task has demonstrated that it is possible to design very low-energy buildings that at the same time have high thermal comfort, good indoor air quality, and low environmental impact.**

Task 13 has shown that it is possible to reduce the total energy consumption to a small fraction of the typical consumption today. The average total projected energy consumption of the evaluated buildings developed in the Task show a reduction of 60% of the typical consumption in residential buildings in the participating countries.

In addition to low energy use, some of the Task 13 design teams also paid special attention to other resource uses. The Canadian house in Waterloo, which makes extensive uses of recycled materials, has reduced purchased water use by 73%, use of ozone-depleting chemicals by 99% and waste sent to landfill by 98%. The other buildings do not have quite as impressive numbers for resource use reduction, but they all demonstrate both reduced energy use and, therefore, reduced environmental impact.

- **The total energy consumption does not differ very much from country to country.**

This is partly because the consumption for water heating, lights, and appliances is relatively independent of climate, but also because the building codes are not. The insulation levels are generally low in countries with wild climates and high in countries with cold climates. The energy consumption per square metre, therefore, does not differ as much as one would expect when looking at the climatic differences.

- **The energy consumption for water heating is becoming a larger fraction of the total energy consumption. The energy consumption for space heating and for lights and appliances are still the largest fractions.**

In fact, in most of the Task 13 buildings, the electricity demand is as large as the thermal energy demand for space heating. This is because the emphasis in many countries has so far been on technologies for reducing space heating consumption and this consumption has therefore been reduced more than the rest. It is now necessary to consider buildings' other energy uses. This actually provides more opportunities for innovation, as can be seen in the Task 13 buildings.

- **It is necessary to consider the total energy use, and not to focus on space and/or water heating alone.**

It is, for example, important to consider both heating and cooling, as several countries found that focusing on one season only could lead to problems during the other season. Also, reducing cooling loads was often a greater challenge than reducing heating loads.

The largest reduction in energy consumption for space heating in the Task 13 buildings is achieved by the use of traditional energy conservation technologies. A further reduction in the energy consumption for space heating, and a reduction in the consumption for water heating, lights, and appliances, required the use of solar technologies. As the objective of the Task was to reduce the total consumption, all of the Task 13 buildings therefore use both energy conservation and solar technologies.

- **It is necessary to consider the building as a system, where the different technologies used are integral parts of the whole.**

The order in which the technologies are introduced into the design appears to be quite important. Generally, energy-conservation technologies are considered first, passive solar second, and active solar third. In most cases all of these technologies are used, often in combined systems. The emphasis in Task 13 has therefore been on developing whole-building concepts rather than on developing specific technologies.

- **Energy conservation, using high levels of insulation and super-windows, should be the first option considered.**

High levels of insulation are beneficial in all climates, including those where cooling is the major issue. Super-windows, i.e. windows with multiple layers, low-E coatings, and gas fillings, are also always beneficial. They proved to be a better option than windows with transparent insulation. Such super-windows render orientation less critical, allowing the use of larger glazing areas on non-south orientations.

- **Mechanical ventilation systems appear to be essential in low-energy buildings, but their use should be challenged.**

All the Task 13 buildings use some form of mechanical ventilation, and many also use heat recovery on ventilation air. The need for such systems should be challenged, however, both as they use electricity and as they add complexity. Parasitic power consumption was found to be unreasonably high in many of the buildings.

The problems of mechanical ventilation systems should be, and were, addressed during the design phase. It is clear that there is a need for further development, especially in the area of low-energy fans and low-pressure heat exchangers.

- **Passive solar gains can make a major contribution to space heating in all climates and do not lead to overheating if proper solar protection is used.**

Passive solar cooling also proved to work. In both the heating and the cooling situations it was necessary to include thermal mass in the direct gain passive solar designs, as that extended the usability of the systems by increasing the time constant and slowing down heat build-up in the summer.

Phase-change storage materials did not function properly in the cases where the technology was tried, however. PCMs were tested in the Japanese house and simulated for the Netherlands' house. In both cases the results were quite negative, indicating that this particular technology is not yet mature.

- **Passive solar systems can offer several benefits and give solar houses a market advantage.**

Sunspaces and daylighting systems add amenity value to the buildings and contribute to their energy performances when properly designed. Sunspaces are especially attractive in dwellings, where they can reduce space heating when used to preheat ventilation air. The amenity benefits have clearly helped market these houses.

- **Solar DHW is an effective way to reduce to water heating requirements.**

After conservation, solar heating of domestic hot water was found to be one of the most effective technologies. It is therefore used in many of the buildings. In the Canadian Waterloo house, it proved to be the most cost-effective way of further reducing consumption. This is because the water-heating load is relatively constant, making it possible to take advantage also of the high solar gains in summer.

- **Active solar space heating is technically feasible but not cost-effective.**

With increasing levels of insulation, the heating season becomes shorter, and heat demand is concentrated in mid-winter, when there is little solar gain. Seasonal storage is therefore a logical solution. The storage can be reduced as the insulation levels increase, but the investment in storage components may not be cost-effective with very small heating loads.

- **Photovoltaics are not, presently, cost-effective for general use, but PV systems that operate other solar equipment may make sense.**

A few of the Task 13 houses use grid-connected photovoltaic systems that supply general power. None of these are cost-effective. Cost-effectiveness may be achieved, however, in cases where the system is used to operate solar equipment, such as shading devices or pumps for solar thermal collectors, or where the cost of connecting to the grid is high, such as may be the case for outdoor lighting.

The Canadian house in Waterloo and the Danish row houses in Kolding have a PV-powered pump for the solar water collector that is more reliable and costs less than an ordinary controller and pump. The use of PV as a heat source for preheating ventilation air, as in the Japanese house, is, on the other hand, technically feasible but not cost-effective.

- **Designing new, innovative building concepts requires a multi-disciplinary design team.**

The extensive use of solar technologies, which often are integral parts of the design, makes the design process somewhat different from the traditional one. It requires the energy aspects to be considered at the early design stage, and also that the architects and engineers work together from the start. The fruitfulness of the Task 13 workshops, where architects, engineers, physicists, and materials scientists worked together to develop schematic designs, shows this quite clearly.

- **There is a lack of advanced calculation tools for integrated design.**

Integrated designs require the use of tools that can evaluate total building concepts with a number of different energy conservation and solar technologies. As such tools were not available when the Task started, the Task participants had to work with several tools in combination, thereby making it harder to evaluate the buildings as total concepts. Most of the models that were available were not user-friendly.

- **Simulation can be reasonably accurate and give a good indication of how the building will perform, before it is built.**

All Task 13 design teams used hourly simulation programs to guide design decisions. Such hourly simulation provided an insight into the building performance not otherwise available using more conventional calculation tools. The simulation of building and system performance was also useful for designing the monitoring programs used in evaluating the performance in practice.

The monitored results available show that the actual energy consumption in almost all cases only slightly exceeds predictions. This is partly a result of the fact that the users do not behave quite as expected. Typically they do not optimise their behaviour from an energy point of view. Also, the builders do not always build as airtight or as exactly as prescribed. The monitored results are therefore somewhat poorer than what is predicted in the idealised situation created on the computer.

- **Training of builders and on-site supervision is particularly important in low-energy buildings.**

In very low-energy buildings, the energy consumption is more strongly influenced by construction practices and by user behaviour than in conventional buildings. For instance, air tightness and the avoidance of thermal bridges is much more important in a well-insulated building than in a traditional building, and the tightness of the ductwork is more critical as these buildings have more equipment.

- **In many countries, the acceptance of new technologies by the trades is difficult.**

More innovative features were simulated and tested in the laboratory than were actually built.

Some of the more innovative ideas developed did show an extremely high energy-saving potential. However, insufficient funding for the extra cost of prototype solutions, as well as liability problems for the builders, limited the number of new systems actually used in the buildings. Many of the very innovative ideas in the original designs had to be replaced by today's more suitable and commercially available products.

- **Laboratory testing of new and innovative components is necessary and useful.**

A number of the prototypes designed required testing in the laboratory first. This turned out to be a very useful activity, helping the design teams to gain confidence in their solutions and to refine further some of the components and systems. A few solutions were abandoned because of negative test results. In some cases, a laboratory in one country was also used for testing the components developed by another country, thereby illustrating the usefulness of international co-operation.

- **The Task 13 buildings provided motivation to experiment with new technologies.**

The Task created a forum for a very fruitful exchange of ideas. The experiences, the contexts, and the climates of the participating countries differ. Therefore, the participants all had something to learn and something to contribute to the development of each of the experimental buildings presented.

- **There is a market for low-energy buildings, at least in some of the countries.**

In many countries, the Task 13 buildings have provided good examples and have resulted in a noticeable demand for similar types of buildings. In some cases, the driving force appears to be energy savings, while in Canada the driving force appears to be environmental impact and amenity value. Obviously, buildings that can demonstrate low energy use, low environmental impact, and high amenity value, as many of the Task 13 buildings do, should therefore be attractive options in the future market in any country.

5

Monitoring results

Bart Poel (editor), Kirsten Engelund Thomsen, Jørgen M. Schultz

5.1 Introduction

This chapter considers the monitoring results. The underlying information was collected in the form of questionnaires. These were formulated in such a way that participants were provided with a uniform layout for filling in their results. Thus, it was possible to compare the measured, calculated or predicted performances for the different houses. The questionnaires were compiled in the working document 'Monitoring Results of IEA SHC Task 13 projects'.

Table 5.1.1 lists the Task 13 projects by type of project and whether monitoring results were submitted.

The monitoring results are based on the returned questionnaires. No results were available regarding the US Yosemite house. The Canadian Brampton house and the US Grand Canyon house were dealt with in a written report and not by the questionnaires. Only some results are included from these projects and only if the information fitted the uniform description.

No monitoring results have come from Italy because no project was realised. Only some short-term experiences were available from the Japanese house. No evaluation results were received regarding the Swiss house, but the house was dealt with in IEA SHC Task 28: Solar Sustainable Housing. No monitoring results were available from Sweden; however, they used their experiences with the test cell in a three-storey family house. Norway did not provide reliable monitoring results as several developments were beyond the control of the monitoring team. Norway, however, did contribute several of the lessons learned.

The climate and locations for the different countries are listed in Table 5.1.2.

Chapter 5.2 describes the evaluation approaches used, while the energy performance of buildings and components are considered in chapters 5.3 and 5.4, respectively. The evaluation results concerning thermal comfort are presented in chapter 5.5, and user aspects and conclusions are presented in chapters 5.6 and 5.7, respectively. Finally, literature and references are listed in chapter 5.8.

The Working Group would like to thank all participants for their co-operation in collecting the monitoring results.

It should be kept in mind that Task 13 focuses on the future and expands on unconventional solutions. Some of the components referred to are prototypes that are not yet commercially available. In that sense,

Table 5.1.1. List of IEA SHC Task 13 projects

Country	Project	Type of project			Monitoring results submitted
		Building	Test cell	Design	
USA	Grand Canyon house	x			(x)
USA	Yosemite at El Portal	x			
Belgium	Pleiade	x			x
Canada	Advanced House in Brampton	x			(x)
Canada	Green Home in Waterloo	x			x
Denmark	Low-energy row houses, A and B	x			x
Finland	IEA5 Solar Houses in Pietersaari	x			x
Germany	Zero Heating Energy House, Berlin	x			x
Germany	Ultra House in Rottweil	x			x
Italy	Prototype apartment building			x	
Japan	WISH House	x			x
Netherlands	Urban Villa called Licht en Groen	x			x
Norway	Row houses at Hamar	x			x
Sweden	Low-cost house in Röskär		x		
Switzerland	Swiss duplex in Gelterkinden	x			

Table 5.1.2. Climate and location of different countries that completed questionnaires

Location	Climate
Belgium, Louvain-la-Neuve	Maritime, 51° N latitude
Canada, Waterloo, Ontario (80 km west of Toronto)	Cold continental, 43.5° N latitude
Denmark, Vonsild (a few km south of Kolding, Jutland)	Maritime, 55° N latitude
Finland, Pietarsaari	Cold, 62° N latitude
Germany, Berlin	Temperate, 52° N latitude
Germany, Rottweil (Southern Germany)	Cool, 48° N latitude
Japan, Iwaki (180 km north of Tokyo)	Warm, 37° N latitude
Netherlands, Amstelveen (10 km south of Amsterdam)	Maritime, 52° N latitude
Norway, Hamar (120 km north of Oslo)	Cool, moderate continental, 61° N latitude

the economic constraints should only be considered important insofar as the energy approach could potentially be feasible in the future.

The evaluation results should, therefore, be judged from this point of view. Task 13 offers a rather experimental approach and should not be regarded as a demonstration of proven technology. Problems encountered in the projects should, therefore, not be looked upon as failures but as important lessons for future developments.

5.2 Evaluation approach

INTRODUCTION

In order to verify their expected performance, most Task 13 houses were monitored for varying periods after they had been built. The evaluation approach varied from project to project with respect to the level of detail and items focused on.

The main objective for all the Task 13 projects was to design and build houses displaying considerably lower total energy consumption than ordinary houses and, in most cases, making extensive use of passive and active solar energy for heating domestic hot water, space heating and, in a few cases, for generating electricity.

Thermal comfort, especially with respect to avoiding overheating during the summer, was an important objective in all projects.

Special attention was also paid to electricity consumption, especially with respect to the 'parasitic energy consumption' resulting from the increased use of electric equipment required for running energy saving installations (e.g. mechanical ventilation systems with heat recovery, control systems, pumps, etc.).

A major objective for one of the Canadian houses was to minimize the overall environmental impact of the building through a careful choice of materials and extensive use of recycled materials.

MONITORING PROGRAMMES

Monitoring programmes were drawn up for evaluating the overall performance of the houses as well as for more detailed investigations of equipment and systems. Three types of measurements were generally presented, i.e. continuous measurement of overall performance and component tests in the unoccupied house, overall performance during occupation and spot measurements before and during occupation. In most cases spot measurements of air tightness, adjustment of airflow, etc. were made before occupation. Longer monitoring periods were available before occupation only in Belgium.

The minimum set of values measured were:

- solar radiation
- wind speed (and direction)
- outdoor temperature
- indoor temperatures
- purchased energy (space heating, domestic hot water, and lighting and appliances)
- active solar energy (electricity and heating).

Computerized data acquisition had been set up in all houses for continuous measurements. Manual readings of integrating systems such as kWh meters, gas meters and water meters were used to inspect the values acquired by the computer systems.

In most cases, several additional measurements were made to evaluate the performance of systems and components, as well as indoor air quality and thermal comfort.

ENERGY CONSUMPTION FOR SPACE HEATING

In general, energy consumption for space heating was measured as the energy supplied to the heating system. The energy losses of district heating systems (outside the dwelling) were not taken into account, but losses from boilers and storage tanks were included.

Measured and calculated space heating demands were compared on the basis of the registered climatic and indoor conditions, as well as the actual use of the house (if registered). In the German houses, a detailed registration of opening and closing of windows formed a well-defined basis for estimating the air changes by natural ventilation.

THERMAL COMFORT

Keeping a comfortable indoor temperature during the winter does not generally present a problem and the evaluation of the indoor thermal comfort focuses on overheating problems. As an indicator for summer comfort, the general approach is to compare indoor air temperatures in the living areas with outdoor temperatures. The objective is to achieve indoor air temperatures that are lower than the outdoor temperature during warm periods. It should be noted that widespread use was made of passive cooling techniques. However, both Canadian houses offered the possibility of mechanical cooling; the Waterloo house by means of water circulated in pipes buried below the basement floor slab and the Brampton house by water cooled by the evaporator of a heat pump for domestic hot water and space heating. The Japanese house, too, was equipped with an active cooling system.

ELECTRICITY CONSUMPTION

Electricity consumption for household, lighting and technical installations was monitored in all houses. In general, electricity consumption was divided into three groups: appliances, lighting and wall sockets, and technical equipment for running the energy systems in the house (also called parasitic energy consumption – see chapter 4). In one of the Canadian houses, however, the plug loads and outdoors loads were measured individually, leading to a more detailed analysis of consumption.

In the PLEIADE house, the electricity consumption was monitored in detail with individual measurements of lighting, wall sockets, control, ventilation, stand-by, refrigerator, stove, washer, dryer and dishwasher.

DOMESTIC HOT WATER

The energy required for domestic hot water is significant compared with the heat losses in highly insulated buildings. This was the object of energy saving measures through the application of active solar systems and/or water saving devices. The energy consumption for domestic hot water was monitored in all the projects except in the US Grand Canyon house. Where water was heated in a separate tank, the heat loss from the tank was included; in other projects, energy meters were used to measure the energy used for heating the water.

AIR QUALITY

Air quality is, in general, considered satisfactory when the air change rate meets the design values.

Only Belgium and Canada evaluated indoor air quality by measuring specific chemical compounds. The CO_2 concentration was monitored continuously in the Belgian house and in both the Canadian houses. In Canada, the indoor air quality was investigated further by carrying out single measurements of formaldehyde, radon, volatile organic compounds (VOCs) and particles both before and after occupation. The concentrations of carbon monoxide (CO), nitrogen dioxide (NO_2) and sulphur dioxide (SO_2), as well as the relative humidity and the electromagnetic radiation, were measured after the house had been occupied.

SYSTEMS AND COMPONENTS

The values reported at equipment and system levels give a good idea of the main interests and/or uncertainties for the various projects.

The air tightness of the building plays an important role with regard to infiltration losses and the efficiency of the mechanical ventilation system with heat recovery. Consequently, pressurization tests were conducted in most houses as well as tests of the thermal efficiency of the heat recovery units. Belgium and the Netherlands also measured the airflow rates in the balanced ventilation systems. The air tightness of the ventilation ducts was also tested in the Netherlands.

Solar collectors were generally subject to special investigation.

Belgium and the Netherlands monitored the overall efficiency of their gas boilers for space heating and domestic hot water. In Canada, a specially designed integrated heating and ventilation system was used; its overall performance was monitored. In Finland, special measurements were conducted on the heat pump for space heating and domestic hot water, as well as on the PV system.

At a construction level, Belgium, Finland and Germany measured the actual U-values of the windows; the vacant monitoring period in the Belgium house offered an excellent opportunity to measure the U-value of all the constructions (walls, floor, roof, etc.). In the US Grand Canyon house, a short-term test was performed that resulted in the determination of the overall heat loss coefficient, the thermal capacity and, in case of favourable climatic conditions, the effective solar gain coefficient.

USER ASPECTS

The user aspects were dealt with very differently, ranging from no investigation at all to a very close interaction between the occupants and the project group. Canada, Finland and the Netherlands offered the new tenants a thorough introduction to the house and how it should be used. In the Netherlands, this was followed up by questionnaires and interviews during the monitoring period. Planned questionnaires/interviews were also provided at the end of the monitoring period in Belgium, Finland and the USA.

However, this does not mean that there was no contact between occupants and project groups in the other projects. Both Canada and Denmark reported the comments collected during the project period.

5.3 Energy performance of the buildings

INTRODUCTION

The energy performance of the Task 13 houses has been evaluated and the results are considered very positive. In some cases, the performance was not in line with expectations but, even then, the project is a great source of information. The measured energy consumption for space heating, domestic hot water, cooling and domestic electricity are compared below with the simulation results.

This comparison is followed by an overview of the experiences with the heating system, the building design, ventilation, internal and solar gains, and the indoor air temperature.

SPACE HEATING

During the design process, the energy consumption for space heating was predicted by simulation under standard conditions. In many cases, these original simulations did not match the actual building realised nor did they reflect the occupant behaviour and the outdoor conditions during the measuring period. A list of the simulation tools used is given in Appendix A, Table A.1. As part of the analyses during the evaluation, the simulations were adapted to the actual situation in many projects. Figure 5.3.1 shows the results obtained from the questionnaires (CDN = Canada, FIN = Finland, J = Japan, N = Norway, NL = Netherlands). For each project, the available results given for space heating are derived from the original simulations, the adjusted simulations and the measurements. In the absence of adjusted simulations, the values for the adjusted simulations are considered the same as the original simulation results.

For the two German projects, the measured values of two heating seasons were averaged. The contribution of active solar gains for the Berlin project was subtracted from the space heating demand to make it comparable with the values from the other projects. The Japanese (J) and Norwegian (N) projects did not provide monitoring results for space heating.

The key issues relating to space heating for each of the projects are explained below.

At the **Belgian Pleiade House** (B), the measured energy demand for space heating during the first winter was substantially higher than predicted. There are several reasons for this difference: the actual level of internal gains was lower than the level assumed in the calculations, the mechanical ventilation system was not used as it was supposed to be, and there were some minor differences in the construction of the building. The adjusted simulation results, which take the malfunctioning heat recovery unit into account, were much higher – namely 27 kWh/m^2 per year. Simulations of the actual conditions gave a net energy demand of 36 kWh/m^2 per year, which corresponded to the measurements. During the 1997/98 winter, the measured energy demand was 24.3 kWh/m^2. The decrease is due to improvements to the heating system, fewer occupants and a milder climate.

In the **Canadian Green Home** (CDN-W), the difference between the predicted and actual energy consumption is due to higher than expected use of natural gas for space and water heating. The primary

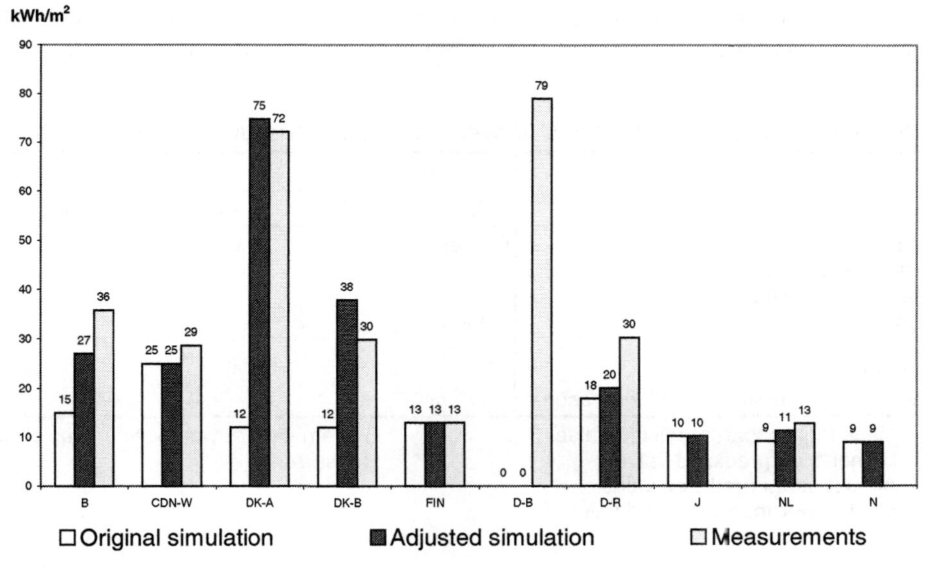

Figure 5.3.1 Energy use for space heating of the Task 13 house

reason appears to be the integrated heating ventilating appliance (IHVA), which displayed a lower efficiency than expected.

The **Danish project** has two different houses: house A (DK-A), which has North–South façades, and house B (DK-B) with East–West façades. The measurements from house A differed considerably from those of house B. Energy consumption, too, was much higher than predicted. This is attributed to the following:

- The indoor temperature measured was higher than predicted.
- The heat recovery system was turned off during the measuring period.
- The weather data used in the calculation differed from those in the measuring period.

Afterwards, the calculations were adjusted to the actual conditions. This led to smaller differences from the measured results.

The **Finnish project** (FIN) shows a perfect match between the calculations and the measurements.

For the **German projects**, the measured values for two heating periods were averaged. It appears that the space heating for the Zero Energy House (D-B) was very high. This was due to higher transmission and ventilation losses. Moreover, the heating system turned out to be inefficient, as 40% of the heating energy supplied was lost in both heating periods. The Ultra House (D-R) also displayed higher energy consumption. This is mainly due to higher ventilation losses.

The original predicted energy consumption for space heating for the **Dutch apartments** of Urban Villa (NL) was 9 kWh/m^2 per year. However, the efficiencies and the influence of thermal bridges had not been taken into account. After adjustments to the simulation, the energy consumption appeared to be 11 kWh/m^2 per year. The measured energy consumption over the first year was 23–68 kWh/m^2 per year, with an average of 37 kWh/m^2 per year. The differences can be explained as follows:

- Around 7 kWh/m^2 per year was used to dry the building materials.
- The heating season was colder than usual.
- Most of the occupants set the thermostat somewhat higher than expected.

The measured results were corrected for these effects in order to make a proper comparison with the simulations. Annual space heating consumption was then 13 kWh/m^2 per year. This is just slightly higher than the simulations and is given as the measured result in Figure 5.3.1.

Overview: space heating

It is of interest to compare the results of the simulations with the actual measured data. This can only be done for the projects where all these results are available. This means that the Japanese and Norwegian projects cannot be taken into account due to the lack of measured data. Also excluded are the Danish A (DK-A) project and the German Berlin (D-B) project. This is because the functioning of these systems during the measurements was not in accordance with the concept. In the Danish project, the heat recovery unit was turned off by the occupants. In the Berlin house, the heat recovery system did not function and the project suffered from the dwelling's substantially higher heat losses and inefficient heating system.

The average results of the remaining six projects from Figure 5.3.1 are:

Original simulation	14 kWh/m^2
Adjusted simulation	22 kWh/m^2
Measurements	25 kWh/m^2

This shows that the original simulation results are on average 39% lower than the measured space heating consumption, whereas the adjusted simulations deviate about 12% from the measured values. Once the experimental character of the projects and the use of prototype components are taken into account, these results are satisfactory and provide considerable insight into the possibilities of solar low energy concepts.

Table 5.3.1 summarizes the most important causes of deviations between predicted energy use and energy use measured for space heating.

Table 5.3.1. Causes of deviations between predicted energy use and measured energy use for space heating

	Project						
	B	CND-W	DK	FIN	D-B	D-R	NL
Actual climate	Colder		Colder				Colder
System efficiency		--			-	-	
Heat recovery efficiency	-*		--**		--*	-*	
Infiltration	++				++	+	
Internal gains	--				+	+	
Solar gains					-		
Indoor air temperature (heating season)			++		++		+

++ Much higher than predicted value
\+ Higher than predicted value
Almost equal predicted value
-- Much lower than predicted value

* System did not work well or was shut off
** House A

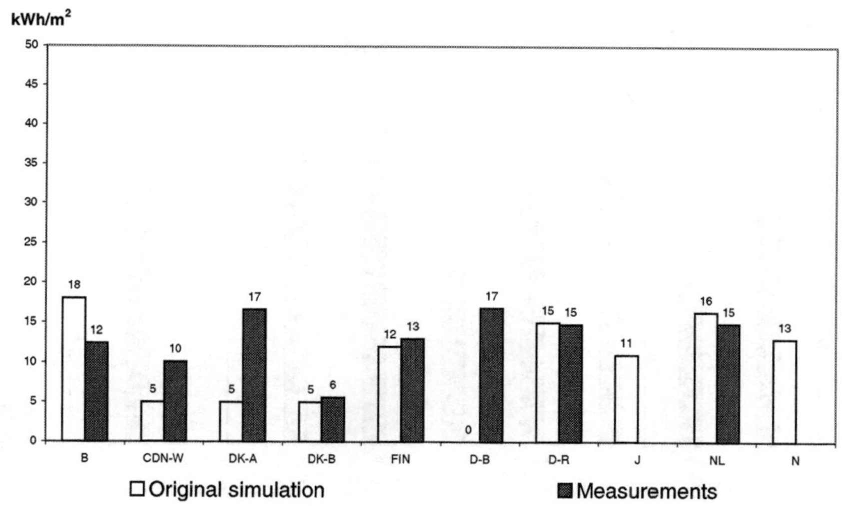

Figure 5.3.2. Energy use for domestic hot water in the Task 13 houses

COOLING

Of the monitored projects, active cooling was only practised in the Canadian and the Japanese houses. The (predicted) energy use for cooling was very low: 0.9 kWh/m^2 per year for the Canadian Waterloo house and 5.2 kWh/m^2 per year for the Japanese WISH house. There were no measured results.

DOMESTIC HOT WATER

The energy use predicted for domestic hot water (DHW) by the original simulations is compared with the measured values in Figure 5.3.2. No measured results were available for the Japanese and Norwegian projects.

The predictions of energy use for DHW came close to the measured results for the Danish house B, the Finnish project, the German Ultra House and the Dutch Urban Villa. In other cases, the difference between the predicted and measured energy use was large, e.g. in the Berlin Zero Energy House. In this case, the heat losses from the storage tank and the distribution system were high. Constant circulation water caused losses that accounted for 30% of the energy demand for DHW. Moreover, in the German Ultra House, approximately 45% of the energy for DHW was lost through storage tank thermal transmissions and circulation. But, in this case, the consumption of DHW was much lower than predicted.

The DHW backup system of the Finnish so-called IEA5 Solar House was poorly designed. The heat pump, however, used less energy for DHW than expected.

In the Danish row house A, there was a considerable difference between simulated and measured energy use for DHW; the solar gains were only 86 kWh compared with the expected value of 1000 kWh due to a valve closed because of a failure. House B, on the contrary, shows a small difference.

Like space heating, the values for both German projects were averaged for two seasons. Part of the solar gains for the Berlin project was subtracted from the energy demand for DHW.

OVERVIEW: DOMESTIC HOT WATER

On average, the predicted and measured values match very well; average values are 11 kWh/m^2 calculated and 12 kWh/m^2 measured.

As in the case of space heating, the Japanese (J) and the Norwegian (N) projects are not taken into account because there are no measured results. The Danish A (DK-A) project and the German Berlin (D-B) project are disregarded because that the system did not operate to reflect the concept envisaged.

The most important reasons for the discrepancy between the predicted and measured performance for domestic hot water are summarized in Table 5.3.2.

Table 5.3.2. Reasons for the discrepancy between energy use predicted and measured for domestic hot water

	Project						
	B*	CND-W	DK	FIN	D-B	D-R	NL
Consumption						--	
System heat losses					--	--	
System efficiency		--		-			
Solar heating			--***		-		-**

++ Much higher than predicted value
+ Higher than predicted value
Almost equal with predicted value
- Lower than predicted value
-- Much lower than predicted value

* No explanation
** System did not work well or was shut off
*** House A

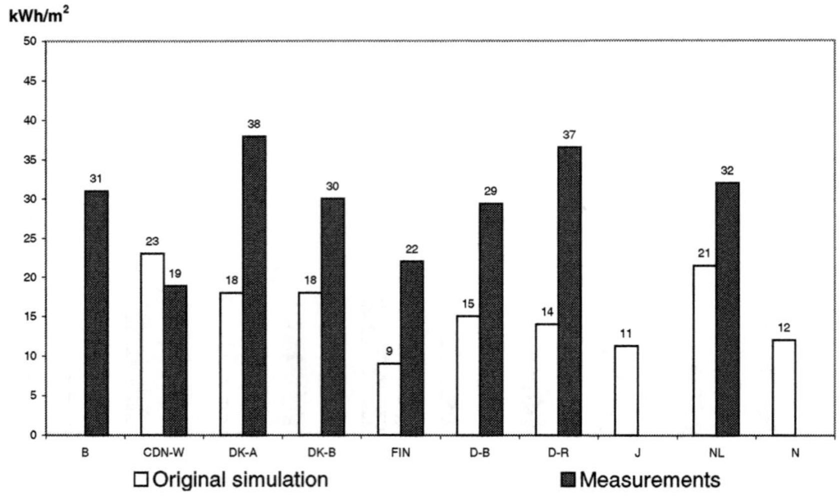

Figure 5.3.3. Total electricity demand (other than space heating or DHW) in the Task 13 houses

Table 5.3.3. Causes for deviations between predicted and measured electricity use

Higher/lower energy use	Project						
	B**	CND-W	DK	FIN	D-B	D-R	NL
Lights and appliances		+	+		+		
Pumps, fans, control systems		-		+		+	+

\+ Higher electricity use - Lower electricity use ** No explanation

ELECTRICITY

The electricity use is the total demand for light and appliances, fans and pumps. Figure 5.3.3 shows the results of the simulations and measurements. It should be noted that PV production has been subtracted from the electricity consumption.

In general, we can see that the actual demand is higher than the predicted demand. In some cases, the difference is large (Denmark house A, Finland, German Ultra House). The predicted value of the Belgian project was not based on simulations, but was a 'first guess' and not, therefore, taken into account in Figure 5.3.3.

Although the total electrical energy use of the Canadian Green Home came close to the predicted values, there appear to be significant variations within specific categories. Specifically, the monitored fan energy use was only 50% of the predicted value. The energy requirements of the various electrical devices, even those that are considered to be off but are actually in a stand-by mode (e.g., VCR digital displays, instant-on TVs) add up to a significant electrical load. The sump pump, which was installed as a rarely-to-be-used safety feature, consumed several hundred kilowatt hours annually.

Electricity demand for pumps, fans and control systems in the Urban Villa was expected to be high. The measured auxiliary electricity consumption was even higher; this is attributed to higher capacities and longer operation periods.

OVERVIEW: ELECTRICITY CONSUMPTION

On average, the measured electricity consumption for fans, pumps, control systems, lights and household appliances exceeded the predicted values by 30% (Belgium, Japan and Norway excluded). The electricity consumption is largely determined by the occupants (lights and appliances) and only a smaller portion is related to the building and its systems, and thus related to the energy concept. The reasons for deviations between predicted and measured electricity consumption are summarized in Table 5.3.3.

TOTAL ANNUAL ENERGY USE

Total energy use covers space heating, DHW, cooling and total electricity use. PV production and solar gains had already been subtracted from energy consumption.

Those projects for which both simulation results and measured results are available for the total energy consumption are shown in Figure 5.3.4. Simulated and measured values are available for total energy consumption for seven projects. The Belgian project provides no simulation results for electricity use, and the Norwegian and Japanese projects provide no measured results. For both the Danish-A and the German Berlin projects, the building and its operation differed so much from the original concept that a straightforward comparison between the simulated and measured values is not relevant.

A comparison based on the remaining five projects (CDN-W, DK-B, FIN, D-R and NL) show the following average total energy consumption:

Original simulations	44 kWh/m² heated floor area
Measurements	69 kWh/m² heated floor area.

This means an underestimation of 35% and, given the experimental character of Task 13, is considered a fairly good result.

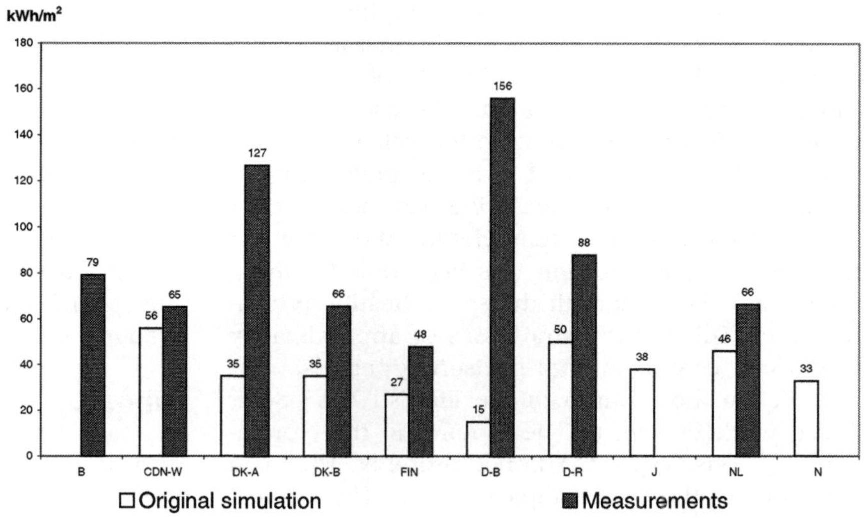

Figure 5.3.4. Total annual energy consumption per m² heated floor area for the Task 13 houses

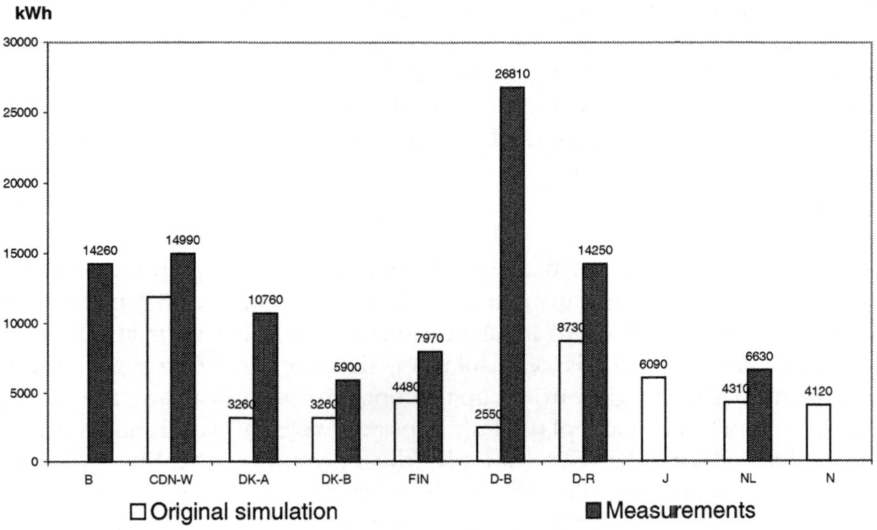

Figure 5.3.5. Total annual energy consumption per dwelling in kWh for the Task 13 houses

Figure 5.3.5 shows the total energy consumption per dwelling in kWh rather than per m² heated floor area.

SPECIAL ISSUES CONCERNING SPACE HEATING

This section summarizes the experiences relating to important issues affecting the energy demand for space heating.

Heating system

The heating systems of the Belgian, Danish, Finnish and Dutch projects worked well. Other cases call for specific comments.

For example, the gas furnace of the Canadian Green Home displayed a predicted energy-efficiency of 85% for the IHVA but, in practice, the average monitored combined efficiency was only 65%. The 20% difference in efficiency corresponds to a 30% increase in energy consumption.

The heating of the Zero Energy House in Berlin was provided by a gas boiler and hot water from the storage tank of the solar system (see Figure 5.3.6). The system as a whole proved to be inefficient because, in both heating periods, 40% of the heating energy supplied was lost.

Figure 5.3.6. Storage tank in German Zero Energy House

A local generator power plant supplied heating energy to the German Ultra House in the form of hot water. In the heating periods, 30–37% of the energy purchased was lost before it could be used owing to poor insulation of the piping in the cellar. The heat losses to the cellar resulted in an average temperature of 25 °C during the measuring period. For better control of the indoor air temperature, the hot water intended for space heating was first cooled using a plate heat exchanger with the space heating returning water. This led to extra losses of approximately 4 kWh/m^2 per year in both measuring periods.

While the heat pump in the Finnish IEA5 Solar House worked well, the heat pump in the Norwegian row house never functioned properly. The main problem was the very complex system. The original ground source system was a coiled pipe buried in sand. When tested, it was found that the coil had been damaged, probably by heavy traffic. The coil was then replaced with a shorter ground loop, closer to the surface, with less thermal storage capacity.

Building design

The building envelope in the Task 13 buildings displays a very high level of insulation. This implies that attention should be paid to insulation-related issues.

Due to the high levels of insulation, the energy transport to the neighbouring apartments through dividing floors and walls played an important role in the Dutch Urban Villa. For example, the consumption of the apartment with the highest energy use was 13 kWh/m^2 per year higher due to heat loss to the apartment above. Calculations show that, for a centrally placed apartment that is 1°C warmer than its surrounding apartments, the annual energy consumption can be as much as 22 kWh/m^2 per year higher. This effect has to be taken into account, especially in low-energy houses. Insulating the dividing walls was considered, but this was not carried out because of the relatively high costs involved. Furthermore, the energy consumption for space heating for the whole building is not affected by insulation of the partitions.

Thermal bridges were an important issue in many projects. Disregarding them could easily have doubled the transmission losses through the envelope. The special attention paid to thermal bridging leads to sound solutions.

Infiltration

Most houses performed well with regard to air tightness. An airtight design of the joints and quality control during construction are crucial for good performance. In the Belgian Pleiade house, however, the air change rate was 2 per hour measured at an air pressure difference between the inside and outside of 50 Pa. The target was an air change rate of 1 per hour.

Moreover, in the German Zero Heating Energy House, the infiltration losses were much higher than planned. Measurements showed an air change rate of 3.3 per hour at 50 Pa, whereas the aim was 1 per hour at 50 Pa. The main reason for the high ventilation rate is the presence of the ventilation openings above the storage tank, which could not be closed airtight. The openings were meant to prevent overheating in the house during the summer. Moreover, the air barrier in the construction component under the solar collectors appeared to be defective and could only be repaired at enormous cost because the collectors would have to be removed.

In the German Ultra House (see Figure 5.3.7), the ventilation losses are significantly higher due to the high rate of natural ventilation from opening win-

Figure 5.3.7. Interior of the German Ultra House

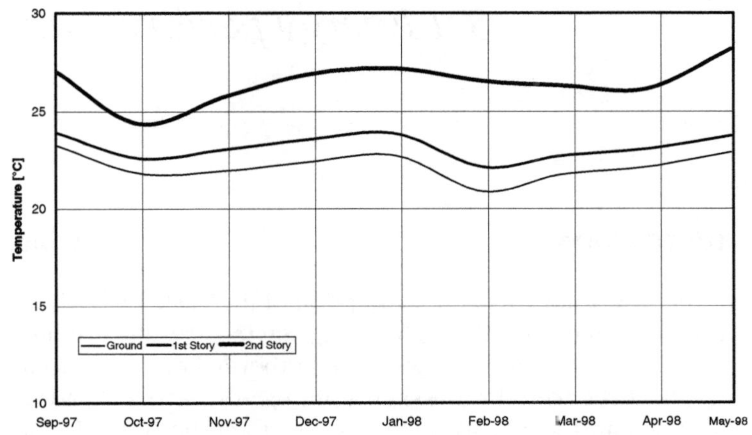

Figure 5.3.8. Average monthly temperatures in the German zero energy house

dows and the inefficient use of the heat recovery system. This meant an average air change of 1.5 per hour. The house itself was fairly airtight, displaying an air change of 1.6 per hour at 50 Pa.

Ventilation

The heat recovery system worked well in the Dutch project. There were no comments on the Finnish heat recovery system. Malfunctioning or low efficiency occurred at the Danish low-energy row house, the German Zero Heating Energy House and the German Ultra House. The ventilation system of the Belgian project was wrongly dimensioned and showed air leaks.

The heat recovery system of the German Zero Heating Energy House was not used due to noise problems and the high infiltration rate.

The heat recovery system of the German Ultra House also exhibited noise problems. Therefore, the heat recovery system was not used continuously as planned. When in operation, there were only minimal air volume flows (90 m^3/hour) through the unit and the windows were opened frequently. For all these reasons, the unit's efficiency was reduced considerably.

Internal gains

Energy use for space heating in low energy houses can be very sensitive to changes in the internal gains. In the Belgian project, for example, the calculations were based on the reference value according to Walloon regulations of 5.42 W/m^2. In practice, however, the internal gains in this building were around 2.2 W/m^2, which had a major influence on the space heating required.

On the contrary, the internal gains of the German projects proved to be higher than predicted.

Solar gains

In several cases, the active and/or passive solar gains are in line with expectations. At the German Ultra House, the passive solar gains were approximately the same as predicted, despite the fact that the measured solar radiation was higher than in the test reference year. In the Finnish IEA5 Solar House, the PV energy was only a little lower than expected. The solar collectors furnished approximately the amount of energy predicted.

The solar gains of the German Zero Heating Energy House were lower than predicted, mainly because trees along the southern boundary of the section blocked out approximately 20% of the solar radiation. It was originally planned to build the house 6 metres further away from the trees.

In some projects, there were some problems with active solar systems. In the Danish house A, the active solar heating system furnished much less energy than assumed. For the PV system of the Norwegian row house, problems arose at the A.C. inverter system and control equipment, which kept the PV system out of operation for long periods. In the Dutch Urban Villa, the solar domestic hot water system did not work properly during monitoring due to a poor design. Once this problem was solved, the contribution of solar gains seems to be as expected. Newer exact measured results are not available. To allow the values to be compared, an adjusted calculation based on the measured consumption of DHW was carried out.

Indoor air temperature

The indoor air temperature during the heating season was higher than assumed in the Danish houses, the German Zero Energy House and the Dutch Urban Villa. This was mainly due to occupant behaviour, which caused higher transmission and ventilation losses. In the German Ultra House, the indoor air temperature was lower than expected. This led to lower transmission losses.

Figure 5.3.8 gives the average monthly temperatures for the German Zero Energy House over the heating period 1997/98. The thermostat is situated on the ground floor and there is an open connection with the top floor. This results in a constant heat demand, the warm air from the ground floor rising to the top floor.

5.4 Energy performance of components

INTRODUCTION

The very low total energy consumption of the Task 13 houses was achieved through energy savings resulting from high insulation levels, airtight thermal envelopes, heat recovery from exhaust air, use of passive solar gains and active solar energy for domestic hot water. Experimental buildings permit the evaluation of the actual performance of various systems and components. The results can be used to improve the systems and components and/or to obtain better input values for the simulation tools used to predict the energy consumption and indoor thermal comfort.

Only values obtained from the questionnaires are shown in the following section. However, more information can probably be obtained from the individual participants.

THERMAL ENVELOPE

Construction

The performance of the thermal envelope can be characterized by its U-value and air tightness. Belgium and the Netherlands are the only participants that measured the U-value of building construction. Table 5.4.1 shows that the measured results are almost identical with the calculated values.

In the USA, a short-term energy performance test was performed from which the overall heat loss coefficient can be derived. The test used a theoretical model of the building to calculate the thermal behaviour under well-defined conditions. The calculated values are compared with the values measured and the main parameters in the model are adjusted by a least square method until the best fit is reached. The results are presented in Table 5.4.2. The floor and Trombe wall heat loss coefficients are not included in the results as they are considered to be well-known.

These experiments show a good agreement between the calculated and measured performance of the envelope.

Windows

The windows play an important role in that they account for the largest heat loss coefficients but, on the other hand, offer the possibility of direct solar gains. Windows are complex constructions consisting of frame, glazing and glazing edge sealing, all of which influence the thermal performance of the window. Heat loss coefficients for windows are often judged on a calculated centre U-value of the glazing and the frame U-value. The total U-value of the windows was measured in the Finnish and the German projects. The results are shown in Table 5.4.3.

The results in Table 5.4.3 show that, even if the centre U-value of the glazing in these cases is below 0.5 W/m²K, the total U-value is considerably higher due to the relatively poor insulating frames and the spacer in the glazing edge sealing.

The Netherlands also measured the solar gain factor of the glazing. The value of 0.49 given by the manufacturer agreed with the measured transmittance of 0.49.

Air tightness

Air tightness of the thermal envelope is important to avoid unnecessary ventilation losses. Table 5.4.4 shows the design and measured values of air tightness obtained by pressurization tests. It appears that there are large variations in the air tightness obtained compared with the design values (see the explanation in chapter 5.3.7 under infiltration). This leads to the conclusion that rendering houses airtight is not an easy job and is one that requires carefully designed joints and supervision during construction.

Table 5.4.1. Heat loss coefficients of constructions (windows excluded)

Country	Component	Calculated (W/m²K)	Measured (W/m²K)
Belgium	Floor	0.20	0.21
	North façade	0.12	0.13
	South façade	0.16	0.17
	Roof	0.10	0.11
Netherlands	South façade	0.80	0.90

Table 5.4.2. Total building heat loss coefficient including windows and infiltration

Country	Component	Calculated (W/m²K)	Measured (W/m²K)
USA	Total house (except floor and Trombe wall)	69	79

Table 5.4.3. Heat loss coefficient of windows

Country	Description	Calculated (W/m²K)	Measured (W/m²K)
Finland	Triple, krypton-filled + storm glazing	–	0.7
Germany, Berlin	Triple, xenon-filled	0.68–0.88	1.0–1.1
Germany, Rottweil	Triple, xenon-filled	–	0.75

Table 5.4.4. Air tightness of thermal envelope expressed as air change rate

Country, house	Design (per hour)	Measured (per hour)
Belgium, Pleiade	1.00	2.00
Canada, Waterloo	1.50	0.8–0.95
Canada, Brampton	0.75	–
Finland	1.00	0.80
Germany, Berlin	1.00	3.30
Germany, Rottweil	–	1.6
Netherlands, Licht en Groen	0.38	0.40
USA, Grand Canyon	0.04	0.02

VENTILATION SYSTEMS

This section covers only results that are linked directly to ventilation systems, i.e. reported heat recovery, air distribution and air tightness of the ductwork. In many of the projects, the ventilation systems form part of an integrated system for space heating, domestic hot water and ventilation. In these cases, it is difficult to distinguish the different parts from each other and such systems are described by their combined performance.

Of special concern is the energy required for the operation of the fans. This value was not reported separately, but is generally included in a combined value for energy consumption by pumps, fans and controls.

Air tightness of ductwork

The Belgian and Dutch teams evaluated the air tightness of the ductwork. Leakage in the ducts will reduce heat recovery for the ventilation air.

In the Belgian house, air leakage was evaluated to account for an 8% increase in heating energy consumption. The air tightness of the ventilation system in the Dutch project came up to the requirements envisaged.

Heat recovery efficiency

A crucial point in many strategies for achieving a low-energy demand for space heating is the efficiency of the heat recovery unit. The values supplied by the manufacturer were achieved under certain laboratory conditions and are probably not applicable for practical use. It was planned to measure the heat exchanger efficiency in many of the projects but, in the German and the Danish projects, the ventilation was stopped or misused, and no results are available. The available measured results are given in Table 5.4.5.

The general assumption that heat exchangers show significantly lower efficiencies after a few months of operation cannot be verified by the results shown in Table 5.4.5. The results from the Netherlands show a higher efficiency than expected, but this can be attributed to a somewhat pessimistic forecast of the expected performance. The efficiency measured during the design phase at the Technical University in Delft was in the range of 79% to 88%.

Table 5.4.5. Temperature efficiency of heat recovery units in balanced ventilation systems

Country, house	Expected (%)	Measured (%)
Denmark, House B	80–85	83*
Canada, Brampton	–	74
Finland	80	80
Netherlands	75	76–82

* Only in operation for 3 months

Adjustment of ventilation rates

In all projects, ventilation rates were probably adjusted with mechanical ventilation systems, but detailed results are only available from the Belgian and the Dutch houses (see Table 5.4.6).

The results and comments from Belgium and the Netherlands show that the design values are difficult to reach even if the installer knows that the system will be subject to detailed monitoring. The most important reason seems to be insufficient knowledge by practitioners and lack of quality control.

HEAT GENERATION UNITS

Efficient space heating and heating of domestic hot water not covered by solar systems is one of the main topics for the Task 13 houses. The very low heating demand calls for systems displaying high efficiencies, particularly at small loads.

Different kinds of measurements were performed reflecting the area of interest in the different projects ranging from the specific performance of, say, gas

Table 5.4.6. Airflow rates in the Belgian and Dutch houses

	Room	Design (m^3/h)	After installation and correction (m^3/h)
BELGIUM			
Supply	Bedroom SW	39	42
	Bedroom SE	34	51
	Office	30	33
	Bedroom NW	34	34
	Bedroom NE	34	16
Supply		171	176
Extraction	WC ground floor	25	23
	Kitchen	50	31
	WC 1st floor	25	30
	Bathroom	50	54
	Shower	50	43
Total		200	181
NETHERLANDS			
Supply	Living room S	69	70
	Bedroom S	75	94
	Bedroom N	38	41
	Kitchen N	68	68
Supply		250	273
Extraction	Kitchen N	150	130
	Bathroom	50	60
	WC	25	28
	Storage room	25	25
Total		250	242

Table 5.4.7. Heat pump performance

Country, house	Expected COP	Measured COP
Canada, Brampton	2.5–3.0	1.7–2.1
Finland	3.3	2.4

boilers and heat pumps to the overall performance of the entire system.

Component performance

The Belgian and Dutch project groups reported the efficiency of gas boilers to be 78% and 75%, respectively.

Heat pumps were used in the Canadian Brampton house, the Finnish, Norwegian and US houses. However, results as to the coefficient of performance (COP) are only available from Canada and Finland (see Table 5.4.7). The results clearly show that the actual performance values measured were far less than those expected. In the Canadian system, some of the discrepancies were due to the unexpectedly large energy consumption of the pumps circulating the glycol. Another reason was that, in reality, the heat pumps did not run at optimal temperature levels.

Integrated systems

An integrated system is a system with combined functions such as heating, cooling, ventilation and heating domestic hot water. Detailed analyses of the performance have been made for the two Canadian houses – the Waterloo house using gas and the Brampton house using an electrically driven heat pump.

The system performance factor (SPF) for the heating season was calculated for both systems and is given by the ratio:

$$SPF = \frac{\text{energy supplied for space heating (and DHW)}}{\text{total amount of energy supplied to the system}}$$

For the Waterloo house, the measured SPF is 65%, i.e. well below the expected value of 85%. Heating domestic hot water does not form part of the integrated system. There is no explanation for the discrepancy between the predicted and measured performance.

For the integrated system in the Brampton house, the performance measured across the heating system was 1.54 while the predicted value was as high as 2.0. The values cannot be compared with the results from the Waterloo house due to the different heat sources. The discrepancy can mainly be attributed to a lower than expected COP for the heat pump and to higher electricity consumption by fans and pumps.

Overall system performance

The overall system performance is the ratio of the net energy demand to the amount of energy purchased. This value is available from Germany and the Netherlands (see Table 5.4.8).

Table 5.4.8. Overall heating system efficiency (space heating and domestic hot water)

Country, house	Expected (%)	Measured (%)
Germany, Berlin	–	60
Germany, Rottweil	–	70
Netherlands	57.5	80

In the Berlin house, 40% of the energy was lost due to inexpedient location of the utility room under the roof with a high air change rate. This meant that the heat losses from the system did not contribute to heating the house.

The Rottweil house displayed a higher system efficiency than the Berlin house, but there was still a loss of 30%. The energy for space heating and domestic hot water is furnished as hot water (60°C) from a local power plant through pipes in the cellar. The energy is transferred to the system of the house via heat exchangers. The main reason for the waste of 30% energy is poor insulation of the piping in the basement, losses in the heat exchangers and circulation of the domestic hot water.

The overall system efficiency in the Dutch house is 80%, which is much higher than expected. The pessimistic expectations were based on a general estimation of domestic hot water consumption and expected heat losses due to piping and water tanks that turned out to be much smaller.

SOLAR ENERGY USE

Maximum utilization of solar energy was a key issue in all the projects and use was made of both active and passive techniques. The passive techniques cover optimal use of direct solar gains through careful planning of window size and orientation, sunspaces for preheating ventilation fresh air, and Trombe walls for indirect space heating. The active systems are solar collectors for heating domestic hot water, and grid-connected and directly coupled PV systems.

Direct solar gains

Direct solar gains were quantified by measurements in the two German houses and in the US Grand Canyon house. The results are shown in Table 5.4.9.

The US house showed a large difference between the measured and predicted values. A short-term energy test was performed, applying physical conditions on the building and measuring the energy performance. The overall solar heat gain coefficient is expressed as a fictive equivalent glazed opening of 2.9 m^2 with 100% transmission. The actual total glazed

Table 5.4.9. Direct solar gains

Country, house	Expected (kWh/year)	Measured (kWh/year)
Germany, Berlin	4210	3300
Germany, Rottweil	6480	6700
USA, Grand Canyon	6890	2683

area is 8.9 m^2, so the effective solar energy transmittance calculated is 2.9/8.9 = 0.33. According to the manufacturer, the value for the used glazing is 0.56. However, shadows from unexpected vegetation also played a part.

In the Berlin house, vegetation was also the reason for the discrepancy observed between the predicted and measured solar gains (the house was built closer to a row of trees than planned). The Belgian measured results indicate that about 45% of the total annual heating demand was covered by solar gains.

Trombe wall

The Grand Canyon house in the USA was the only one using a Trombe wall. The design of the wall delayed the penetration of the solar energy into the house until the direct gains had disappeared.

The measured efficiency of the Trombe wall was approximately 20%, which was much lower than expected. The reason seems to be the effective solar energy transmittance of the glass cover, in that only 44% of the incident solar energy was absorbed on the outside of the wall surface. The glass cover should have a solar energy transmittance of 85%.

Active solar systems for domestic hot water

Active solar systems for heating domestic hot water were used in several of the projects. The predicted performance is generally slightly higher than the measured performance (see Table 5.4.10). No specific explanations are available.

Table 5.4.10. Performance of active solar systems

Country, house	Predicted (kWh/year)	Measured (kWh/year)
Canada, Waterloo	1387	1314
Denmark, House B	812	703
Finland	1992	1992 *
Germany, Berlin	–	5160
Netherlands	1630	1220 **

* 3151 kWh/year measured, of which 1992 kWh/year was usable.
** Malfunctioning of the system during monitoring.

Photovoltaics

The only available results concerning the performance of PV systems were from the Finnish house. The expected output was 2320 kWh/year, which was approximately 25% higher than the measured value of 1826 kWh/year. There is no explanation for the difference.

PARASITIC ENERGY CONSUMPTION

Parasitic energy consumption is understood to be the energy consumed by the components associated with the installed systems for heating, cooling and ventilation such as pumps, fans and control systems.

The predicted and measured values available are given in Table 5.4.11. The measured values are generally considerably higher than expected except for the Brampton house. In this case, the fan turned out to be far more efficient than expected, which was mainly due to the very efficient electronically commutated motor (a different type of motor to the one foreseen).

The measured parasitic energy relative to the total energy consumption ranges from 5% to 27% (see Table 5.4.11). However, most of the parasitic energy contributes to heating the house.

ENERGY CONSUMPTION OF APPLIANCES

Available information on the energy consumption of appliances is difficult to compare. The Danish measurements, for instance, represent only one specific refrigeration and freezing unit, whereas the results from the Canadian Waterloo house include all common appliances. The results are shown in Table 5.4.12.

There is a surprisingly large difference between the two Canadian houses even if the predicted energy consumption of the freezer (420 kWh/year) is subtracted from the values for the Waterloo house. The gas stove seems to be the weak point in the Waterloo house in that it had a predicted energy consumption of 1050 kWh/year. This is three times the value for the electric Brampton stove. However, it should be noted that the values are very dependent on user behaviour.

Table 5.4.11. Energy consumption of pumps, fans and control systems

Country, house	Predicted (kWh/year)	Measured (kWh/year)	Percentage of total energy consumption
Belgium, Pleiade		1400	10
Canada, Waterloo	1845	822	5
Canada, Brampton		3750	19
Denmark, house A	425	595	6
Denmark, house B	450	564	10
Finland	1162	2656	27
Germany, Berlin*		1680	7
Germany, Rottweil**		1250	8
Netherlands	450	600	14

* Calculated from the value of 1265 kWh measured over 9 months.
** Calculated from the value of 940 kWh measured over 9 months.

Table 5.4.12. Energy consumption of appliances

Country, house	Predicted (kWh/year)	Measured (kWh/year)
Canada, Waterloo*	2786	3128
Canada, Brampton**	1608	1502
Denmark, house A***	394	419

* Freezer, refrigerator, dishwasher, stove, washer, dryer
** Refrigerator, washer, dryer, dishwasher, stove
*** Low energy refrigeration and freezer unit

OTHER SYSTEMS OR COMPONENTS

A greywater heat exchanger was installed in the Canadian Brampton house, with recovery obtained using a heat pump. Its recovery rate was measured at an average value of 63% of the water-heating load.

EXPERIENCES WITH SPECIFIC SYSTEMS AND COMPONENTS

Ground loop cooling

Circulating cistern water through plastic tubing buried below and around the foundation, and through a coil in the air-handling system provided mechanical cooling for the Canadian Green Home. Installation of the ground loop presented several problems. First, constant supervision was required to ensure proper installation because the contractors were not familiar with the system. Secondly, the piping was ruptured (and subsequently repaired) several times during construction. When the system was commissioned, it was discovered that one of the three cooling loops was broken and had to be isolated. The monitoring showed that the cooling output was modest (2 kW) and did little to lower the house temperature. In conclusion, ground loop cooling is not recommended, but passive cooling techniques appear to be effective.

Sunspaces for ventilation air preheating

Sunspaces are widely used for ventilation air preheating, as in the Canadian Brampton house. However, the efficiency in this house was significantly reduced owing to leaky partition doors between the sunspace and the rest of the house. In fact, 55% of the air sucked into the sunspace came from the house itself. The partition doors were protected from wind, rain and snow by the sunspace. Thus, from this point of view, no specific requirements needed to be met. However, using the sunspace for ventilation air preheating requires doors with the air tightness of normal outside doors.

The Dutch apartment building was provided with a central sunspace. Ventilation air was preheated by the sunspace before entering the heat recovery unit of the dwellings. Both provisions in series reduced the ventilation losses by about 90%. The contribution of the sunspace was about 10% and the heat recovery unit had an efficiency of about 80%. The investment of the sunspace in the Dutch project cannot be justified by the energy saving alone; other features like functionality and architecture should be taken into account.

5.5 Thermal comfort

INTRODUCTION

Thermal comfort, especially with respect to avoiding overheating during the summer, is an important objective. Four countries (Belgium, Denmark, Germany and the Netherlands) offered evaluation results concerning thermal comfort. Finland measured the indoor air temperatures during the summer; these never exceeded 26°C. Belgium and Canada evaluated indoor air quality.

BELGIUM

Summer thermal comfort

Monitoring showed that the Belgian house provided comfortable indoor conditions even during periods of extremely warm Belgian weather. Figure 5.5.1 shows the outdoor and average indoor temperature during the monitoring period. It is remarkable that, thanks to the application of the overheating prevention strategy (solar control and night-time ventilation), the mean indoor temperature remained below 25°C, whereas the daytime outdoor peak temperature kept rising and reached 30°C on the last day of the data set. This shows that proper application of intensive night-time ventilation can be an effective strategy in Belgium for achieving summer comfort in residential buildings when used in combination with other conditions such as accessible thermal mass and sun protection.

Indoor air quality

As long as the ventilation system was operating under normal conditions, indoor air quality was acceptable. CO_2 concentrations were not above the recommended value of 1500 ppm, but most of the time the ventilation supply was switched off (extraction only) due to draught problems in the bedrooms. In this case, the CO_2 concentration sometimes exceeded the recommended value, but only in the bedrooms.

Comments

From the indoor temperatures found, it may be concluded that very comfortable indoor conditions were maintained in the Belgian house in spite of a severe heat wave (it is commonly admitted that temperature levels ≤25°C are acceptable in the summer).

CANADA

Formaldehyde, radon, VOCs and particulate concentrations were measured in November 1993 during the demonstration period in the Green Home in Waterloo. Formaldehyde levels were 0.03 ppm – well below the Canadian guideline value of 0.1 ppm. The radon concentration was measured at 0.003 working level units (WLU) – again well below the conservative US guideline value of 0.02 WLU. The VOC concentration was 0.18 mg/m³, approximately half the European guideline value of 0.3 mg/m³. These readings were among the lowest values in all Canadian advanced houses. The airborne particulate reading was 92 µg/m³ – higher than some of the other Canadian advanced houses. This was attributed to the high outdoor particulate levels generated by nearby building construction.

The indoor air quality measurements were repeated in August 1994 after the homeowners had moved into the house. The formaldehyde reading tripled to 0.09 ppm, but remained below the Canadian guide-

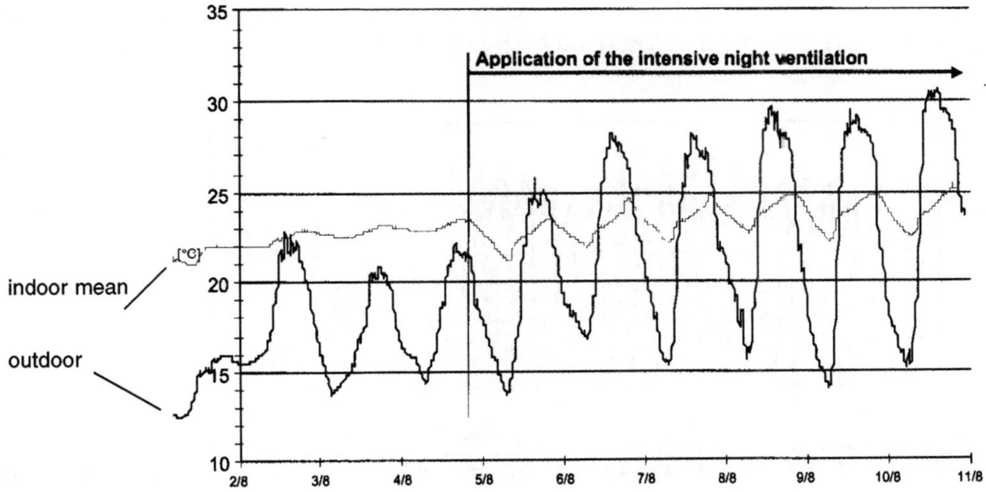

Figure 5.5.1. Outdoor and average indoor temperatures during the monitoring period

line value. The higher reading was attributed to the replacement of display furnishings by the homeowner's own furnishings, some of which were manufactured from particle board. Likewise, the VOC concentration was 0.26 µg/m³ – 45% higher than the previous reading. The radon concentration was only slightly higher at 0.004 WLU. The airborne particulate reading decreased by 75% to 22 µg/m³, presumably due to less nearby construction activity.

CO_2 was measured continuously in the Waterloo house from mid-January 1995 to the end of August 1995. CO_2 is produced by people and can provide an indirect measure of ventilation effectiveness. The monthly average values ranged from a high of 685 ppm in January to a low of 218 ppm in July. The maximum daily average reading was 869 ppm. ASHRAE Standard 62 recommends a limit of 1000 ppm to satisfy comfort criteria (i.e. to limit odour buildup); thus, the ventilation system as designed and operated in the Waterloo house provided adequate fresh air.

Detector tube tests were conducted to measure CO, SO_2 and NO_2. The tests were performed in January 1994. No detectable levels of any of these gases were found at the supply registers while the furnace was operating.

The technical requirements of the Canadian advanced houses mandated a minimum indoor relative humidity of 30%. A humidifier was to be installed if required. In the Waterloo house, the IHVA transfers moisture as well as heat from the stale exhaust air. Indoor humidity was measured continuously in the stale air stream of the heat recovery ventilator (HRV). The monthly average relative humidity was >40% all winter and the minimum daily average humidity was 35%. Thus, the IHVA appeared to be effective in maintaining adequate humidity levels in the house.

Electromagnetic radiation was measured in the Waterloo house in October 1993 using an electric field measurement Gauss meter (see Table 5.5.1). The recommended safe levels of electromagnetic radiation cover a wide range from 3 mG to 2 mG, depending on the authority cited. Average measurements in typical housing today are approximately 2–3 mG.

Throughout the Green Home, the vast majority of measurements were well below 1 mG. The electronic ballasts used in the fluorescent fixtures and the low currents required for the energy-efficient appliances were two reasons for the low readings. There were a few locations where the readings were >3 mG. Generally, these were not areas where people spent extended periods. Table 5.5.1 summarizes the electromagnetic field readings in the Waterloo house.

Table 5.5.1. Electromagnetic field readings in the Waterloo house

Location	Reading when touching the device (mG)	Reading 300 mm from the device (mG)
Living spaces	<1	<1
Microwave oven	1,200	70
Compact fluorescent	90	0.2
Electric panel	76	3–10
Damper motor*	30,000	10

*Damper since replaced by one that does not need a motor.

DENMARK

House A: comfort in the summer

Figure 5.5.2 shows that the indoor air temperatures in Danish house B rose far too much during summer. The average temperature in August was 28.2°C (maximum 32.6°C, minimum 24.8°C). The measured indoor temperatures are much higher than ones calculated during the design phase. The most obvious reason for this deviation is a presumed air change rate of 3 per hour when the indoor air temperature exceeds 23°C. This was probably not achieved in practice; according to the inhabitants, it was difficult to achieve natural cross-ventilation in the building.

House B: comfort in the summer

The indoor air temperatures in Danish house B followed the outdoor temperature and were even below it during the warmest periods. But, as can be seen in Figure 5.5.3, the decrease in indoor air temperature led slowly to a high average indoor temperature. The average temperature in August was 25.9°C (maximum 29.1°C, minimum 23.6°C). Once the high outdoor temperatures are taken into account, the measured indoor temperatures are reasonable.

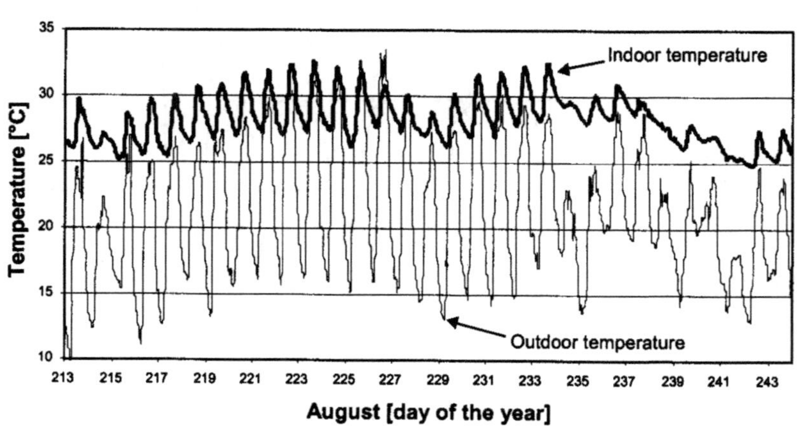

Figure 5.5.2. Indoor and outdoor air temperatures in summer in house A

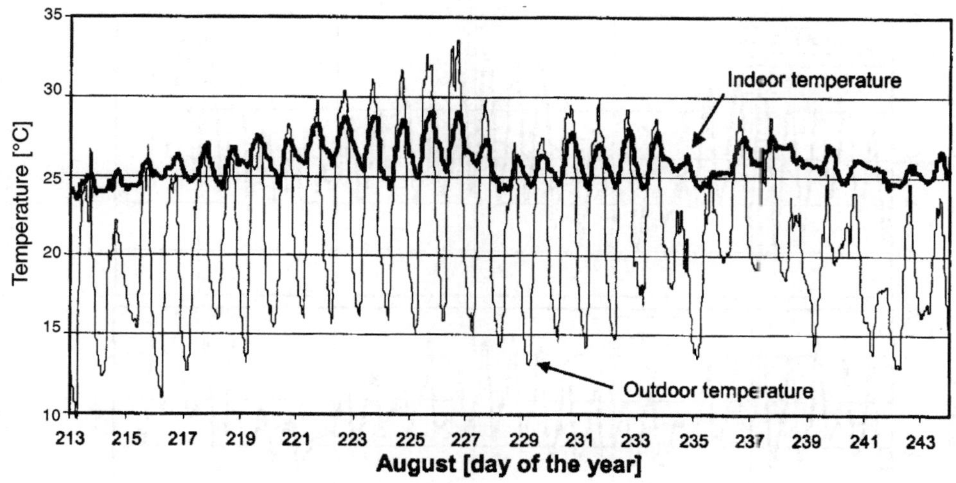

Figure 5.5.3. Indoor and outdoor air temperatures in summer in house B

However, it should have been possible to cool the house overnight. The reason why this did not happen is probably a presumed air change rate of 3 per hour when the indoor air temperature exceeded 23°C. This was probably not achieved in practice and the actual air change rate was not recorded.

Comments

It seems obvious to assume a certain ventilation rate during the design phase, e.g. one controlled by the indoor temperature. However, caution is recommended, as the occupants will not be at home for much of the time to react to increasing temperatures. Furthermore, planning of the building should include windows that can be opened at both façades in order to create cross-ventilation.

If the indoor thermal comfort is based on increased ventilation to prevent overheating, automatic opening of windows is required (or increased mechanical ventilation that is controlled automatically).

GERMANY

Zero Heating Energy House in Berlin, Spandau, 52°N

Figure 5.5.4 shows the results of continuous temperature measurements across the three levels of the house from the ground floor to the top floor compared with the outside temperature for August 1997. For the ground floor, the living room temperature is shown during the warmest period of the measurements. The maximum temperature was 27.7°C, despite outdoor temperatures reaching over 30°C for 10 days. On the next level, the highest temperatures were measured

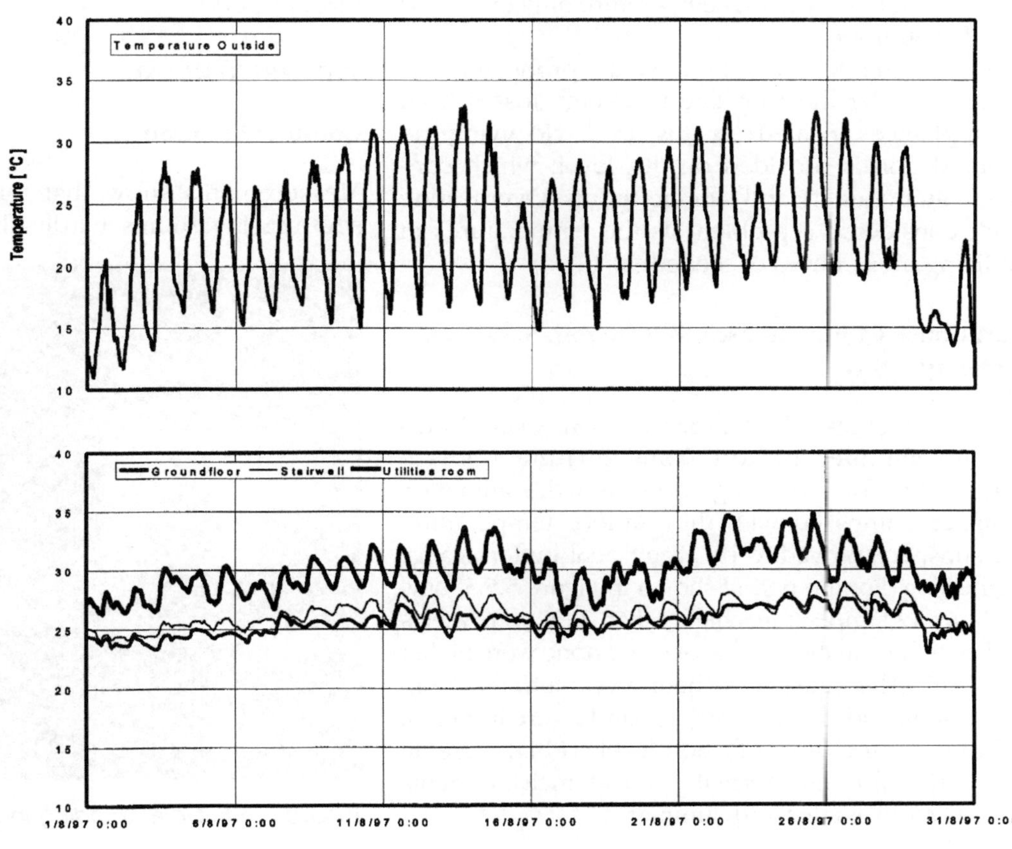

Figure 5.5.4. Temperatures in summer in the Berlin house

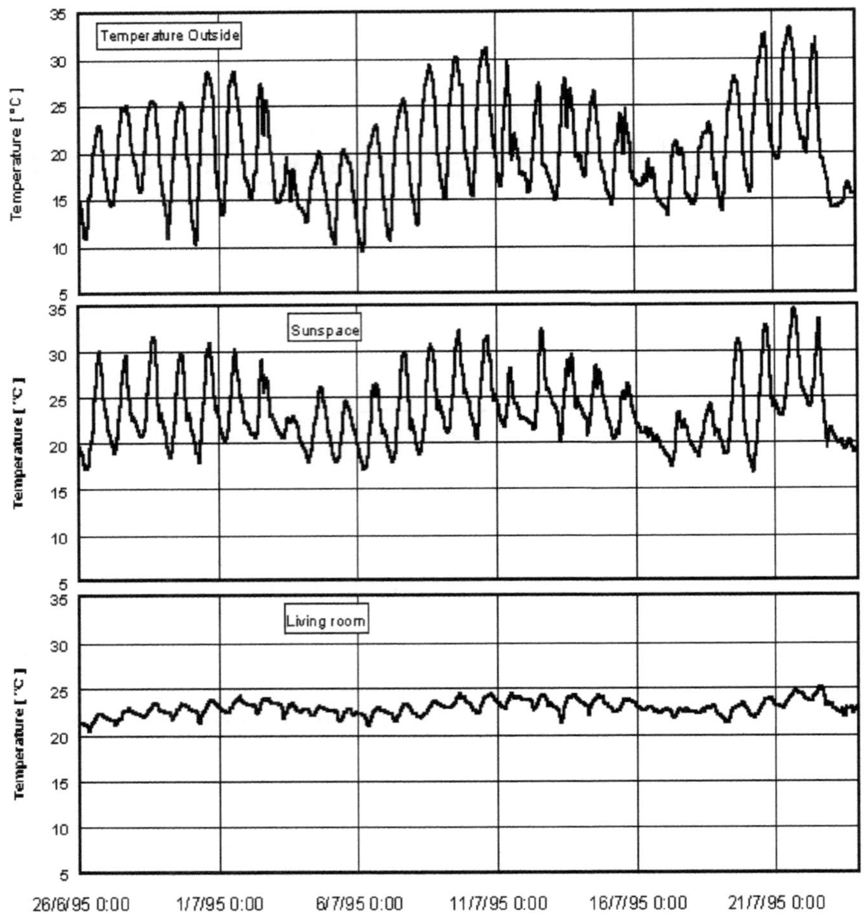

Figure 5.5.5. Temperatures in summer in the Rottweil house

in the stairwell, which is open to the sunspace. Despite being located near the storage tank, the maximum temperature was 28.9°C. In the south-facing bedroom on this level, the temperature only exceeded 28°C for 30 hours.

Overheating proved to be the main problem in the top level under the roof due to energy losses from the appliances located on this level. No living accommodation is provided on this level, which contains only the utility and storage rooms. Overheating is not considered a problem as the temperatures in the living areas never exceeded 30°C.

Ultra house (3 litre house) in Rottweil, Southern Germany, 48°N

Figure 5.5.5 shows the temperatures measured during the summer period from 26 June 1995 to 23 July 1995 for the living room and the adjoining sunspace compared with the outdoor temperature. The sunspace proved to be an efficient buffer against overheating. For the period shown in Figure 5.5.5, only five hours of temperatures higher than 26°C were measured anywhere in the living space and none were higher than 28°C. The sunspaces themselves suffered from overheating and, at a later stage, blinds were installed.

The predicted dissatisfaction levels (PPD) were all below 10% for the thermal comfort measurements carried out in winter and summer.

Moisture and poor air quality were not a problem due to the HRV and high window opening times; the average air change rate was 1.5 per hour during the heating period.

THE NETHERLANDS

Comfort in summer

Measurements show that the temperatures in the south-facing rooms hardly fluctuated, even during

Figure 5.5.6. Solar shading in the Urban Villa

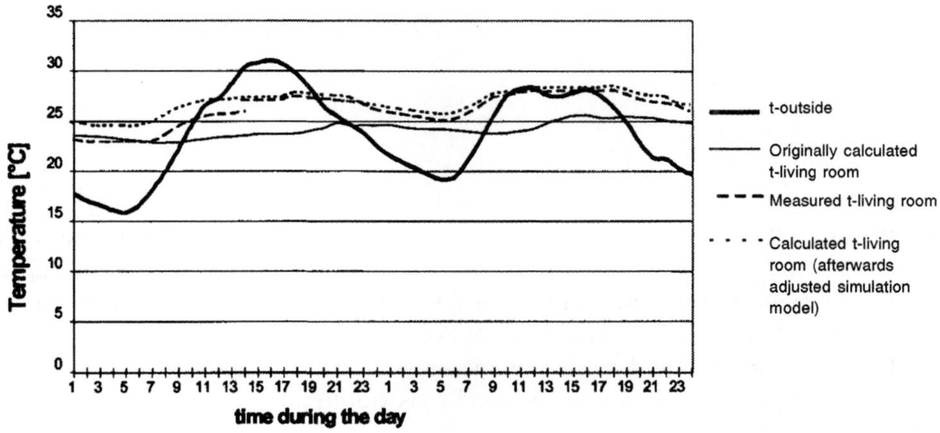

Figure 5.5.7. Temperatures in summer in the Dutch house

days with intensive solar radiation. This phenomenon was due to building mass, solar shading (see Figure 5.5.6) and the functionality of the ventilation facilities (passive cooling). The effects on the indoor temperature were not determined separately.

Figure 5.5.7 shows the variation in measured temperatures, the ones calculated originally and the temperatures calculated using an adjusted behaviour pattern. The calculated and measured values are almost the same during the nights; in the daytime, the difference was approximately 2°C. In the summer, the indoor temperature in the bedroom is only around 25°C, whereas the outdoor temperature is 30–35°C.

The temperature in the atrium was registered on four levels (ground floor, 1st, 2nd and 3rd floors) at the eastern and western side. The values exceeded the outside temperature by around 2°C, which was considered satisfactory.

Comments

The concept was shown to guarantee a pleasant indoor climate, even under extreme conditions as in the very hot summer of 1995. Most occupants were also satisfied with the temperature control in other seasons.

Floors, windows and stairs inside the atrium often show some condensation on the inside. The occupants expected the temperatures to be lower during summer and to be higher in the winter.

General remarks

In Northern and Central European countries, it is possible to prevent overheating in warm periods by applying passive cooling techniques.

The Belgian and the Dutch houses had indoor temperatures ≥25°C, even during very warm days with intensive solar radiation. This is mainly due to thermal mass, sun protection and night-time ventilation.

The Berlin house in Germany displayed rather high indoor temperatures during the summer, with considerable variations in outdoor temperatures. It should be possible to cool the house overnight and make use of more thermal mass.

Danish house A displayed far higher indoor temperatures than predicted. This was mainly because it was difficult to create cross-ventilation in the house and that is the reason why the air change rate became much lower than predicted. Furthermore, the increased ventilation was based on manual opening of windows, which relied on the occupants being at home. If indoor thermal comfort is to be based on increased ventilation, then automatically opening windows should be provided.

5.6 User aspects

INTRODUCTION

User aspects were monitored in four projects, namely the Belgian Pleiade house, the Finnish Solar House, the Dutch Urban Villa and the Norwegian row house. The (very positive) opinion of the owner is quoted for the Canadian project.

No specific investigation was conducted in the Danish project, but some comments can be made based on contacts with the users during monitoring. No comments were received regarding the user aspects of the Japanese WISH House, as it had to be demolished because it was not built in accordance with local regulations. There are no general comments on the two German projects; in these houses, noise problems caused by the heat recovery units occurred.

APPRECIATION

Most of the projects were valued positively, but for different reasons. The architecture, the house and the installations were valued positively, especially in the Belgian and the Canadian projects. The Belgian house, moreover, displayed a very good indoor climate. In the Finnish project, the indoor climate 'met the expectations' of the occupants. In the Dutch project, the indoor climate was regarded as quite good, but the control system was found to be too 'complex'.

The two projects with overheating problems were in Denmark and Norway. Overheating problems were rather serious in Danish house A and the neighbours even moved out because of their problems with overheating. In the Norwegian row house, overheating occurred only in the sunspace and not in the living room, but only as long as the doors were kept shut.

BELGIUM

After the Belgian Pleiade row house had been occupied for five years, its occupants were still very positive with regard to the house and the installations. The only exception was the mechanical air ventilation system. Sometimes there were draught problems in the bedrooms. The occupants were involved in the design and were positive about the architecture: the details and the choice of the furniture form part of those findings. They mentioned the quality of the different ambiences during different moments of the day and the seasons.

Thus, the various objectives of the project were completely attained, i.e. thermal comfort in winter, thermal comfort in summer and daylight in all the places.

The occupants recognized the importance of the domotic system, especially with regard to the regulation of the shading system and the artificial light system.

CANADA

Quote from the owner of the Green Home in Waterloo, Rolf Theisen:

> "What has it been like to live in the Green Home? Wonderful! The vaulted ceilings add an extra spaciousness to the house. The large, south-facing windows make for a very bright interior. The layout and architectural lines are sensible and aesthetically pleasing. The combination furnace and heat recovery ventilator keep the air fresh and cosy warm."

DENMARK

No specific investigations on the user aspects were conducted in the two Danish low-energy row houses. However, please note the following comments. In general *(for both houses)* the fresh air inlet near the dining table results in an uncomfortable draught, even though the fresh air passes a high efficiency counterflow heat exchanger. The draught made the occupants shut off the air supply fan (and thus the heat recovery). In house A the indoor temperature becomes very high due to sunshine (also during the winter). It was an important reason why *some occupants* moved out. The pumps and ventilation system cause minor noise problems. It is difficult to control the floor heating system due to the large time constant, especially when the occupants are not aware of how the building physics work. In house B the indoor temperature on the first floor becomes too high during the summer and serious noise problems occur from the ventilation system, depending on the wind direction. There is only one thermostat for controlling the floor heating system in four different rooms. Finally it is considered inconvenient that the central ventilation stopped between 22:00 and 7:00. This meant that neither the kitchen nor the bathroom could be aired.

FINLAND

User satisfaction with the Finnish IEA5 Solar House was high; in particular, the indoor climate met the expectations of the occupants. They had a great influence on the architecture, structural solutions and building services in that they played an active role in the pre-design and construction phases.

THE NETHERLANDS

At the time of purchase, the 16 apartments of the Urban Villa were designated as apartments with a high-quality indoor climate, meeting good technical and living standards. The low energy consumption hardly played a role when the dwellings were being bought.

The occupants were instructed about their new apartments and installations both before delivery and during the first year of occupation. The instructions given were evaluated as sufficient, but at certain points as too complicated. The majority of occupants, even after one year, regarded the control system as complex.

It seems that there are two groups of occupants: the intensive users and the marginal users. However, it turned out that the concept itself guaranteed a high-quality of indoor climate without intensive interference from the occupants. This is mainly due to the high insulation level, the programmed ventilation system and the automatically controlled outside solar shading. The designers and occupants generally considered the concept to be a success.

Unfortunately, the failure of the active solar system for domestic hot water during the introduction period detracted from the many positive aspects of the concept. The occupants were very dissatisfied as regards the slow and inadequate solution of problems and complaints. Other appliances appeared to work properly and contributed to the project's goals.

In general, the occupants were satisfied with the architectural quality; in particular, the view from the rooms and the atrium were valued positively. The access of sunlight to the kitchen and the small bedroom was regarded as inadequate, as these areas were oriented towards the atrium.

NORWAY

When interviewed, the occupants of the Norwegian row house reported that the hydronic heating system, fed by the heat pump, was hardly ever used, except in the sunspace/staircase (see Figure 5.6.1). The sunspace was overheated at times in summer, even when the vents were open. The living room next to the sunspace remained comfortable as long as the doors to the sunspace were kept shut. The heat pump compressor caused audible noise in the living room, but not in the downstairs bedrooms. The levels of daylight and the ventilation were judged satisfactory.

Figure 5.6.1. Sunspace/staircase of the Norwegian row house

5.7 Conclusions

GENERAL CONCLUSIONS

Many of the monitored projects are successful. In some cases, the performance of the dwellings was not in line with expectations, but even those projects turned out to be a great source of information. Although many of the concepts include prototype components and the energy concepts were stretched beyond common practice, the results are considered very positive.

The IEA SHC Task 13 project has contributed significantly to the further development of solar low energy design. Figure 5.7.1a shows the measured energy use for the IEA SHC Task 13 houses that contributed their evaluation results. The results of both the Danish A project and the German Berlin project are not included in this value because essential components of the energy concept were turned off during the evaluation period (making the results unrepresentative of the concept envisaged). For reference, the predicted annual energy consumption based on the design is presented as an average value for all the projects in Figure 5.7.1b and average energy use in typical houses in all the participating countries in Figure 5.7.1c.

A very ambitious target was originally set to achieve a saving of approximately 75% on the energy use of the typical houses. The energy saving measured turned out to be 60% of the energy use of the typical houses. This is a fairly good result, considering the experimental character of Task 13.

The difference between the target and the actual energy saving is due to space heating and electricity consumption. The two main reasons for this are:

- The performance of some of the building components did not meet expectations (air tightness of the envelope, efficiencies of system components).
- The occupants used more energy than assumed originally (higher indoor air temperatures, higher electricity use for household appliances).

Furthermore, the results show that:

- In the participating countries, there is still a substantial energy saving potential in new housing and that technical facilities are available to achieve this reduction in energy consumption.
- A comfortable indoor climate in solar low energy houses can be reached by simple means. The use of shading devices and night ventilation are powerful techniques to prevent overheating in regions with a moderate climate such as Western Europe.
- In general, there should be a close interaction between the building physicists, architects and system engineers during the design process. Of particular importance are the first phases in which ideas and know-how are exchanged. Discrepancies in the design have a negative effect on the concept and on the application of important energy saving techniques. It is difficult and sometimes even impossible to rectify changes afterwards.
- As far as energy is concerned, a three step approach (the so-called *trias energetica*) is recommended: (1) reducing the energy losses in order to lower the energy demand, (2) trying to use renewable energy sources to fulfil this energy demand, and (3) covering the remaining demand for energy as efficiently as possible with fossil fuel.
- Quality control during the construction phase turned out to be very important to reach the performance required. Special attention needs to be paid to air tightness of the envelope and the ventilation systems. A good description of equipment in terms of the performance and testing methods required is crucial for quality control. Moreover, a sound system of warranties will enhance the overall quality of the building.
- The functionality of the concept should be simple and easy to explain to the occupants; a good instruction and user manual is helpful. Efforts should also be made to design a control system that is simple and easy to operate.

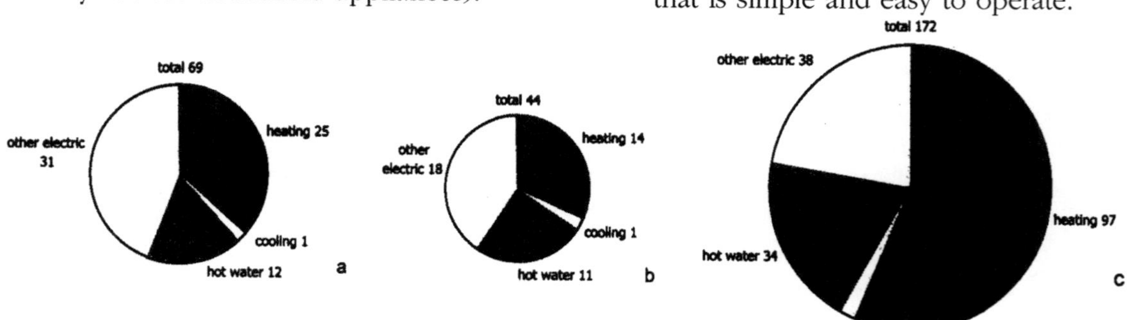

Figure 5.7.1. Energy use measured in six houses (a), as originally predicted by simulations (b) and in typical houses (c), expressed in kWh per m² heated floor area

Simulation turned out to be a powerful tool to predict the energy performance of the projects. The calculated results are close to reality provided the input is adjusted to the actual indoor conditions and properties of the building components.

CONCLUSIONS PER ASPECT

The experiences with the energy concepts have given rise to negative as well as positive comments. They provide lessons for us all. In this section, we describe the most important 'lessons learned'.

Heating system

In general, the heating systems for space heating and domestic hot water worked well. It appears that attention is necessary to every detail of the design of installations. Moreover, the costs of maintenance need to be considered. The message is that the simpler the system, the less that can go wrong.

There should be optimal co-operation among building physicists, architects and engineers. Another important aspect is the instruction of the occupants (see the comments on the complexity of the systems from Denmark, Finland, Germany and the Netherlands).

From a point of view of energy efficiency, it is important to reduce heat losses from burners and heat storage as much as possible. The efficiency can be further improved by locating these components in an area where the heat loss to a certain extent contributes to space heating. In the German Zero Energy House, nearly 40% of the total heating load was lost because the utility room was located under the roof in an area with a high air change rate and no heat recovery. More attention should also be paid to back-up heating systems.

The Norwegian project showed that further development of small heat pump systems is necessary as this kind of system did not perform satisfactorily in the Norwegian project.

Building design

In general, it is possible to obtain a high level of insulation of the building envelope. It is important that thermal bridges are avoided. In the Belgian Pleiade house, for example, there is a good agreement between the theoretically calculated U-value of a building and the physical behaviour of the constructed building.

The Green Home wall system, composed of wooden I-beams, has twice the insulation level of conventional houses, but uses one-third the amount of wood. In addition, the high-performance windows reduce the transmission losses and contribute considerably to the indoor climate. Heating devices do not necessarily need to be placed below the windows to prevent cold draughts.

In apartment buildings with a high level of thermal insulation, special consideration should be given to the energy losses through partitions between apartments. Insulation between the apartments will avoid those losses, but increase the cost. This internal insulation has no effect on overall energy consumption of the apartment building as a whole.

High-performance envelope components are safer investments than technical energy systems. Better insulation has a longer lifetime. Moreover, it requires less maintenance and servicing than complex mechanical systems. Priority should, therefore, always be given to measures for improving the energy performance of the envelope.

Ventilation and infiltration

The necessary air tightness of the houses was achieved in most projects. In some cases, however, (Belgian Pleiade row house, German Zero Energy House) no high level of air tightness was attained. To create a proper air barrier, special attention should be paid to ensuring that all joints are properly sealed and, in particular, around windows, external doors, the ventilation system and the roof construction. This is particularly important when a heat recovery system is used.

In some projects, it appeared to be difficult for installers to meet the ventilation standard requirements. A combination of more performance-oriented standards/legislation and field controls could improve the situation significantly. Likewise, it would probably be advisable for the ductwork to satisfy special air tightness and insulation requirements.

For various reasons not all heat recovery systems apparently worked properly (Belgium, Denmark, both German projects). Some of the problems were eventually dealt with. In the case of the Dutch projects, however, the heat recovery system was found to produce satisfactory results.

Once the calculations are adjusted to the real situation, it was possible to match the simulated results to the measured ones (Belgium, Denmark, the Netherlands).

Comfort

In a moderate climate, it is possible to prevent overheating in warm seasons by applying mainly passive cooling techniques (Belgium, the Netherlands). This implies a reduction in solar gains and making use of ventilation and thermal mass. Active cooling is needed in a particularly warm climate (e.g. Japanese WISH house).

It is important that the assumed control strategy during the design phase should also work in actual practice. In the Green Home, for example, an enhanced ventilation rate was assumed to prevail when indoor temperatures exceeded a certain level. In reality, this assumption depended on the occupants being at home and opening the windows. An automatic window opening system would have led to a better performance.

In the building design, account should be taken of the influence on the indoor thermal comfort. For instance, two-storey rooms make it difficult for the temperature to be kept down in the adjacent rooms on the upper floor (Denmark).

More consideration needs be given to reducing noise emanating from heat recovery and ventilation systems (German projects, the Netherlands). Shorter duct routes and silencers would contribute to making these systems more acceptable to the occupants.

The high efficiency of the heat recovery system does not always guarantee an air supply without draught (the Netherlands, Denmark). The installer should pay more attention to adjusting the balanced ventilation system.

Other environmental issues

In many projects, attention was paid to other environmental issues, e.g. the choice of building materials, reduction of the demand for water, etc.

Special efforts were made in the Canadian Green Home project, e.g. reducing waste by using recycled or re-used materials. Those materials were found to be of the same or superior quality, durability and appearance as conventional construction materials.

Major reductions in water use are possible. The Canadian monitoring showed that a reduction of at least 73% was achieved compared with the water use of a typical four-person family and 85% compared with that of an average five-person family.

Indoor air quality measured results show that pollutant concentrations in the Green Home were far below the most stringent regulations in the world.

CONCLUSIONS PER PROJECT

Specific conclusions for many of the evaluated projects are given below.

Belgium

The Pleiade dwelling exemplifies how very low energy consumption can be achieved along with good comfort in winter and summer. The main characteristic is the integrated approach developed, i.e. interaction between architecture, building, energy, systems and controls.

The project has highlighted the need to go beyond the concept of average thermal insulation level and/ or the concept of the net heating demand. Optimal solar control, energy efficient heating and ventilation systems, summer comfort, etc. are crucial challenges for future buildings.

There is, therefore, need for additional requirements in Belgium. An energy performance regulation like the one under preparation in the Flemish Region of Belgium may offer an attractive way of creating an assessment procedure that takes account of these various issues.

Canada

The Waterloo Region Green Home was designed to be energy efficient and environmentally responsible, and to meet the requirements of the advanced houses programme. The house was expected to require only 30% of the energy and water of a conventional new house.

Tests and monitoring showed that the building envelope met or exceeded design requirements. The air tightness was measured at an air change rate of 0.8 per hour at 50 Pa, almost half the design requirement of 1.5 per hour. The building space heating load met the design predictions.

In the Canadian Green Home, the use of natural gas-fired equipment and appliances was preferred to the use of electric devices due to the lower CO_2 emissions of natural gas. Mechanical cooling was provided by circulating cistern water through plastic tubing buried below and around the foundation. Installation of the ground loop presented several problems. In conclusion, ground loop cooling is not recommended.

Formaldehyde, radon and VOC concentrations were significantly below the most stringent guidelines in the world.

Co-operation between the public and the private sectors resulted in a successful and high profile demonstration project. Some of the benefits are already apparent. The requirements of the R-2000 program were upgraded to include some of the advanced house requirements. Many leading edge builders are incorporating advanced house technologies into their housing, e.g. engineered wood products are becoming the floor system of choice, fibreglass windows are seeing dramatic increases in market share and wall-mounted water heaters are available for lease from natural gas utilities.

Denmark

House B had a heating demand for space heating of as low as 30 kWh/m^2 – about 45% of the Danish energy standard. The many technical problems in house A, however, led to a demand of 72 kWh/m^2. These figures show both the potential and the difficulties in building energy efficient houses.

Despite the low energy demand for house B, the measured energy consumption is higher than the predicted value. The reasons for this difference only relate to the concept to some extent as the reasons were a higher indoor air temperature, the heat recovery system was turned off and electricity consumption was higher than anticipated. Moreover, the active solar system for domestic hot water did not work well in house A.

Both houses had problems with the thermal comfort. In house A, the indoor air temperature became too high on sunny days (both in winter and summer). In house B, high temperatures were mainly a problem on the first floor during the summer. Automatic opening of windows or automatic solar shading devices are recommended in order to create a comfortable indoor climate. Furthermore, the ventilation system caused a draught in the living area and was noisy. An improved system design can solve these inconveniences.

It is clear that the technical aspects have to be improved in future low-energy houses. Furthermore, it will be useful to give more information to the occupants on how the technical equipment works with respect to the building physics. They will thus be able to adjust more easily to the low-energy concept involved.

Finland

The heating energy consumption of a typical Finnish detached house can be reduced by 50% by decreasing the house's heat losses. By introducing a solar design, energy consumption can even be reduced to 25% compared with that of typical buildings. The architecture of the house was found to be pleasant; the appearance of the house does not remind one of a typical solar house.

The project itself had no great influence on the building practice in Finland, but it served as a good example of how environmental issues and user satisfaction may be combined.

Germany: Zero Heating Energy House

The German Zero Heating Energy House in itself cannot be considered a success in that the original goal of the project was not achieved (i.e. building a house that did not require fossil fuels for heating energy). This failure was mainly due to changes in the construction of the house that were not originally calculated and dramatically altered the outcome of the results. The concept, however, was successful. It has been further developed by the leading supplier of prefabricated housing in Germany and is now commercially available.

Germany: German Ultra House

The concept of the German Ultra House proved to be worthwhile and has been put into practice, in various forms, by many builders throughout Germany. The Ultra House itself was subject to unduly high ventilation and heat losses, which caused energy consumption to be higher than predicted.

Japan

The WISH house was designed and constructed to reduce the energy consumption to half the average energy consumption of recently built Japanese single-family houses. For this reason, use was made of innovative technologies such as a PV thermal hybrid collector system and PCM storage panels. As the house was demolished, the results could not be properly evaluated.

The Netherlands

The Urban Villa was evaluated as very successful and provided a good example for similar follow-up projects or for application of energy-efficient technologies in existing or newly built houses. It showed how a low-energy concept can be combined with passive cooling techniques, with very good results indeed.

Moreover, the Urban Villa exemplifies how a strict energy performance requirement can be achieved for houses built after 2000. The Urban Villa and similar projects form the starting points for the commercial production of certain energy-efficient technologies that are too costly at present.

Norway

The experiment has shown that it is quite easy to develop windows displaying a substantially better performance than normal by applying well-proven technology with low-E glazing and inert gas filling. Energy-wise super windows will perform just as well as very thick and much more expensive glazing with transparent insulation materials and, at the same time, offer a clear visual contact. Therefore, the building code requirement for the U-value of windows should be stricter.

The failures in the two major energy systems of the Norwegian row house underline the conclusion that such systems should be designed, installed and serviced under one contract by one company. When several parties are involved, there will always be discussions about responsibility for problems and repairs.

5.8 Bibliography

BELGIUM

[1] *PLEIADE: conclusions* (1998). Ministère de la Région wallonne.
[2] *Study and monitoring of the PLEIADE dwelling. Final report 1994-1996* (1997). Belgian Building Research Institute with the collaboration of Architecture et Climat UCL).
[3] *Study and monitoring of the PLEIADE dwelling. Final report 1996-1998.* (1998). Belgian Building Research Institute.
[4] Vanderick L and Deprêter Laborelec A (1996-1997-1998). *PLEIADE, Consommations énergétiques.*
[5] Vanderick L and Deprêter Laborelec A (1997). *PLEIADE, Gestion des charges électriques par un système domotique.*
[6] De Wit K (1996). *Warm waterproduktie in de PLEIADE woning.* ARGB.
[7] *La maison PLEIADE: chaude l'hiver, fraîche l'été* (Spring 2000). CSTC Magazine.

CANADA

[1] *Design and performance of the Waterloo Region Green Home* (1996). Prepared for CANMET, Ottawa, Ontario, by Enermodal Engineering Limited, Kitchener, Ontario, Canada.
[2] *Green Home – planning and building the environmentally advanced house* (1993). Prepared by Wayne Grady, Camden East, Ontario, Canada.
[3] *Waterloo Region Green Home* (1993). Prepared for CANMET, Ottawa, Ontario, by Enermodal Engineering Limited, Kitchener, Ontario, Canada.

DENMARK

[1] Engelund Thomsen K, Jensen S Østergaard and Saxhof B (1992). 'Optimization of a Danish third generation low-energy house concept', in *Proceedings 5th International Conference: North Sun'92 – Solar Energy at High Latitudes, Trondheim,* pp. 65–71.
[2] Engelund Thomsen K, Wittchen KB and Saxhof B (1992). 'Advanced solar low-energy buildings – Danish work within IEA Task 13', in *Proceedings of ASHRAE/DOE/BTECC Conference on the Thermal Performance of the Exterior Envelope of Buildings – Thermal Envelopes V, Clearwater Beach, Florida,* pp. 614–620.
[3] Saxhof B, Schultz JM and Engelund Thomsen K (1993). 'Two Danish Task 13 low-energy houses – designs and parametric studies', in *Proceedings of IEA and CIB W67 International Symposium on Energy Efficient Buildings, Stuttgart,* pp. 211–225.
[4] Saxhof B, Schultz JM and Engelund Thomsen K (1993). 'Thermal analyses of Danish low energy row houses for IEA SHC Task 13 Advanced Solar Low Energy Buildings', in *Innovative Housing '93, Vancouver.*
[5] Saxhof B, Schultz JM and Engelund Thomsen K (1993). 'Two- and three-dimensional heat losses in superinsulated buildings', in B Saxhof (ed.), *Proceedings 3rd Symposium on Building Physics in the Nordic Countries: Building Physics '93.* Laboratoriet for Varmeisolering, Lyngby. ISBN 87-984610-0-1 (Vol. 1). pp 109–116.
[6] Engelund Thomsen K, Saxhof B and Schultz J M (1994). *Two Danish Task 13 low-energy houses – designs and parametric studies.* ISBN 0905-1511.
[7] Saxhof B and Engelund Thomsen K (1996). 'Two different row houses in Kolding: the effect of orientation', in *Proceedings of EuroSun'96, Freiburg.*

FINLAND

[1] Nieminen J and Kouhia I (1993). 'Low-energy residential housing concepts for northern climate', in *Innovative Housing Conference, Vancouver.*
[2] Nieminen J and Kouhia I (1994). 'Solar House IEA 5. The Finnish IEA Task 13 demonstration house', in *International Conference on Cold Climate: HVAC'94, Rovaniemi.*
[3] Nieminen J and Kouhia I (1994). 'Low-energy residential housing concepts for northern climate', in *International Symposium on Cold Region Development: ISCORD'94, Espoo.*
[4] Nieminen J and Kouhia I (1994). 'Solar House IEA 5. The Finnish IEA Task 13 demonstration house', in *North Sun'94, Glasgow.*
[5] Nieminen J (1996). 'A single family house in Pietarsaari: combining active solar and PV', in *Proceedings of EuroSun'96, Freiburg.*
[6] Nieminen J and Kouhia I (1997). 'The Finnish IEA Task 13 experimental building in Pietarsaari: completed results', in *Proceedings of North Sun'97, Espoo.*
[7] Kouhia I and Nieminen J (1995). *IEA 5 Solar House.* VTT Research Notes 1674.

GERMANY

[1] *Null-Heizenergie-Haus Berlin* (1996). Brochure, Fraunhofer Institüt für Bauphysik, Stuttgart.
[2] Trautner A (1998). *Validierung der Energiekonzepte für das Null-Heizenergie-Haus Berlin-Spandau.* Degree thesis, Fraunhofer Institüt für Bauphysik, Stuttgart.
[3] DGS Sonnenenergie Verlags GmbH (1996). 'Europe's first zero-heating energy house built by a public proprietor in Berlin', in *Proceedings of EuroSun'96, Freiburg.*
[4] *Ultra- and low-energy house Rottweil* (1998). Brochure in English and German, Fraunhofer Institüt für Bauphysik, Stuttgart.

[5] Schlechte M (1997). *Validierung der Energiekonzepte für das Niedrigenergiehaus und das Ultraniedrigenergiehaus in Rottweil.* Degree thesis, Fraunhofer Institüt für Bauphysik, Stuttgart.

JAPAN

[1] Kimura K, Udagawa M and Ishida K (1996). 'A house for three generations: using a hybrid PV collector and PCMs', in *Proceedings of EuroSun'96, Freiburg.*

[2] Hestnes AG, Hastings R and Saxhof B (1997). *Solar energy houses,* James & James (Science Publishers) Ltd, London.

THE NETHERLANDS

[1] Winter R (1996). *1 Jaar Urban Villa Amstelveen.* RIGO, Amsterdam.

[2] Schiphouwer H (1996). *Evaluatiemetingen Urban Villa Amstelveen.* TNO Apeldoorn.

[3] Poel A, Bax E and de Vries G (1997). *Evaluatie Urban Villa Amstelveen.* Damen Consultants, Arnhem.

[4] Poel A and Eijdems HHEW (1996). *Urban Villa Amstelveen, energie- en comfortberekeningen.* Damen Consultants, Arnhem.

[5] Liem SH and van Paassen AHC (1995). 'Digital control systems for extreme low energy apartments', in *Paper from 19th IIR/IIF Congress,* Delft University of Technology, The Netherlands.

[6] Hestnes AG, Hastings R and Saxhof B (1997). *Solar energy houses,* James & James (Science Publishers) Ltd, London.

[7] *Praktijkonderzoek warm tapwater* (1997). Energiened, Gasunie, Kiwa, Vewin.

NORWAY

[1] Aschehoug Ø, Hestnes AG, Jacobsen T, Raaen H and Marit T (1990). *Parametric studies for a Norwegian IEA Task 13 house.* Report STF62 A90018, SINTEF, Trondheim.

[2] Aschehoug Ø (1993). 'Anvendelse av solenergi i boliger. Snekkerstua lavenergiboliger på Hamar', in *Seminar Lavenergiboliger, Bygg Reis Deg, Oslo, September 1993.*

[3] Aschehoug Ø (1994). 'Low energy timber frame row houses in Hamar', in *Proceedings of Eureka Technology Conference – a sustainable approach to housing qualities, Lillehammer.*

[4] Aschehoug Ø, Hestnes AG *et al.* (1994). 'Snekkerstua low-energy dwellings', in *Proceedings of North Sun'94, Glasgow.*

[5] Brattset O (1995). *Snekkerstua lavenergiboliger. Delrapport 1. Bygningsutforming og bygningsutførelse.* Report STF62 A95021, SINTEF, Trondheim.

[6] Lien AG, Hestnes AG and Øyvind A (1995). 'The use of transparent insulation in low-energy dwellings in cold climates', in *Proceedings of ISES World Congress, Harare.*

[7] Winther, BN (1998). *Energibelastninger ved lavenergiboliger.* Thesis, Building Technology, NTNU. Trondheim.

Appendix A

Building energy-analysis tools

A.1. INTRODUCTION

Building Energy Analysis Simulation computer programs (BEA) were used as a design aid to calculate the viability of the new energy-saving concepts in Task 13. Some of the advanced technologies considered within the Task could not be directly modelled with the available simulation programs. Unconventional use of existing computer models was required for some parameter studies in the analysis of the houses. There was therefore a need for support in modelling these new technologies, as well as a need to validate these partially newly developed models. In addition, software tools to optimize building units were used (e.g. FRAME, VISION, WINDOW).

A Simulation Support Group (SSG) was formed and set up within Subtask A to provide the required information. The approach the Simulation Support Group took was divided into two stages:

Two standard cases were calculated in detail, using the simulation models available to the experts, and the results compared, the so-called 'base-case comparison' approach. The objective was to arrive at a common basis for which there was sufficient confidence in the models to enable the simulations to be carried out within the Task. Within IEA SHC Task 8 a thorough design-tool evaluation was carried out. In order to benefit from the work in Task 8, two suitable Benchmark Test cases were selected from the Task. The two base cases were simulated with the aid of each of the models used in Task 13. It should be stressed that this activity was not intended to evaluate thoroughly the models for their physical quality. After a first calculation process, the results were exchanged, and after a discussion about possible model errors, or input errors, a second calculation process was carried out.

This second stage of the activity was directed at the new technologies. As a first step, the participants established which technologies needed support. For each of these technologies a 'Technology Simulation Set' was developed to serve as a support for modelling that technology.

A.2. SIMULATION MODELS

Table A.1 shows which simulation models were used by the participating countries as well as a contact person for that work. The SUNCODE model used by three of the countries has nearly identical codes while each of the TRNSYS models used contained different, specially developed subroutines.

Apache

Apache is designed to operate on microcomputers (MS-DOS operating system) and workstations (UNIX operating system). It is a dynamic thermal simulation program which is particularly versatile in the treatment of plant and control. The software is interactive, with numeric input, and has the possibility of scanning in architectural drawings. Up to 600 zones can be simulated. Quantities that can be calculated are heating load, cooling load, room-temperature profiles, solar fractions, U-values, heat recovery, heat pumps, hot-water rates, and loads due to lighting. A newer version of the program, used in Task 13, can calculate surface temperatures.

The weather data required on an hourly basis is ambient temperature, relative humidity, direct insolation, and indirect diffuse insolation, with wind speed, pressure, and cloud cover as optional. The software is available commercially and is designed for use by engineers, technicians, and research analysts.

Table A.1. Simulation models used in Task 13

Country	Model	Contact person	Address
Belgium	MBDS	Magali Bodart	Université Catholique de Louvain
Canada	Enerpass	Stephen Carpenter	Enermodal Engineering Ltd., Kitchener
Denmark	tsbi3	Kirsten Engelund Thomsen	DTU, Lyngby
Finland	TRNSYS / vtt	Ismo Heimonen	VTT, Espoo
	TASE	Tapio Haapala	University of Technology, Tampere
Germany	TRNSYS / itw	Norbert Fisch	ITW, University of Stuttgart
	SUNCODE	Hans Erhorn	IBP - Fraunhofer Institute, Stuttgart
Japan	EESLISM	Mitsuhiro Udagawa	Kogakuin University, Tokyo
The Netherlands	TRNSYS	Rob Kouffeld	W & MT - Technological University of Delft
	ESPr	Leo Bakker	TNO, Delft
	SUNCODE	Bart Poel	Damen Consultants, Arnhem
Norway	SUNCODE	Marit Thyholt	SINTEF, Trondheim
	TRNSYS	Inger Andresen	SINTEF, Trondheim
Sweden	tsbi3	Gudni Johannesson	KTH, Stockholm
Switzerland	SUNCODE	Patrick Steineman	ETH - Hönggerberg, Zürich
UK	Apache	John Littler	University of Westminster, London
USA	Custom	Doug Balcomb	NREL, Golden, Colorado

Custom-made programs

An alternative to using a packaged program is to develop a custom program written in a language such as Basic or Fortran. This is more work but allows great flexibility to incorporate algorithms that describe special features of the design. While this approach provides the greatest opportunity to simulate the entire system accurately, it requires knowledge and programming skills normally beyond those possessed by the building designers.

Custom-made programs were used to simulate the American houses. These programs use a thermal network representation of heat flow (similar to SUNCODE). Hourly weather data were developed for the two locations based on local long-term averages and hourly data from a nearby site. Detailed simulation models were developed to characterize the performance of the integrated mechanical systems for each of the four operating modes. Natural ventilation was also simulated.

EESLISM

EESLISM is an overall simulation program for predicting room thermal environment and energy used in buildings. The program is designed to accept any type of building with passive, active, or combined types of environmental control systems.

The main features include:

- simulation of building thermal environment expressed by room air temperature and room surface temperature
- space heating and cooling loads
- energy simulation of heating and cooling primary and secondary systems including domestic hot-water heating.

The building thermal model is based on the unsteady-state model using the implicit type of finite-difference method. As the multi-room model is used, the effects of the room-temperature variations of the adjacent spaces can be considered. Radiative and convective heat transfers on room surfaces are calculated separately. The simulation of radiant heating and cooling panels is included in the building thermal model. Use of thermal sensation indices for both evaluation and control of thermal environment are also included.

The mechanical components included in the simulation are boilers, refrigerators, heat pump units, heating and cooling coils, heat exchangers, heat storage tanks, solar collectors, pipes, ducts, pumps, and fans. Each component is expressed by one or more component equation(s) describing the relationship between inlet and outlet fluid temperatures passing through the components.

EESLISM has been developed and used on the Unix workstations. The ANSI C computer language is used for the program, which was first developed in 1988. In the earlier version, linear sub-system models simulating several kinds of typical combinations of components were used. However, in the current version the method is extended to simulate any combination of system components. The software is not available commercially.

Enerpass

The Enerpass program, developed by Enermodal Engineering Limited of Canada, is a detailed total-building energy analysis program. The program determines the energy consumption and cost for space heating, cooling, water heating, pumps/fans, lights, and appliances. The program does a complete loads and systems calculation each hour based on hourly values of measured weather data. The program is capable of analyzing the performance of most residential and commercial buildings. It includes routines for many air handling systems and equipment and allows for heat and mass transfer between zones. The program can analyse up to seven zones in a building.

The software is available commercially. It is designed to run on an IBM PC with 512K of memory, using MS-DOS version 2.10 or higher. It requires a weather file with hourly ambient temperatures, relative humidity, and global insolation on a horizontal surface with wind speed as an optional extra. It is designed to be used by engineers, consultants, and research analysts.

ESPr

The ESPr version used in the Task 13 BEA comparisons is not available commercially. There is, however, a commercial version of ESPr available. It runs on a SUN-SPARC workstation under the Unix operating system. It uses a dynamic state model for computation with the possibility of defining around 27 zones for calculation. The input is interactive in a numeric form.

Quantities that can be calculated on an hourly basis include heating load, cooling load, room temperature profiles, surface temperatures, solar fraction, heat-transfer coefficients, and infiltration-ventilation fluxes.

The weather data required on an hourly basis is ambient temperature, relative humidity, direct insolation, indirect diffuse insolation, and wind direction with wind speed as optional. The software is designed for use by engineers, architects, technicians, and research analysts.

MBDS

The MBDS Computer program was developed by the University of Liège. It was derived from the 'TYPE 46' of the program TRNSYS, and designed for providing full information on the dynamic behaviour of a multi-zone building under the effect of cooling and heating loads, solar radiation and infrared losses, outdoor temperature changes, and free gains such as occupancy and lighting. The available daylighting, leading to hourly lighting loads, cannot be accounted for directly. Possible outputs per zone from MBDS contain detailed hourly information as to what is happening physically (e.g. net heat and cooling demands, surface temperatures, the gains through convection with adjacent zones, the radiation absorbed by each wall). If any heating or cooling power is insufficient for the satisfaction of a prescribed set-point temperature, the program switches automatically to free floating temperature calculation in the zone under consideration.

The meteorological hourly data used in the MBDS program is a typical mean year. Irradiation data and the sunshine period integrated over the day to calculate infrared losses are available.

SUNCODE and SERI/RES

SUNCODE is the PC version of the mainframe building energy analysis program SERI/RES; the calculation modules are essentially identical in the two programs and have the same capabilities. The computer should be IBM-compatible with the operating system MS-DOS 2.0 or greater. A new version is currently being developed to operate under Microsoft Windows. The programs calculate hourly heating and cooling thermal loads for multizone buildings and are appropriate for analysis of residential and small commercial buildings.

SUNCODE and SERI/RES were developed to provide detailed, dynamic analysis of buildings that incorporate passive solar systems. SUNCODE consists of three modules. EDITS is a building file editor and preprocessor that provides an interactive, menu-driven building-description input environment. LOADS is the calculation module; it uses a thermal balance technique to calculate hourly heating and cooling loads. VIEW is a graphical postprocessor that allows the viewer to examine simulation results.

The load calculation is based on a thermal resistance network analysis. Explicit models are provided for many passive solar systems, including Trombe walls, rock beds, and phase-change-material storage systems. Using the multizone (maximum ten) features of the program allows explicit simulation of more complex passive systems such as envelope-integrated air collectors.

Weather data required on an hourly basis are ambient temperature, dew point temperature, direct insolation, and indirect diffuse insolation with wind speed as optional. The software is available commercially and is designed for use by engineers, technicians, research analysts, and architects.

TASE

The design tool TASE was developed by Tampere University of Technology for its own research projects and is not available commercially. It operates on workstations or on a main frame computer. VAX, APOLLO, and STARDENT are the computer types used, with operating system VMS or Unix. It is a dynamic simulation program with the emphasis on energy analysis. The input is numeric, and up to ten zones can be simulated. Quantities that can be calculated are heating load, cooling load, room temperature profiles, surface temperatures, solar fractions, U-values, and surface heat fluxes.

Weather data required on an hourly basis are ambient temperature, direct insolation, indirect diffuse insolation, and cloud cover. The software is designed for use by research analysts.

TRNSYS

TRNSYS, the 'transient system simulation program', is a comprehensive tool for simulating the energy performance of buildings. It was originally developed to allow detailed analysis of buildings with active solar technologies, and has incorporated a range of conventional heating and cooling equipment. TRNSYS is a modular program that performs calculations of solar system performance and auxiliary energy requirements.

The design tool operates on a PC, a workstation, or on a mainframe computer. Operating systems that can be used are MS-DOS, Unix and VMS. A Fortran compiler is required with the Fortran code included. TRNSYS uses a graphical preprocessor (PRESIM) and a system description input language which allows the user to specify the links between the program modules. A library of active and passive solar and conventional HVAC equipment and control component models is provided. In addition, there are component models for single and multizone buildings that generate the loads seen by the solar and conventional systems and for processing the meteorological data used in the simulation.

The weather data required on an hourly basis are ambient temperature, relative humidity, direct insolation, and indirect diffuse insolation, with wind speed as optional. The software is designed for use by engineers and research analysts.

TRNSYS can analyse up to 25 building zones. However, the TRNSYS/VTT version has a zone model included which allows an unlimited number of zones to be analysed. While TRNSYS is available commercially, the zone model used by the VVT is not. The software is designed to be used by engineers, technicians, and research analysts.

tsbi3

The computer program tsbi3 provides analysis of the energy use and indoor climate conditions in buildings on an hourly basis: time steps down to 4 seconds may be used for detailed analyses. It integrates thermal load and mechanical system calculations, accounting for operation of the control system. The program is modular to facilitate addition of new techniques. It has a user-friendly interface based on Microsoft Windows 3.1 or Microsoft DOS, version 3.30 or later, and menus that are accessed using the keyboard and mouse. It is designed to operate on an IBM-compatible PC with a 80286, 386, or 486 microprocessor. The number of zones in a building that can be analyzed is limited only by the computer's RAM capacity.

The output of the program includes indoor temperature and humidity, as well as heating and cooling energy requirements based on an hourly, daily, weekly, monthly, or yearly basis. Detailed energy balance information is also available, including solar and internal gains, ventilation, infiltration, and conduction gains and losses. Weather data required on an hourly basis are ambient temperature, relative humidity, cloud cover index, direct insolation, and indirect diffuse insolation, with wind speed as optional. The software is designed for use by engineers and research analysts.

A.3 TECHNOLOGY SIMULATION SETS

An important feature of the Simulation Support Group's work was to assist in the accurate modelling of the new technologies. The following technologies were divided among the participating countries:

- multi-layered glazing
- atria
- solar hybrid systems
- phase change materials
- transparent insulation materials
- PV systems
- dynamic insulation
- heat-recovery ventilators
- ground heat loss

Initially, three or four levels of modelling the technology were described. Each successive level increases the complication of the model in an effort to improve the accuracy. Normally, the first level was used in the first design phase, as it could give the quickest results, i.e. minimum time for developing the model, and it needed the least CPU time. The increasing levels of modelling not only increased the accuracy but usually supplied extra information. Modelling at the highest level also required a higher level of experience in modelling. In many cases the highest level of modelling was not

Figure A.1.Comparison of the calculated heating loads for four test cases used by the Simulation Support Group

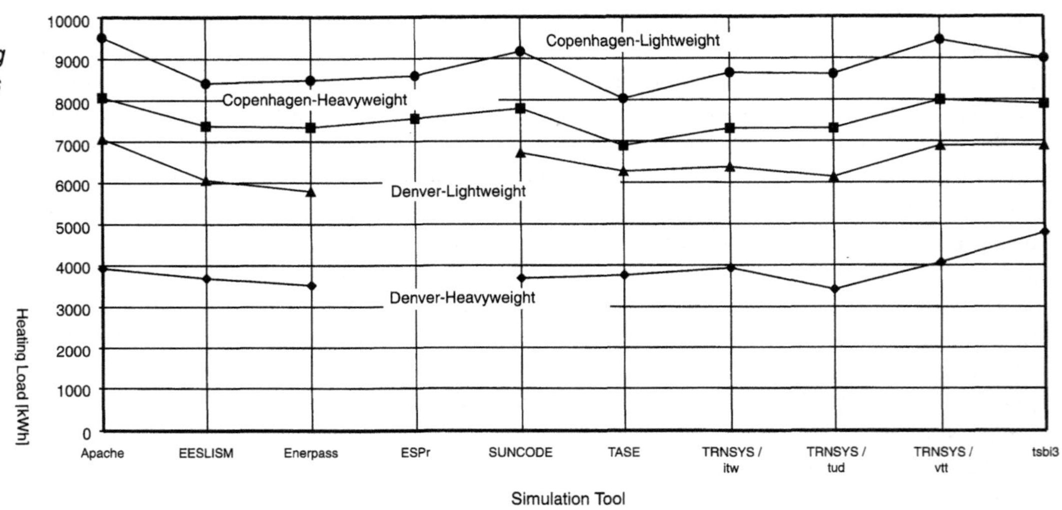

Figure A.2. Yearly mean and range for inside air temperatures for the Copenhagen lightweight case

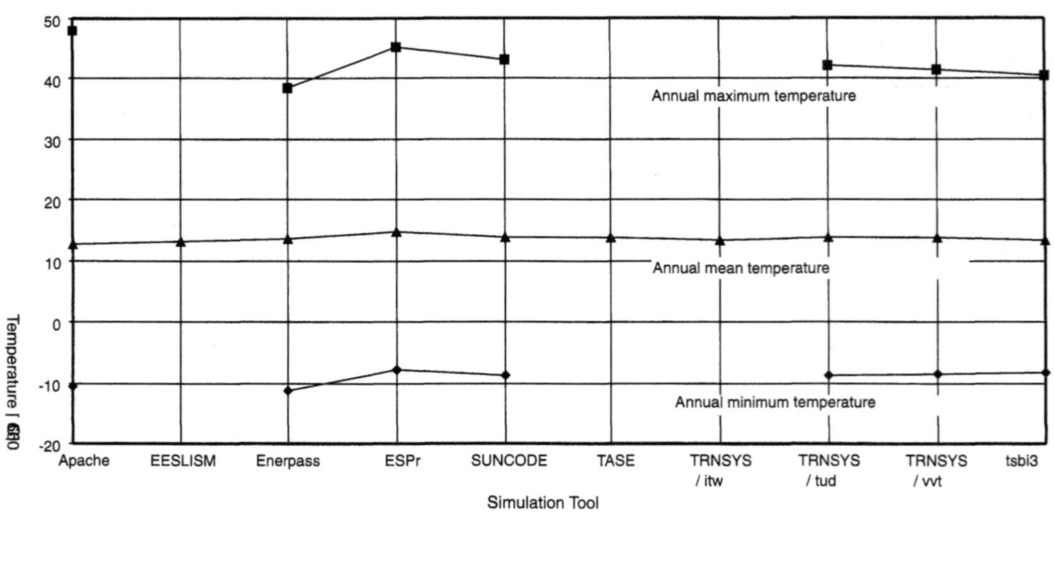

achieved with the simulation program that was used. The models were also validated by other members of the group, who recalculated the same case studies.

A.4 CONCLUSION

In order to compare the simulation programs used in Task 13, all the participants modeled four benchmark cases: Two locations with different climates, Copenhagen and Denver, were selected. The building to be modeled was a simplified (Shoebox) building of two construction types (lightweight and heavyweight).

Generally the models showed comparable results. Examples of the comparisons can be seen in Figures A.1 and A.2.

Comparing the simulation starting point and results proved very helpful, in that there was a significant converging of simulation approaches.

FURTHER READING

Base Case Comparison: Eijdems H (1993). IEA SHC Task 13 *'Advanced Solar Low Energy Buildings'*, Working Document; Damen Consultants, Arnhem, The Netherlands.

Technology Simulations Sets: Poel A (ed.) (1994). Technical *Report IEA SHC Task 13 Advanced Solar Low Energy Buildings*; report number IEA/SHC/T.13.A.1. Damen Consultants, Arnhem, The Netherlands.

Passive Solar Commercial and Institutional Buildings: Hastings S R (ed.) (1994). A Sourcebook of Examples and Design Insights; IEA SHC Task 11 *Passive Solar Commercial Buildings.* ETH Zürich.

Appendix B

The International Energy Agency

The International Energy Agency (IEA), with its headquarters in Paris, was founded in 1974 as an autonomous body within the framework of the Organization for Economic Cooperation and Development (OECD) to co-ordinate the energy policies of its members. The 23 member countries seek to create the conditions in which the energy sectors of their economies can make the fullest possible contribution to sustainable economic development and the well-being of their people and the environment.

The policy goals in the IEA countries include diversity, efficiency, and flexibility within the energy sector, the ability to respond promptly and flexibly to energy emergencies, the environmentally sustainable provision and use of energy, more environmentally acceptable energy sources, improved energy efficiency, research, development, and market deployment of new and improved energy technologies, and co-operation among all energy market participants.

These goals are addressed in part through a program of collaboration in research, development, and demonstration of new energy technologies consisting of about 40 Implementing Agreements. The IEA's R&D activities are headed by the Committee on Energy Research and Technology (CERT), which is supported by a small Secretariat staff in Paris. In addition, four Working Parties (in Conservation, Fossil Fuels, Renewable Energy, and Fusion) are charged with monitoring the various collaborative agreements, identifying new areas for cooperation and advising the CERT on policy matters.

IEA SOLAR HEATING AND COOLING PROGRAMME

The Solar Heating and Cooling Programme was one of the first collaborative R&D agreements to be established within the IEA, and, since 1977, its participants have been conducting a variety of joint projects in passive solar, active solar, and photovoltaic technologies, primarily for building applications. The twenty members are:

Australia	Japan
Austria	The Netherlands
Belgium	New Zealand
Canada	Norway
Denmark	Spain
European Union	Sweden
Finland	Switzerland
France	Turkey
Germany	UK
Italy	USA

A total of 22 projects or 'Tasks' have been undertaken since the beginning of the Solar Heating and Cooling Programme. The overall Programme is monitored by an Executive Committee consisting of one representative from each of the member countries. The leadership and management of the individual Tasks are the responsibility of Operating Agents.

The Tasks and their respective Operating Agents are listed below (an asterisk indicates that the task has been completed).

- Task 1:* Investigation of the Performance of Solar Heating and Cooling Systems, Denmark
- Task 2:* Coordination of Research and Development on Solar Heating and Cooling, Japan
- Task 3:* Performance Testing of Solar Collectors, Germany/UK
- Task 4:* Development of an Insulation Handbook and Instrument Package, USA
- Task 5:* Use of Existing Meteorological Information for Solar Energy Application, Sweden
- Task 6:* Solar Systems Using Evacuated Collectors, USA
- Task 7:* Central Solar Heating Plants with Seasonal Storage, Sweden
- Task 8:* Passive and Hybrid Solar Low Energy Buildings, USA
- Task 9:* Solar Radiation and Pyranometry Studies, Canada/Germany
- Task 10:* Material Research and Development, Japan
- Task 11:* Passive and Hybrid Solar Commercial Buildings, Switzerland
- Task 12:* Building Energy Analysis and Design Tools for Solar Applications, USA
- Task 13*: Advanced Solar Low Energy Buildings, Norway
- Task 14:* Advanced Active Solar Systems, Canada
- Task 15: Not initiated
- Task 16:* Photovoltaics in Buildings, Germany
- Task 17:* Measuring and Modelling Spectral Radiation, Germany
- Task 18: Advanced Glazing Materials, UK
- Task 19: Solar Air Systems, Switzerland
- Task 20: Solar Energy in Building Renovation, Sweden
- Task 21: Daylighting in Buildings, Denmark
- Task 22: Solar Building Energy Analysis Tools, USA

TASK 13: ADVANCED SOLAR LOW-ENERGY BUILDINGS

The energy consumption for space heating has in many countries been greatly reduced over the last few years.

This is mostly achieved by the use of traditional energy conservation and solar technologies. The total energy consumption, especially in residential buildings, is, however, still large and warrants considerable effort. To obtain a significant further reduction in the energy consumption for heating, and also to reduce the consumption for cooling, ventilation and lighting, it has become necessary to develop new building concepts. Such concepts require the use of new materials, components and systems.

Task 13 was started for this purpose. Its official objective is to advance solar building technologies through the identification, development, and testing of new and innovative concepts that have the potential for eliminating or minimizing the use of purchased energy in residential buildings while maintaining acceptable comfort levels.

The focus of the Task is the application of passive and/or active solar technologies for space heating of single family and multi/family residential buildings. The use of passive and active solar concepts for cooling, ventilation, and lighting is also addressed, as well as advanced energy conservation measures to reduce heating and cooling loads.

Since the emphasis is on innovation and long-range (after the year 2000) cost-effectiveness, the materials, components, concepts, and systems considered need not be currently feasible, economical, or on the mass market today.

In order to accomplish the foregoing objective, the Participants undertake work in three subtask areas:

- Subtask A: development and evaluation of concepts (Leader: Hans Erhorn, Germany)
- Subtask B: testing and data analysis (Leader: Bjarne Saxhof, Denmark)
- Subtask C: synthesis and documentation (Leader: Robert Hastings, Switzerland)

To facilitate the effective planning and implementation of the program of work, a feasibility phase was conducted prior to initiation of the research phase of the Task. The primary result of the feasibility phase is a detailed work plan for the research phase. The research phase was started on 1 September 1989 and is scheduled to last until 1 September 1996.

OTHER TASK 13 PUBLICATIONS

Poel B (ed.) (1994) *Technology Simulation Sets.* Distributed through Damen Consultants, P.O.Box 694, 6800 AR Arnhem, The Netherlands.

Saxhof B (ed.) (1994) *Component and System Testing.* Distributed through Department of Buildings and Energy, DTU, 2800 Lyngby, Denmark.

Hastings S R (ed.) (1995). *Solar Low Energy Houses of IEA Task 13.* James & James (Science Publishers) Ltd, London

National Contact Persons for Task 13

Operating agent
Anne Grete Hestnes
Faculty of Architcture, NTNU
N-7034 Trondheim

Austria
Helmut Krapmeier
Inst. für Sinnvollen Energieeinsatz
Bahnhofstrasse 19, Postbox 51
A-6851 Dornbirn

Belgium
André De Herde
Université Catholique de Louvain
Place de Levant 1
B-1398 Louvain-la-Neuve

Canada
Stephen Carpenter
Enermodal Engineering
650 Riverbend Dr.,
Kitchener
Ontario N2K 3SR

Denmark
Bjarne Saxhof
Department of Buildings and Energy
Building 118, DTU
DK-2800 Lyngby

EC
Peter Wouters
Belgian Building Research Institute
Violetstraat 21-23
B-1000 Brussels

Finland
Jyri Nieminen
VTT, Building Technology
P.O. Box 18011
FIN-02044 VTT

Germany
Hans Erhorn
Fraunhofer Institut für Bauphysik
Nobelstrasse 12
D-70569 Stuttgart

Italy
Aldo Lauritano
Dipartimento di Energetica ed Applicazioni di Fisica
Università di Palermo
Viale delle Scienze
90128 Palermo

Japan
Ken-ichi Kimura
Department of Architecture
Waseda University
1-24-2 Nishi-Shijuku, Tokyo 160

The Netherlands
Bart Poel
Damen Consultants
Postbox 694
NL-6800 AR Arnhem

Norway
Øyvind Aschehoug
Faculty of Architecture, NTNU
N-7034 Trondheim

Sweden
Gudni Johannesson
Dept. of Building Technology
KTH
S-10044 Stockholm

Switzerland
Robert Hastings
Architecture, Energy & Environment GmbH
Kirchstrasse 1,
CH-8304 Wallisellen

UK
John Littler
Research in Building Group
University of Westminster
35 Marylebone Road
London NW1 5LS

USA
J. Douglas Balcomb
National Renewable Energy Lab.
1617 Cole Boulevard
Golden, CO 80401-3393

For Product Safety Concerns and Information please contact our EU representative GPSR@taylorandfrancis.com Taylor & Francis Verlag GmbH, Kaufingerstraße 24, 80331 München, Germany

Batch number: 10731598

Printed by Printforce, the Netherlands